Energiepolitik und Klimaschutz. Energy Policy and Climate Protection

Series Editors

Lutz Mez, Berlin Centre for Caspian Region Studies, Freie Universität Berlin, Berlin, Germany

Achim Brunnengräber, Environmental Policy Research Centre, Freie Universität Berlin, Berlin, Germany

Diese Buchreihe beschäftigt sich mit den globalen Verteilungskämpfen um knappe Energieressourcen, mit dem Klimawandel und seinen Auswirkungen sowie mit den globalen, nationalen, regionalen und lokalen Herausforderungen der umkämpften Energiewende. Die Beiträge der Reihe zielen auf eine nachhaltige Energie- und Klimapolitik sowie die wirtschaftlichen Interessen, Machtverhältnisse und Pfadabhängigkeiten, die sich dabei als hohe Hindernisse erweisen. Weitere Themen sind die internationale und europäische Liberalisierung der Energiemärkte, die Klimapolitik der Vereinten Nationen (UN), Anpassungsmaßnahmen an den Klimawandel in den Entwicklungs-, Schwellen- und Industrieländern, Strategien zur Dekarbonisierung sowie der Ausstieg aus der Kernenergie und der Umgang mit den nuklearen Hinterlassenschaften.

Die Reihe bietet ein Forum für empirisch angeleitete, quantitative und international vergleichende Arbeiten, für Untersuchungen von grenzüberschreitenden Transformations-, Mehrebenen- und Governance-Prozessen oder von nationalen „best practice"-Beispielen. Ebenso ist sie offen für theoriegeleitete, qualitative Untersuchungen, die sich mit den grundlegenden Fragen des gesellschaftlichen Wandels in der Energiepolitik, bei der Energiewende und beim Klimaschutz beschäftigen.

This book series focuses on global distribution struggles over scarce energy resources, climate change and its impacts, and the global, national, regional and local challenges associated with contested energy transitions. The contributions to the series explore the opportunities to create sustainable energy and climate policies against the backdrop of the obstacles created by strong economic interests, power relations and path dependencies. The series addresses such matters as the international and European liberalization of energy sectors; sustainability and international climate change policy; climate change adaptation measures in the developing, emerging and industrialized countries; strategies toward decarbonization; the problems of nuclear energy and the nuclear legacy. The series includes theory-led, empirically guided, quantitative and qualitative international comparative work, investigations of cross-border transformations, governance and multi-level processes, and national "best practice"-examples. The goal of the series is to better understand societal-ecological transformations for low carbon energy systems, energy transitions and climate protection.

Reihe herausgegeben von
PD Dr. Lutz Mez   PD Dr. Achim Brunnengräber
Freie Universität Berlin   Freie Universität Berlin

Weitere Bände in der Reihe http://www.springer.com/series/12516

Arwen Colell

Alternating Current –
Social Innovation
in Community Energy

 Springer VS

Arwen Colell
Berlin, Germany

This book is based on the author's PhD dissertation "Alternating Current – Social Innovation in Community Energy" as submitted to the School of Public Policy at the Technical University of Munich (date of defense Dec 16, 2019 | ISBN: 978-3-00-066927-9). The dissertation was supported by a research grant of the Heinrich Boell Foundation.

ISSN 2626-2827   ISSN 2626-2835   (electronic)
Energiepolitik und Klimaschutz. Energy Policy and Climate Protection
ISBN 978-3-658-32306-6   ISBN 978-3-658-32307-3   (eBook)
https://doi.org/10.1007/978-3-658-32307-3

Responsible Editor: Stefanie Eggert
This Springer VS imprint is published by the registered company Springer Fachmedien Wiesbaden GmbH part of Springer Nature.
The registered company address is: Abraham-Lincoln-Str. 46, 65189 Wiesbaden, Germany

Acknowledgements

It has been an inspiration and a great pleasure to work with all the community energy projects that form the heart of this research project. To the interview and workshop partners in the communities of Samsø, Schönau, Oldenburg, Fintry and Mull: my research would not have been possible without you generously and patiently sharing your time and knowledge (and enduring a distinctly pregnant researcher and occasionally carsick children). I am deeply grateful for your kindness and support, and truly appreciate the time all of you gave on top of your community work and very often your day jobs in alternative professions. Thank you! I would also like to express my thanks to interview partners in community energy projects and associations in Bristol, Edinburgh, Yorkshire and Berlin, which provided additional insights.

This book is based on my doctoral dissertation, which was submitted to the School of Governance at the Technische Universität München in June 2019 and defended in Munich on December 16, 2019. I am deeply grateful for the continued support and thoughtful mentoring of my *Doktormutter* Prof. Dr. Miranda A. Schreurs. For comments and conversations on presentations and draft chapters and for your encouragement – thank you! I am also very grateful for the support and encouragement, professionally and personally, of my second supervisor Prof. Dr. Anita Engels. Vielen Dank! For discussions on drafts and ideas, thank you to Karen Litfin, Catherine Mitchell and Rafael Ziegler. I am grateful to the administrative offices at Freie Universität Berlin and Hochschule für Politik TUM; special thanks to Daphne Stelter and Dr. Stefanie Lernbecher. I am exceedingly grateful for the support of a PhD scholarship of the Heinrich Boell Foundation as part of the research cluster "Transformation", which offered the freedom to pursue my research interests without financial insecurity and provided space for reflection and inspiration. I am deeply grateful for the support of my writing group of fellow "Cluster PhDs". If there is one thing I learnt, it is that really no one should be doing such a thing by themselves. Very special thanks, for everything!, to Dr. Angela Pohlmann. Very special thanks also to my mentor Dr. Janis Baumert, for timely support in every way. For comments on presentations and draft chapters, general encouragement, help when the storm hit and just coffee, thank you to Dr. Dörte Ohlhorst and to Liv Colell.

It truly took several villages to write this dissertation, and I am most deeply indebted to my own. A number of teachers and mentors have challenged and supported me over the years, and helped me grow. Special thanks to Dr. Friederike Meinel for having a voice like the sea and helping me to find my own, and for reminding me that a special kind of freedom lies in the right blend of discipline and serendipity. To Stefanie Schmoeckel and Volker Bohnsack. Thank you to my family and friends, who support and inspire me – dissertation or no dissertation. I hope all of you know how important your support is! Very special thanks to Prof. Dr. Georg Baumert and Dr. Lore Baumert and to Susanne and Björn Colell – for everything. To my sisters, whom I could miraculously take along this ride in various unexpected ways. Very special thanks also to Luise Neumann-Cosel, for political vision, passion and being the best soprano neighbor among other things, to Anne-Ruth Müller, Fides Sochaczewsky, and Nina Fritsche. To my children, who were the most amazing research travel partners and are just in general the most amazing people to get to be around. To my partner in life and love and all things. Meine Jungs, ich danke euch von Herzen!

Contents

List of Figures

List of Tables

Part I
Introduction

Introduction

1

1.1 A Story of Social Innovation

Standing on the ferry in the spring of 2016, I watch as the island of Samsø grows larger on the horizon. I am visiting an exception. As I set foot on the small island in the Kattegat, ten years have passed since the local community achieved a 100% shift to renewable energies, offsetting any remaining fossil fuel consumption with additional renewable capacities. I am visiting an exception that should be the rule. Energy system change is an old idea. Internationally, the academic and political debate on shifting the fuel base away from fossil and towards renewable resources dates back, at least, to the oil price shocks and growing environmental awareness of the 1970s (Lovins 1979 (1977)). Yet, the challenge prevails. The most recent report of the United Nations' Intergovernmental Panel on Climate Change concluded that just a decade remains to limit global warming to 1.5 °C and avoid the direst consequences of climate change (IPCC 2018). Energy consumption accounts for over 80 percent of greenhouse gas emissions (UBA 2016). Providing warmth in winter or light in darkness, energy is inextricably linked to people's everyday lives. How to breathe new life into the old idea of energy system change? Could we all become Samsø—and should we?

Shifting human life systems such as energy infrastructure to perform sustainably requires fundamental change. Increasingly, scientists are emphasizing the socio-cultural parameters underpinning and driving this change. Social scientists use the term 'imaginaries' to describe the values, institutions and narratives people use to imagine their social situation and its prospects. Sheila Jasanoff and

© The Author(s), under exclusive license to Springer Fachmedien Wiesbaden GmbH, part of Springer Nature 2021
A. Colell, *Alternating Current – Social Innovation in Community Energy*, Energiepolitik und Klimaschutz. Energy Policy and Climate Protection, https://doi.org/10.1007/978-3-658-32307-3_1

Sang-Hyun Kim apply this concept to socio-technical systems to explain the connotations of public purposes and common welfare within visions for scientific and technological progress (Jasanoff and Kim 2015). Providing new solutions also includes providing new ideas for their meaning and legitimacy, and related knowledge resources (Göpel 2016, p. 161). New ideas and solutions may shift power relations and improve the capabilities of individuals and communities involved (Nicholls and Ziegler 2017(2015)). Social innovations focus attention on the ideational and material capabilities of individuals and communities to assume power in decision-making and implementation. Against this backdrop, this book provides answers on how social innovations in community energy can explain energy system change: How do social innovations in community energy explain energy system change?

Answers are based on the qualitative analysis of five cases of community energy, the Elektrizitätswerke Schönau (EWS) and olegeno in Germany, the Renewable Energy Island of Samsø in Denmark, and Fintry Development Trust (FDT) and Green Energy Mull (GEM) in Scotland. These five cases provide examples of renewable energy projects majority-owned and managed by citizens, and providing diverse energy services including production, distribution, retail, as well as reduced and efficient consumption. While all three countries include targets for renewable energy generation in their respective climate change mitigation strategies, different policy approaches were chosen to attain them. The analysis included data obtained in semi-structured interviews and workshops, as well as documents and reports provided by projects, national strategy and policy documents. Data were processed with methods of situational analysis (Clarke 2003).

Stories of energy system change often tell tales of civil society engagement. The German 'Energiewende', the shift of the energy sector to a renewable resource base has resulted in a roughly 40% share of renewable energies in electricity generation to date. Citizens' financial and organizational investments account for the lion's share of this (trend:research and Leuphana Universität Lüneburg 2013). Often, this is attributed to the technological features of renewable energy generation, well-suited to small scale, decentralized production units. But renewable energy implementation shows that scalability and centralization of renewable energy production are not only possible but openly desired pathways of technology development by energy utilities and policy makers alike (Grashof 2019). Parameters of technological innovation alone do not explain civil society engagement. Also, citizen-owned projects across the country and beyond the German case share organizational purposes of accountability and legitimacy, and paradigmatic features such as sustainability or civil society control. But they do not necessarily share policy parameters fostering investment or organization, such as the German feed-in tarif frequently invoked when seeking an explanation for the excitement of citizens with photovoltaics (PV) and wind turbines.

What drives citizens to implement system alternatives to energy infrastructures, if it is not technological or economic opportunity? What role is there for political strategy? And what happens within these projects that are implementing alternatives to conventional energy system structures? The academic literature provides many and diverse answers to these questions. A unifying theme in the political economy, political science and sociology literature is the recent call of scholars to recognize people as social beings and their socio-cultural belief systems in explanatory approaches to processes of change (Göpel 2016, p. 158). Stories of social innovation provide additional depth to such approaches by including the study of power dynamics. Social innovation approaches specifically focus on the ability of individuals and communities to alter power relations by developing and implementing new solutions and processes (Nicholls and Ziegler 2017(2015); Ziegler 2017). To understand the relationship between social innovations and energy system change in community settings, this study combines the analysis of social innovations with resource mobilization theory (McCarthy and Zald 1977, 2001, 2017 (1987); Walker and McCarthy 2010).

This study offers empirical, theoretical and methodological contributions to social innovations research. Empirically, it compares community energy cases in five different, international locations, with respect to patterns of resource mobilization and innovation biographies of community engagement. Innovation biographies of community energy develop in distinct ways and create corresponding energy system alternatives. Projects mobilize similar types of resources in response to development stages or challenges. But what can become a resource within these types is closely tied to innovation biographies. This also adds to the understanding of resource mobilization in community setting with respect to the relationship of material and non-material resource types. Theoretically, this study refines concepts of power in social innovation by combining theories of social innovation and framing. It will be argued that processes of framing and frame resonance can account for the interventions to power relations required to implement social innovations. Methodically, this study provides diverse examples of using situational analysis mapping to visualize power relations and processes of framing and frame resonance in community energy projects. Such maps provide a way "into the data" of qualitative case studies (Clarke 2003, p. 570), while offering visualizations that increase and improve traceability and transparency of qualitative analyses in academic and trans-disciplinary contexts.

The story of social innovation in community energy is told in five sections. Section I, to which this introduction provides the first part, continues to introduce the five cases included in this qualitative case study, and the three countries wherein the cases are located. Section II will discuss previous research on community energy actors (2.1) and relate the study of community energy projects to concepts of power (2.2). It then introduces the theoretical framework of this research project, drawing on concepts of social innovation and innovation biographies, as well as resource mobilization theory (2.3–5). An introduction to methods of data collection and analysis (chapter 3) concludes this section. Section III begins the empirical analysis by presenting the innovation biographies of community energy projects in Schönau (chapter 4), on Samsø and Mull (chapter 5), in Fintry and in Oldenburg (chapter 6). It characterizes project development in each case with respect to the actors, institutions and narratives, as well as the technologies involved in project implementation. Section IV continues the empirical analysis by assessing mobilization of material and non-material resources across cases. Patterns of resource mobilization are explained during project emergence and establishment (chapter 7) and project maintenance (chapter 8), as well as in response to challenges (chapter 9). Section V summarizes key findings of the analysis (chapter 10) and discusses these in reference to the literature on community energy projects, framing and social innovation (chapter 11). Chapter 12 presents relevant conclusions and briefly indicates future prospects for the study of social innovations.

This dissertation carries a conversation with the literature on civil society engagement in energy systems. It attempts to bridge conventional readings of community engagement in energy sector change as a means of increasing acceptance for technological innovations, and recent studies on the role of narratives and symbolic resources. By integrating material and non-material resource dimensions in the study of project development across national borders, this project seeks a more balanced understanding of the kinds of social innovations that occur in community led energy system change, and the conditions of their development. The urgency of profound change to the networked infrastructures of human life to remain within its planetary boundaries is undisputed. Maja Göpel speaks of paradigms, people, purpose, process, and planet in building a model of transformations literacy that could kindle the 'great mindshift' required in sustainability transformations (Göpel 2016). Against this backdrop, the understanding of social innovations as an intervention to established power grids, physically and in decision-making, can also contribute to an on-going debate on societal transformations.

1.2   Case Stories

1.2.1   EWS: "The World's Most Democratically Legitimized Electricity Provider"[1]

The story of Elektrizitätswerke Schönau eG (EWS) began with parents' despair in the face of the 1986 nuclear incident of Chernobyl. Seeking independence of nuclear power, these parents mobilized the village of Schönau in the German Black Forest to reduce local consumption and began implementing small scale renewable energy generation (FUSS 2007). In the vertically integrated and not yet liberalized energy market of the early 1990s, the only way to gain control of the village's power sources was through ownership and operation of the local distribution grid, governed by municipally awarded concessions. The local utility, Kraftübertragungswerke Rheinfelden (KWR), refused cooperation with villagers to bring down consumption and phase out nuclear energy. Instead, KWR sought to prematurely prolong the concessionary contract with the municipality by 20 years in 1991, offering a 100,000 Deutsche Mark (DEM)[2] reward to the local council in the process. EWS offered 100.000 DEM should the council refuse. The council ruled in favor of KWR, which EWS fought with a referendum. The initiative succeeded. The municipality could not renew the contract prior to its formal expiration date in 1994, giving villagers enough time to build their own offer for grid operations.

In 1995, a new local council awarded the concession to EWS forcing KWR to sell all assets of the local distribution grid. The utility fought this decision with a second referendum in 1996. An emotional electoral campaign saw a utility seeking to discredit the cooperative as ignorants and dilettantes, and a community energy initiative becoming the symbolic Gaulish village of a national alliance for renewable energies and against nuclear power. In a highly contested vote, the utility was ultimately defeated and the concessionary decision of the council reinforced through local vote (coining the above quoted statement of the most democratically legitimized energy provider). EWS commenced operations as local distribution system operator (DSO) and energy utility on July 1, 1997. The establishment of EWS occurred in an energy sector dominated by a utility with close ties to, and strong support of, the municipal authority. This was set in a non-liberalized, vertically integrated market structure, characterized by trust

---

[1] Elektrizitaetswerke Schönau 2018.
[2] ca. 88,000 EUR today, including inflation.

in technical expertise over citizen participation in utility management, centralization as opposed to decentralization, and still marginalized support for renewable energies in public debate.

Following market liberalization in 1999, EWS began national retail operations. It has since grown to one of Germany's four largest independent providers of renewably sourced electricity (with 200.000 customers in early 2019, EWS 2/27/2019), the only one of which to be owned by citizens. Previously a partnership under the civil code (Gesellschaft bürgerlichen Rechts, GbR), EWS was legally transferred to a citizen owned cooperative in 2009. Under the law, each member has one vote in the General Assembly regardless of the amount of shares they hold (Genossenschaftsgesetz (GenG 2017) BGBl. I S. 2230, § 43). The General Assembly elects a board which then chooses the management, and also decides on strategic cornerstones of operations (ibid.). On Dec 31, 2016, the cooperative had 5.135 members holding 385.485 shares and 38.548.600 EUR in business assets (EWS 2017, p. 65). Retail is immediately connected to extending renewable capacity with a fixed investment in new generating capacity of 0,5 Cent per kilowatt hour (kWh) of electricity or kilowatt hour per cubic meter (kWh/m^3) of natural gas, the so-called Sonnencent (sun cent). The company requires all electricity to be sourced renewably from providers without connections to nuclear or fossil fuel generation. 70% must be sourced from 'new installations' (Neuanlagen), not older than six years at the time of procurement (EWS 2017, pp. 69–70). The cooperative engages in multiple initiatives to reduce energy consumption and increase energy efficiency. At an average consumption of 2.330 kWh p.a. per household, its customers undercut the national average of ca. 3.500 kWh per household significantly (EWS 2017, p. 61).

Locally, EWS has advanced to one of the principal businesses in its town regarding employees and business tax volume. It gained regional prominence and recognition as an energy utility, also by connecting to regional utilities. For example, EWS holds a 40% share in the retail operations of the utility of Stuttgart, the capital of Baden-Wurttemberg (EWS 2017, p. 41). The local energy landscape has changed, EWS advancing to an incumbent position. With a large local market share and the role of a significant local employer, the cooperative has economic potential to realize its political goals and a strong negotiating position vis-à-vis the municipality. Federal regulatory reform enabled EWS to offer energy services at a national level. This reinforced its role as a symbol of a national movement for decentralized renewably sourced energy systems and against nuclear power, connecting political support and customer relations.

1.2.2 olegeno eG: "To be the Alternative."[3]

olegeno was the result of a window of opportunity in administrative processes. In the fall of 2010, the city council of Oldenburg began discussing the expiration of the current concession governing local electricity and gas grid operations. It decided not to investigate potential alternatives, and opened the application process on July 1, 2011. Applicants had to declare themselves within three months. The Energierat (Energy Council) of Oldenburg, a group of local citizens, had been monitoring and consulting to the city in energy-related questions for decades. It saw an opportunity to regain control of the technical, political and financial assets connected to the local power grids through administrative process. The Energierat had been a critical voice for transformation and participation in the previous concessionary process 20 years earlier. It now opened public debate with a view of establishing a "broad, civic alliance" to work on a new configuration of concessionary contracts in "the interest of consumers and the municipality" (olegeno 2019a). As with EWS, nuclear devastation lent urgency to the cause of energy system transformation, with news of the nuclear incident of Fukushima, Japan, reinvigorating members of the Energierat to seek new ways of energy system change locally.

The Energierat commissioned a local law firm to conduct an indicative calculation of municipally owned grid operations, as well as different models of cooperation as alternatives to renewing contracts with the local incumbent, EWE AG.[4] It also reached out to other communities and municipalities to exchange experience and build momentum on the question of grid operations and management. The Energierat was advised to not be discouraged by financial demands of buying back the grid, one supporter recalls a meeting in May 2011, underscoring instead how locals were "democratically legitimized" to engage in energy politics (Djordjevic 2011). Results of the law firm in August 2011 highlighted that operating local power grids in a municipal model would not only make economic sense but held high potential for a redesign of electricity and gas infrastructures and was a cornerstone of the local Energiewende (olegeno 2018a). When the council remained unconvinced, the Energierat decided to take matters into its own hands. On September 28, 2011, olegeno eG was established as a citizen owned

---

[3][sic!] I9: 113.

[4]EWE was formally the abbreviation of Energieversorgung Weser-Ems; the company now operates exclusively under the name of EWE AG.

cooperative just in time to formally apply for the concession—the only challenger to EWE AG.[5] Twenty years and extensive market liberalization regulation had passed since the formation of EWS. Still, the establishment of olegeno as a competitive bidder for the electricity and gas grid concessions occurred in conditions strikingly similar. Again, a local incumbent had strong ties to the local municipality that held both market and symbolic power. Oldenburg was the primary seat of EWE holdings at the time, translating to significant tax income and relevance as a local employer. EWE's position as a major local employer also held symbolic power, as many locals had direct or indirect ties to the company and its subcontractors. Again, both the municipality and the local utility refused citizen attempts to negotiate alternative configurations of local energy services. Again, citizens chose immediate (economic) action as a response to their political interests: The formation of a citizen-owned energy company to realize their political goals for the local energy system.

In the fall of 2011, the newly elected majority of social democrats and green party in the city council decided to commission an independent investigation into grid operations in its coalition agreement (olegeno 2019a). Concessionary applications were set on hold in the meantime. Despite more favorable reviews in this investigation regarding both the political potential of municipal management or the leasing model and its economic implications, the council rejected the investigation's recommendations with a majority of the votes of Social, Christian and Liberal democrats (olegeno 2019a). Indicative bids were requested of the two applicants in June 2013, olegeno submitted a legally binding offer in November 2013 (olegeno 2013). Upon reviewing both offers, the consulting law firm recommended EWE, stating that olegeno had failed to prove its capacity sufficiently. In January 2014, the city council voted in favor of EWE's offer with two political factions withholding their votes in protest of an intransparent assessment process and the re-concessioning of EWE for twenty years without adaptations to the concessionary contract (olegeno 2019a). Following the concessionary decision in 2014, the cooperative entered an extended period of reorganization around their central goal; local energy system change lead by citizens. The cooperative currently has ca. 300 members and holds assets of approximately 42.000 EUR. The cooperative offers renewably sourced electricity and natural gas tariffs to local customers (olegeno 2019e, 2019d) and engages in local renewable energy generation, especially for renting tenants (olegeno 2019f, 2019c), increased energy efficiency and reduced consumption (olegeno 2019b).

---

[5]While initially a third party had declared interest in operations, they dropped out of the application process in early 2012 before entering the period of indicative bids.

1.2.3   Samsø: The Danish Renewable Energy Island

The 'Renewable Energy Island' (REI) of Samsø grew out of a pride of local potatoes. In 1997, the national government announced a competition for a 'lighthouse
region' that would implement a 100% energy system change to renewable sources,
bringing to life Denmark's commitment of the Kyoto Protocol to reduce greenhouse gas emissions by 21%. The 4.000 islanders on Samsø did not think the
government was talking to them. At approximately 12% self-sufficiency in energy,
Samsø was 'no greener' than the rest of Denmark (Papazu 2018). Famous for its
successful agriculture—the island's famous Spring potatoes are nicknamed Samsø
Gold—and abundant marine nature, Samsø, instead, took pride in its farming tradition and summer beauty. But nestled in the Kattegat, Samsø is also part of
'Udkantsdanmark'—the 'Danish outskirts'—the rural periphery of the country
characterized by aging communities, and the steady loss of institutions and jobs
(Papazu 2018, p. 6). A shared sense of vulnerability of community life drove
the island's mayor and the local representative in the national parliament to commission the island's candidature and a master plan that would transfer the local
energy system to 100% renewables within 10 years (Papazu 2018, p. 6). Samsø
won. The competition provided funding for two employees to bring the 10-year
master plan to life, one engineer and one communicator.

By 2007, the island's energy system was transformed (Hermansen 2007). Eleven onshore wind turbines, each at 1 Megawatt (MW) production, cover 100% of
the island's electricity demand. Diverse ownership schemes were devised based
on the principle of "if you can see the wind turbines, you should at least have the
possibility to own a share in it" (I24: 170), resulting in nine turbines being owned
by local farmers or small associations of farmers and two turbines cooperatively
owned by many local shareholders. Another ten windmills were erected offshore.
The 23 MW of electricity generated are equal to the island's fossil fuel demand
in transportation (offsetting respective emissions) and are fed into the national
grid through an underground connection to the mainland. Samsø Municipality
is the largest investor, owning five offshore turbines, with another three turbines
owned by private investors, one cooperatively owned by local shareholders and
one owned by a professional investment fund. Both on and off shore, the turbines
generate income via the Danish feed-in tariff. Four district heating plants were
set up, three fueled by biomass boilers running on locally grown straw and one
a combination of solar heat and locally grown wood pellets. Two are commercially run by the regional energy utility NRGi (Denmark's fourth largest energy
utility, cooperatively owned and based in Jutland, NRGi 2018), one collectively
owned by consumers and one owned by a locally based company. District heating covers approximately two thirds of island homes. Another 250 homes outside

district heating service have invested in solar heating, heat pumps or wood burners. In 2007, islanders sported a negative balance on carbon dioxide emissions (Hermansen 2007).

The transformation of energy generating systems was deeply social in its connections to patterns of ownership and behavior on the island, the creation of local jobs and investment opportunities, and narratives of community life. Changes in heating and electricity infrastructure, for example, require an immediate participation of essentially every island home. The primary task of Søren Hermansen, hired in January 1998 and the project's lead manager and spokesperson to date, was therefore to turn a top down decision into a bottom up endeavor that would ensure widespread participation (interview, May 2016: 34–38). Hermansen was employed to moderate, whereas a second colleague with a background in engineering brought technical expertise. In local public debates, islanders determined the technological and, more controversially and ultimately importantly, organizational changes they were willing to adopt. The master plan included a multitude of potential projects that far exceeded the required technical improvements to achieve 100% renewables (Samsø Energiselskab 1997). Projects of the plan could therefore be dropped if need be without jeopardizing the overall REI target. The technological choices of the first ten years fell in favor of domestically well-established renewable energy solutions, most notably wind energy and district heating, with diverse configurations of individual and joint local ownership and investment (Hermansen 2007). Overall, transition costs amount to approximately 58 million EUR, 70% of which were raised by personal investments of 3700 local citizens. Diverse and extensive infrastructure projects have increased demand for local craftsmen and created numerous jobs in what was a struggling local economy. Tourism and educational work connected to the island's successful energy system transformation add value to the local economy. Samsø becoming Denmark's 'Renewable Energy Island' reinvigorated the sense of locals that they could secure the future of their community (Papazu 2018).

Samsø Energy Academy, the umbrella organization that has succeeded the Energy and Environment Office, besides engaging in international communication and educational work, continues to work on local transformation processes with a view of becoming a fossil free island by 2030 (Kristensen 2015). A new gas fueled ferry entered operations in 2014. In 1999, the municipality leased four electric vehicles, and islanders today have the highest per capita rate of electric cars in Denmark. Driving electric or on biogas are set to enable fossil free transportation, with a local biogas plant planned on the island. In addition, the Academy spurs community action for transformative strategizing more generally, for example in workshops planning a circular local economy (Flemming 2013).

1.2.4 Fintry Development Trust: 'Doing Good Stuff in Fintry'[6]

Scotland's community wind energy pioneers were born as 'FREE' (Fintry Renewable Energy Enterprise). Established by four energy enthusiasts of the village in 2003, FREE's aim was to turn Fintry, about an hour's drive outside of Glasgow, into a carbon neutral community. Although Scottish wind energy was in its infancy at the time, the potential of the sector had caught attention of professional project developers. Beyond significant fossil fuel resources, Scotland holds importance in the UK energy strategy for its renewable energy potential, its wind energy installations accounting for the lion's share of wind resources in the United Kingdom (60%, or 2.5 GW in 2011, Department for Business, Energy & Industrial Strategy 2012, updated 2018). But community owned wind energy installations were not yet established in Scotland. By coincidence, a commercial project developer planned a nearby wind farm, and the village entered negotiations. FREE founders initially envisioned a turbine outside the village, directly supplying local electricity demands. When it became clear that it would not be technically, or indeed economically, advisable to "run a cable down from the hill" (I19: 33), an alternative was devised to use revenues of a community owned wind turbine within the windfarm to fund energy efficiency in the village. While not physically linked to production in their consumption, the "emotional resonance and identification" with nearby installations, founders hoped, would create the necessary incentive for community members to actively engage in local energy system change including changes in installations and behavior at home (FDT 2018a).

Negotiations were completed in 2006, and a contract between FREE and the windfarm developer was signed. FREE did not become an owner of one of the ten turbines planned in the hills outside the village, but instead paid for installation of one turbine in return for guaranteed revenues under the UK feed-in tariff (FIT) over a period of twenty years. FREE founders had sought charitable status early on, but the authorities had refused the organization because of its commercial relations to the windfarm developer. Yet, founders felt strongly about charitable status, and the assurance that money generated by the wind turbine would indeed benefit the entire community. In 2007, Fintry Development Trust (FDT) was therefore established as a community-owned and oriented, charitable organization with FREE as its commercial arm. Membership of the Trust is restricted to locals, with some 150 villagers currently engaged. The Trust manages turbine

---

[6]Slogan of Fintry Development Trust, fintrydt.org.uk/

revenues, and disseminates these the community in its diverse projects. The board of directors is comprised of seven voluntary members, one of them part of the four founding members of FREE. The board of directors includes one member of the Fintry Community Council as an informal agreement between the Trust and the Council. Project planning is overseen by the board, and discussed by members in Annual General Meetings (AGM). The Trust currently has two paid staff members developing and implementing community projects, and resides in the local community sports club.

The steady flow of income generated by the wind turbine has financed implementation of diverse local projects. While residential energy system were the focus point in the first years, the Development Trust providing free loft and cavity wall insulation to all local residents, activities then branched out to include community institutions such as the Sports Club and the school, as well as educational events. It has ventured beyond the energy sector, implementing sustainability projects in the community such as planting an orchard. In 2010, the Trust hired its first staff member, professionalizing and institutionalizing energy advisory services to villagers. As recognition of the Development Trust grew, additional grant money could successfully be acquired that supported on-going advisory and educational efforts of staff and enabled extended project work (FDT 2018e). Past years have seen increasing attention to heating, with the Trust establishing a district heating system for a residential neighborhood suffering from fuel poverty[7] in 2015, and a project to balance local green energy production and community energy use, 'SMART Fintry' (Smith 2018). Its on-going advisory and educational work in the community and with schools and universities remains free of charge (FDT 2018e, 2018d). Extensive online publications on its activities include minutes from board meetings (dating back to 2009) and annual general meetings (dating back to 2007, FDT 2018b, 2018c).

The first to have successfully negotiated a community wind project, FDT today is an important advisor and role model to communities seeking to establish own projects, as well as to professional project developers seeking advice on community participation and acceptance. The Trust has gained national (and international) recognition, including political authorities (FDT 2018e). The Scottish government has committed to 100% renewables in the electricity sector by 2020, its energy targets reaching beyond this to decarbonization of the economy by 2050 (Scottish Government 2017). To this end, Development Trusts have become an

---

[7]Scottish government defines fuel poverty based on heating which should not account for more than 10% of the household income (Scottish Government 2017b). Poor energy efficiency is among the primary causes. On the incorporation of other energy sources and usages in this definition, see Simcock et al. 2016.

important community vehicle to realize energy related projects oriented towards community benefit (van Veelen 2018, p. 659).

1.2.5   Green Energy Mull: "Every Cloud Really Does Have a Silver Lining"[8]

Mull and Iona Community Trust (MICT), established in 1997 as a community oriented, charitable organization to "support local projects aimed at improving the social amenities, and physical and economic infrastructure of the islands" of Mull and Iona[9] (MICT 2018a), had nothing at all to do with energy upon its founding. Mull is the second largest island of the Inner Hebrides, famous as much for their abundant and diverse wildlife as for their strong winds. Unsurprisingly, the Outer and Inner Hebrides hold some of Europe's largest wind and marine power resources. But high costs of construction in the harsh Atlantic Ocean and strong local concerns over the preservation of unique ecosystems often proved important impediments to renewable energy developments in the area (Carrell 2008a, 2008b). At the same time, communities in the Hebrides remind the reader of 'Udkantsdanmark'—remote and often harsh living conditions, poor wages and few jobs, aging communities and high fuel poverty (73% in 2015, Bunting 2015). Following Scotland's Land Reform Act of 2003 (Scottish Parliament 3/25/2003), many communities in the Hebrides bought back the land from previous private owners, and began exploring the potential for community owned renewable energy generation (Bunting 2015). While often small, community energy projects increasingly gain recognition as sustainable, local energy alternatives as well as significant sources of community income. A community owned project on the Isle of Lewis includes 'just' 2.7 MW or three turbines onshore compared to 28 turbines planned offshore by professional developers in 2008 and abandoned following extensive local protest, but could generate community revenues of up to £1mio annually (ibid.).

On Mull, islanders began actively considering renewable energy generation in 2010 following both the interest of local residents in alternative energies and the potential for income this provided to the Community Trust in the face of reduced grant funds following the financial crisis of previous years (I20: 7–14). Islanders

---

[8]I20.

[9]Iona is a very small island located immediately to the South West of Mull. Accessible by ferry from Mull, the island is closely tied to its larger neighbor—for example through the establishment of a joint community trust.

quickly ruled out wind energy, mostly for fear of endangering wildlife—among others, Mull is famous for the Golden Eagle and the White Tailed Sea Eagle. The community finally decided on a 400 kW river-run-off hydropower system near the village of Garmony. Construction began in March 2014. 'Garmony Hydro' began operations in June 2015, once the underground cable connecting to the mainland and enabling energy exports under the UK Feed-In Tariff (FIT) was completed. Set to generate just over 1 gigaWatthour (gWh) annually in conservative estimates, the Garmony Hydro scheme should create an annual income of around £200.000 to the community. Garmony Hydro is a rather modest exploration regarding its size of 500 kW. This even speaks from the installation itself, which nestled into the hills will soon be overgrown, vanishing into the soft moorlands abundant on the island. Garmony Hydro is owned and operated by Green Energy Mull (GEM), a Community Benefit Society. While Moray Finch, a co-initiator of Garmony Hydro and General Manager of MICT, chairs the board of directors, GEM stands independent from MICT and directorship is tied to investment. Funds were raised via community shares, reaching just under £500.000 in March 2015 (significantly exceeding the required £330,000, Finch 2015). Over 200 predominantly local shareholders invested, although non-locals could also invest. Net profits of GEM can be distributed among investors. In 2017, GEM bestowed £25.000 upon The Waterfall Fund', an independent charity established in 2016 which will distribute the money as grants to community projects on the islands (The Waterfall Fund 2018). Dedicated to "supporting transformational change to our islands" (The Waterfall Fund 2018), grants are bound by their community reference not by topic. Projects funded include a Gaelish choir, supplies for first responders, or fencing for a local orchard (The Waterfall Fund 2019). While GEM is organizationally independent of MICT, this indicates its embeddedness in the Trust's strategy of alleviating the diverse vulnerabilities of community life on a remote island.

During planning, grid connection of Garmony Hydro to the mainland (a prerequisite of FIT exports) was uncertain due to transmission constraints (wherein only connections below 50 or 100 kW may proceed rapidly), that despite British plans for Scottish generating capacities affect almost 60% of the Scottish land area (ACCESS 2015). Although grid connection was ultimately resolved, MICT entered a project partnership to explore alternative local system configurations for renewable generation in 2015. ACCESS (Assisting Communities to Connect to Electric Sustainable Sources) seeks to develop a model demonstrating real time balancing of local renewable generation and demand, as well as a corresponding system of local heat tariffs to commercialize services (ACCESS 2015). Project

partners beyond MICT include Community Energy Scotland (the Scottish national association of community energy), SSE Ltd. (the UK's largest provider of
renewable energy), VCharge (who designed the smart appliances set to balance
Garmony Hydro generation and distributed demand), and Element Energy (working on commercial roll-out and system integration for other communities across
Scotland). ACCESS exemplifies a fundamental strategic conflict of energy system change in Scotland and the UK: Remote large scale generation in the North
servicing large scale demand in the British South, as opposed to small scale,
community oriented developments, sensitive to siting conflicts and adjacent to
consumption (John Muir Trust 10/30/2015, 11/2/2015).

1.3   Country Stories

1.3.1   The German 'Energiewende'

Germany's energy system prior to market reform in the late 1990s was organized in regional monopolies, the state holding majority shares in all electricity
providers either directly or indirectly (Mez 2003, p. 201). With high turnovers
and blurred ownership lines between state and non-state actors on municipal,
regional and federal levels, the energy sector wielded considerable economic and
political power (Lauber and Mez 2004, p. 604). Following the 1970s oil crises,
Germany relied heavily on coal and nuclear resources in electricity generation
(Lauber and Mez 2004, p. 615; Jacobsson and Lauber 2006, p. 261). In 1998, the
German Energy Industry Law (Energiewirtschaftsgesetz EnWG, Federal Republic of Germany 7/7/2005) was reorganized following the EU single electricity
market directive of 1996. In 2000, the Erneuerbare Energien Gesetz (Renewable
Energy Law REL, Federal Republic of Germany 3/29/2000), EEG for short, was
introduced. Germany's energy system fundamentally changed.

Market liberalization resulted in a sequence of mergers creating four major
energy utilities, dominating the energy market since: E.ON, RWE, Vattenfall and
EnBW (Kungl 2014, pp. 12–13). Majority ownership of these utilities is private,
except for Vattenfall, which is a full subsidiary of the state owned Swedish Vattenfall AB. Accounting for 82% of electricity production capacity in 2003 and
generating approximately 90% of the country's electricity (Kungl 2014, p. 14),
these utilities particularly dominate the coal and nuclear based power generation.
Reorganization of ownership was not limited to the electricity sector but revealed
a strong political preference of private ownership in utility management generally
(Héritier 2002). Historically, concepts of 'territoriality' strongly shaped German

energy sector design. Prior to privatization, these translated into local multi-utility service partners, so-called Stadtwerke, often covering vertically integrated energy services (production, distribution and retail) as well as water, sewage and waste management, but also public services such as transportation or communication (Julian 2014). Market liberalization allowed free customer choice of energy providers and allowed regional providers to offer their services nationally. Privatization, although resulting in mergers and underwriting the former Stadtwerke landscape with related holding structures, maintained a highly decentralized structure of energy distribution and retail with approximately 900 distribution grid operators and over 1.000 energy suppliers operating on a national level (Julian 2014). Market reform in 1998 also instated environmental safety in energy law.

The REL became core legislation of the German 'Energiewende', a transformation of resources and ownership structures of the energy sector and the phase-out of nuclear power (Kungl 2014, pp. 12–13; Schreurs 2014). While early support schemes for renewable energy generation in Germany date back to the 1970s, these were limited to subsidies for research and development provided by the Ministry of Research and Technology (BMBF). Nuclear and coal fired power generation, instead, found direct and indirect support of the more powerful Ministry of Economic Affairs. Public concerns over nuclear safety and environmental repercussions of fossil fuels grew, forming the social movements founding the Green Party in 1980 and subsequently starting to demand legislation for renewable generation (Jacobsson and Lauber 2006, pp. 261–263). The anti-nuclear movement gained strength after the nuclear devastation of Chernobyl in 1986. In 1989, the BMBF departed from its focus on research and development, launching a '100 MW Program' for wind power in 1989 as well as a '1.000 rooftops Program' for solar. Legislation ensured 0,04 €/kWh of electricity generated by wind turbines, or an investment subsidy of 60%, and investment subsidies for small-scale PV (Ohlhorst 2009, 114, 132; Mautz 2008, p. 54; Jacobsson and Lauber 2006, p. 264). These early windfarms were typically operated by farmers or citizen initiatives—small and independent power producers—founding a renewable energy movement built on citizen resources (Bolinger 2001).

In 1991, Germany introduced a Feed-In Tariff (FIT) for the first time. The Electricity Feed-In Law (Stromeinspeisegesetz, Federal Republic of Germany 12/7/1990) forced utilities to connect independent (renewable) generation to the grid and pay a compensation of 90% of their average electricity tariff (Jacobsson and Lauber 2006, p. 264). This provided security for investments in renewables. The 1991 FIT was a key driver of wind power installations in Germany, which benefitted most from this level of financial support, and supported citizen owned installations and their challenge to the incumbents of the energy system (Ohlhorst

2009, p. 124; Maut 2008, p. 53; Jacobsson and Lauber 2006, p. 272). Additional measures of support included preferential conditions for loans, building and taxes (Ohlhorst 2009, 133–35, 142), as well as individual support programs of federal states (Ohlhorst 2009, 114–15, 133–34). Federal support of solar power was limited to individual households. But citizen pressure on local governments resulted in municipal utilities providing local feed-in tariffs high enough to cover the cost of PV installations, forcing a first move of utilities towards a more renewable energy friendly stance (Jacobsson and Lauber 2006, p. 266). Wind capacity doubled annually in the first half of the 1990s (Bruns et al. 2010). Regional planning authorities became more important to manage an emerging landscape of small scale and decentralized energy generation including rising local conflicts on the immediate impacts (especially noise and optics) of wind installations and resistance from nature conservation advocates (Bruns 2011). An amendment to the German federal building code created an underlying presumption favorable to turbine (and other renewables) installations on federal level, which counterbalanced so-called planning reservations (designed to separate residential and wind zones) municipalities could instate (Bruns et al. 2008).

While large corporate actors dominated the conventional energy landscape, the advent of renewable energy technologies coincided with a widening of the actor base to include non-corporate actors and citizens, often organized in cooperatives (Becker et al. 2013; trend:research and Leuphana Universität Lüneburg 2013), and reinvigorated municipal utilities (Mez 2003, p. 199). Citizen-owned structures were often more responsive to policy needs on a local or regional level, interacting closely with municipalities, as exemplified by regional support schemes beyond the national FIT in the 1990s (Jacobsson and Lauber 2006, p. 266). Collective or cooperative ownership of energy installations was underpinned by a tradition of cooperative self-help in Germany, also apparent in agriculture and food, banking or housing (Bijman and Hanisch 2012). German utilities mostly opposed feed-in tariffs, claiming competitive distortions as well as an imbalance of technologies due to the emphasis on wind created by the level of subsidies (Sijm 2002).

The 2000 REL, introduced by the country's first coalition government of Social Democrats and Green Party, revised these conditions and provided renewed direction and depth to support for renewable energies (Jacobsson and Lauber 2006, p. 267). FITs were disconnected from average tariffs and defined individually for renewable energy technologies, depending on the respective cost structures. The cost was refinanced through a renewables surcharge (so-called EEG Umlage) paid per kWh by consumers. The law also included a nationwide scheme to share cost burdens of grid operators resulting from increased fluctuating capacities. Amendments in 2004 and 2009 (Federal Republic of Germany 7/21/2004,

10/25/2008) further secured grid access while differentiating FITs, finally introducing a breathing cap which adjusts tariff payments in correspondence to overall installed capacity. The 2009 amendment also addressed grid constraints, partly resulting from a rapidly growing offshore wind sector, requiring grid expansion and optimized management structures. Further reform of the Energy Industry Law (Energiewirtschaftsgesetz, EnWG) in 2005, translating the EU directive on increased market liberalization dubbed the acceleration directive, strengthened federal regulation of regional monopolies in grid operations and unbundling of previously vertically integrated utilities.

For renewably sourced retail to increase demand for renewably sourced production, grid access of renewable generating capacities must be secured to ensure that additional capacity can actually be fed into the grid. The energy grid is pivotal to the Energiewende. In Germany, regulation on non-discriminatory grid access importantly aided suppliers of renewable energy to connect capacities to the grid and the REL prioritized renewable sourced energy (§§ 13, 13a, 14, 15, Federal Republic of Germany 10/25/2008). By this ruling, fossil fuel generation would have to be reduced first in times of overproduction. Yet, the insufficient flexibility of fossil fueled generating capacity to reduce production and ambiguous rulings on priority in the grid—still—result in base load provision through fossil fuels. Renewable capacities are shut off if production exceeds consumption and cannot be rerouted via transmission lines, a practice termed re-dispatch. Although generators receive compensation, this decreases effectiveness of renewable capacities in a highly visible manner: Turbines standing still. While, for example, 2015 and 2017 were both years with comparatively strong winds they also saw relatively high re-dispatch measures (BNetzA 2015 (2016), 2017). Reporting that less wind energy capacities had been shut off in 2016 (BNetzA 2016), therefore did not indicate improved grid integration of renewable supplies but a dead calm year.

Critics of renewable capacity increase gained influence in the late 2000s, fueled by inefficiencies in the regulation of the renewables surcharge. In 2009, a new coalition government of conservative and liberal parties reversed nuclear phase-out and began substantial REL revision. 2012 amendments to the EEG reduced feed-in tariffs considerably (EEG 2012 and so-called PV Novelle, Federal Republic of Germany 7/28/2011, 7/14/2012). The rising renewables surcharge, a result of the intended growth in capacity, did not lead to revision of its calculation (Gawel et al. 2015; Tews 2014). Instead, a 2014 amendment extended the breathing-cap-principle to all renewable energy sources. Large scale PV was excluded from feed-in tariff support completely, to be managed via tenders instead (Federal Republic of Germany 7/21/2014). The shutdown of renewable generating capacities could now go without compensation in case of grid-balancing issues.

The 2016/17 amendment departed from a FIT system, switching to tenders, and limited the expansion of renewable generation (Federal Republic of Germany 10/13/2016). Key motivations for the change to an auction-based system included improved cost efficiency of installations through competitive structures, close observation of targets for intended amounts of capacity increase, and limiting the influence of renewable energy advocates on the policy process.

By design of the policies to foster sector development in the 1990s and underscored by a broad civil society movement, ownership of renewable energy installations in Germany was importantly shaped by individuals and communities. In 2012, the share of renewable energies in electricity generation accounted for 22.8%, up from 4.7% in 1998. Only 5% of the overall installed capacity were owned by the 'Big Four', citizens by contrast accounting for 34,4% of all renewable installations (Kungl 2014, p. 15; trend:research and Leuphana Universität Lüneburg 2013, p. 42). Collective citizen ownership accounted for 20,4% of onshore wind capacity in 2012 and another 25,8% were at least partly owned by individuals (trend:research and Leuphana Universität Lüneburg 2013, p. 45). Replication and up-scaling (expansion of number or size of turbines) of citizen owned windfarms followed word of mouth on the benefits of collective installation (Schreuer 2015, 90–92, 103), although citizen ownership typically did not extend to offshore wind installations (Schreuer 2015, pp. 125–28). Solar installations, by contrast, were typically individually owned under FIT support of the 1990s and early 2000s (trend:research and Leuphana Universität Lüneburg 2013, p. 44), although collective and especially cooperative ownership of larger PV installations has increased recently (Schreuer 2015, pp. 93–95; Debor 2014).

The growing sector of citizen and community led renewable energy projects and advocates of an accelerated energy system change heavily criticized the government's decision to switch to an auction-based system for renewable energy deployment. Increased transaction and financing costs of an auction-based system could outweigh the intended benefits of cost-effective installations. Also, calculations for capacity increase could lack reliability as a result of projects not being completed despite their developers' success in tendering (Hauser et al. 2014, p. 1; Kochems et al. 2015; Kreiss et al. 2017). In other countries, such as the UK, these effects had caused governments to depart from auction-based systems at the time the German government was discussing their introduction (Hauser et al. 2014). Moreover, small scale, citizen-owned investments provided considerable financial investments to the German Energiewende. Critics feared that auctions would threaten the diversity of actors in the energy system (Hauser et al. 2014, pp. 4–5). The federal government attempted to address these concerns by creating favorable conditions for citizen energy projects in auctions. Among other things, regulation reduced conditions for planning permissions and financial security while demanding local references and majority representation in decision-making, and offered more money and time if citizen-owned projects were successful in bidding (Federal Republic of Germany 10/13/2016). Auctions showed, however, that these criteria were insufficient to support citizen owned projects and prevent abuse by professional project developers. The federal regulator Bundesnetzagentur (BnetzA) reported that out of 70 windfarms planned nationally more than 65 successful bidders were citizen-owned projects. But a closer look revealed that these projects were in fact led by professional developers that had created proxies to benefit from improved bidding conditions without ultimately delivering citizen-owned projects (Wetzel 2017). Academic analysis of the effects of auctioning on the actor structure of the German energy system confirmed the investor specific adverse effects this policy change had on citizen-owned projects (Grashof 2019).

1.3.2 Fossilfri Danmark

In 1979, three families in a small town outside of Aarhus laid the cornerstone for Danish leadership in community wind energy. Previously jointly owning a snowplough, soaring oil prices now drove them to invest 15,000 EUR each in a 55 kW wind turbine that was set on the intersection of their properties. Denmark's highly import dependent economy was hit hard by the 1970s oil crises (over 90% of energy was based on imported oil in 1973, IRENA 2013). In its subsequent energy plans (Ministry of Economics 1976; Energiministeriet 1981), the country focused on security of supply via domestic coal and renewable resources. Nuclear power was ruled out as a result of joint lobbying of green, leftist and rural movements favoring decentralized structures and strong wind energy expertise catering to these interests (Cumbers 2012), and ultimately banned in 1985 (IRENA 2013). This political turn initiated what to date is characteristic of the Danish energy sector: The prominence of wind energy as an energy resource but also a driver of scientific and industrial innovation and related economies, and the strong role of individual and collective ownership in a highly decentralized energy system.

The 1970s and 80s were characterized by domestic experimentation with often small scale wind turbines by private individuals, implementing new technologies largely without government support but carried by supportive social movements including anti-nuclear groups and movements to support domestic engineers and innovation (Jørgensen and Karnøe 1995; Bolinger 2001). This intensified rapidly with government support policies enacted in the national energy plans, quickly resulting in larger, more powerful and effective turbine designs (IRENA 2013).

The dynamics of domestic sector development were underscored by international market developments in the early 1980s offering attractive conditions for export and in turn bolstering technological learning (Danish Wind Industry Association 2000). The 'Danish Concept' for three-bladed turbine design now dominates international markets (Olesen et al. 2002; Maegaard 2013). The 1976 energy plan introduced a 40% investment subsidy for construction and limited ownership to within a 3 km radius of the turbine, ensuring involvement of locals in each installation. In 1983, earnings from the first 7 MWh of production were exempted from income tax and 1984 saw the first feed-in tariff with utilities required to pay turbine operators 85% of the price currently paid by large electricity customers (Danish Energy Agency, n.d.). Again, those three families outside of Aarhus led change, as the 1984 decision was also a result of the families' suing their uncooperative grid operator for grid connection.

Early wind energy cooperatives in Denmark primarily served as a means of self-supply. One share typically equaled the production of 1.000 kWh/year and members would acquire shares equivalent to their annual consumption. Revenues beyond this enjoyed tax exemptions if less than 940 EUR p.a. and simple tax revenue forms created easy investment conditions. Tax exemptions were tied to a number between 10 and 20 shares; otherwise, the cooperative would be treated as a company. Early developments were typically small, turbines often at around 30 kW, enabling comparatively swift, affordable and consensual roll-out because repercussions to the immediate environment were negligible. Besides fostering deployment, this enabled technology learning for both developers and the public and fostered acceptance of the energy system shift (Toke et al. 2008). Technologically, feedback and ideas of early citizen adopters created a strategic advantage for Danish wind energy developers (Danielsen 1995; Garud and Karnø 2003; Ornetzeder and Rohracher 2013). Financially, citizens' investments spurred wind energy deployment since the early 1970s, a dynamic that intensified when citizens joined in so-called guilds (preferable to cooperatives at the time for tax reasons), investment risks were reduced politically, and installations were conditioned to regional investment (Tranaes n.d.). Politically, the strong advocacy base in favor of a profound shift of the energy system has supported deployment of strategies directed at savings and efficiency as well as creating a support base for cost intensive measures.

The 1985 prohibition of nuclear power was accompanied by a '100 MW Agreement' to be achieved by 1990 which effectively called for a doubling of capacity at the time (Lauber 2005). Grid capacity was increased, and independent investors received priority access in partnerships with communities. This policy design fostered sector growth while at same time ensuring reliability of the grid systems, as well as encouraging grass-roots involvement and reducing investment risks. In 1989, investment subsidies were abolished. However, this did not significantly impede sector development because retail and operation based incentives created stable investment conditions. By 1990, over 3.000 cooperatives were operating turbines, managing investments of over 150.000 shareholders (Danish Energy Agency, n.d.). 1990 saw another agreement to increase capacity by an additional 100 MW by 1995 as well as a long-term target of 1.500 MW by 2005 or 10% of the country's electricity generation. In 1991, a price premium of 3.6 Cent/kWh was introduced. Committing to ambitious greenhouse gas emissions reductions in the Kyoto Protocol, Denmark created a national competition for a showcase community to implement a transition to 100 renewable energies within 10 years in 1997 (Miljø- og Fødevareministeriet 2005).

Early policy design explicitly focused on citizen and cooperative involvement. Electricity was seen as a commodity to serve the social good of individuals and communities (Hadjilambrinos 2000; Lucas 1985). This changed in the 1990s. In 1992, reform of planning laws favored commercial developers by requiring upfront investments for siting, a difficult condition for citizen owned cooperatives without upfront investment structures. Planning authorities began recommending joint installations of at least three to six turbines together, which in turn necessitated an Environmental Impact Assessment for siting and further drove upfront costs of development. In addition, turbine sizes increased from an average between 55 and 300 kW between 1975 and 1995 (turbines typically no higher than 60 meters realized by 20 to 40 households). After 1995, installations typically ranged between 500 kW and 1 MW. Consequently, between 1996 and 2000, deployment of turbines was dominated by large investors (Olesen et al. 2002). Remaining citizen projects were typically individual farmers rather than cooperatives. Through local tax revenues and positive effects on local value creation, this still served the political target of local communities benefiting from installations even if not directly involved in the ownership. Zones for wind energy deployment had to be specified involving local administrative and non-governmental actors as well as utilities (Toke et al. 2008).

In 1999, the new center-right government terminated the FIT and introduced Renewable Portfolio Standards (RPS) and emissions trading, instead (IRENA 2013). By 2003, all grid connections of turbines were realized through RPS as a combination of market prices with a fixed market premium of 1.3 Cent/kWh for 20 years. The country exceeded its 1990 targets for capacity increase by 2003, achieving double the planned capacity (Meyer, 2007). Yet, sector effects of new legislation were drastic. The wind energy market stagnated, with only 42 MW additional installed between 2004 and 2008 (as opposed to previous *annual* increases of several 100 MW in earlier years, Energistyrelsen 2019). Turbine installations no longer required local ownership, favoring remote investments and adversely affecting local tax revenues and value creation. The 2004 market liberalization reorganized the electricity sector (Maegaard 2009). Low voltage energy distribution was tasked to local non-profit cooperatives, municipalities or companies awarded a concession by municipal authorities. Energinet, a new state-owned company, should handle transmission over 60 kV. Responsibility for electricity generation was split between (1) DONG, a 76% state-owned company operating central power plants; (2) power plants owned by Vattenfall and E.ON; (3) local or municipally owned combined heat and power plants; and (4) wind power developments with an 85% share of ownership by so-called Independent Power Producers (Maegaard 2009). However, local involvement was still not required so that investments could remain remote with little or no direct or indirect benefit to respective communities. In 2007, without departing from market-based policy instruments, a target of 100% renewables by 2050 was phrased. Yet, sector dynamics did not pick up. The 2008 Promotion of Renewable Energy Act reintroduced localized investment conditions, now requiring a minimum of 20% to be held by locals within a 4.5 km radius, as well as support measures for re-powered wind turbines, necessitated in part by the installations of the early 1990s reaching the end of their lifespan. Its effects on citizen held capacities were mixed, as re-powering frequently resulted in larger turbines which few cooperatives had the financial or organizational resources to implement (Maegaard 2013).

In 2009, the Danish Government returned to FITs for turbines of all sizes as part of the Promotion of Renewable Energies Act (Danish Government 12/27/2008). In 2010, wind power covered 20% of total electricity consumption—hitting this target seven years late (Energistyrelsen 2019). In 2011, a new government–whose electoral campaign had centered on plans for a fossil free economy–published its Energy Strategy 2050 (Danish Government February 2011). Targets included a 30% share of renewables in overall energy consumption (50% electricity generation served by wind) by 2020, 100% renewable electricity and heat sectors by 2035, and overall independence of fossil fuels to be achieved by 2050. Implementation is detailed in the 2012 Energy Agreement (Danish Ministry of Climate, Energy and Building 3/22/2012). This set sector targets and instruments for increasing the share of renewables in electricity and heating, a strategy of 'intelligent energy systems' including smart grids and flexible mechanisms of demand and supply management, and charged numerous institutions with fostering energy savings and efficiency in various sectors and actor groups. To maintain and spur public support for this transition, the Agreement included initiatives on transparency in strategies and pricing, for example publishing the estimated added costs per household and potential gains through savings and efficiency. In 2012, regulation on net metering created exemptions from tariffs, duties and value-added tax for self-produced energy, limited to consumer-owned, on-site production as well as by size.

The consecutive policy changes to accelerate sector transformation after 2009 resulted in Denmark returning to annual increases of the share of renewables, setting in back on track for its 2020 target of 50% of electricity sourced renewably (Energistyrelsen 2019). In 2015, the share amounted to 45% already, although 2016 saw a drop in production. This was attributed to atypically low winds throughout the year, as opposed to 2015 which had recorded the windiest year since 1994 (Weston 2017). In 2016, the government decided to phase out Public Service Obligations (PSO), a levy applied to retail taxes creating funds for renewable energy support schemes. Financial support for small-scale solar and onshore wind is set to expire. A 2017 report of the International Energy Agency saw Denmark on track for its long-term energy sector targets (IEA 2017). Phase-out of the PSO as the primary funding source between 2017 and 2020 was reviewed critically, however, as sector effects and alternative sources of funding sector transformation out of the General Budget are yet unclear (IEA 2017, p. 79).

1.3.3   An Energy Strategy for Scotland

The devolution of government power from Westminster to Scotland, Wales and Northern Ireland in the late 1990s created a stronger role for regional authorities in energy politics (Smith 2007a). The 1998 Scotland Act reserved overall powers in energy policy and regulation to the UK Department of Technology and Industry, as well as the UK regulator Ofgem (HM Government 1998). But the Scottish government could promote renewable energies and energy efficiency. Therefore, both Westminster and Holyrood shape the UK energy system.

Following privatization in 1989, five generating companies, two transmission grid operators, and 14 regionally focused distribution companies came to dominate the UK electricity sector (transmission and distribution grid operations were unbundled in 1998). The energy market centers on large scale, inflexible power generation (Mitchell and Connor 2004). To diversify share ownership across society, privatization was embedded in a political design ensuring minimum risk and stable return. Rates of return were allocated based on asset valuation at the time of privatization (the so-called regulatory asset base, RAB), and subsequent capital expenditures. While some regulatory changes were made since, this structure of incomes has not fundamentally changed and continues to incentivize expansion of capital assets and discourage utility interests in reduced consumption (Kern et al. 2015; Mitchell 2015).

Westminster: Market based support and slow deployment
The UK government, almost traditionally, favors market based energy policy instruments. Two policy instruments stand out with respect to RE deployment: The 1989 Non-Fossil Fuel Obligation, or NFFO, and the 2000 Renewables Obligation, or RO (Mitchell and Connor 2004). Directed at nuclear power generation but extended to renewable energies to comply with EU legislation, the NFFO was a levy raised on fossil fuels to refinance investment in non-fossil fuels. NFFO premiums were low, however, which frequently resulted in under-financed bids that could then not be built, and in combination with organizational requirements meant that only wind projects could compete (Mitchell 1996). While attempted reforms included differentiating technology support and extending time periods, the fundamental flaw of its price caps remained unresolved. Deployment was poor.

In 2000, the RO replaced the NFFO to accelerate deployment and broaden the technology base (Mitchell and Connor 2004). Originally designed as an RPS, electricity suppliers were required to purchase electricity from renewable sources for defined (and increasing) shares of their supply. Non-technology specific support and higher premiums increased the number of applications for planning permissions. More market oriented than the NFFO, the RO, however, shifted risk to RE developers, who negotiated prices with suppliers which again impeded deployment (Mitchell et al. 2006). At the same time, electricity suppliers falling short of the required RE share could pay a penalty charge per MWh, creating a de facto price cap for renewably sourced power. In addition, a newly established fund pooled buyout fees, which could then be redistributed among suppliers according to their compliance. While designed to compensate the cost of RO compliance to suppliers, strategic planning for limited compliance instead redistributed funds among suppliers while the investment of RO compliance could be passed on to customers (Mitchell and Connor 2004, p. 1939). Installed capacity (approximately 3.000 MW in 2000 upon introduction of the RO), increased only incrementally and was still below 5.000 MW in 2005 (Department for Business, Energy & Industrial Strategy 2016, p. 158).

Both the NFFO and the RO favor existing players of the energy market, rather than fostering a diversification of the actor base. A combination of short time frames and low premiums under the NFFO discouraged market newcomers with limited resources and experience, favoring utility incumbents instead (Mitchell and Connor 2004, pp. 1936–1937). The RO lifted the key impediment to RE deployment—the tight cost cap (Mitchel and Connor 2004, p. 1941). Still, the larger risks to be assumed by RE developers in combination with ambiguous and complicated regulatory frameworks for calculating premiums and estimated returns of installations further dissuaded newcomers and smaller players from entering the market (Mitchel et al. 2006, p. 305). In direct connection with impeding diversification of the actor base in renewable energies, technology diversification also suffered from adverse policy design. The NFFO cost cap was so low that only renewable technologies already close to market could perform accordingly. This did not change with the introduction of the RO, despite its non-technology specific design, which had been drafted to counter this flaw of the NFFO. By not differentiating between specific technologies in its initial version, the strong market orientation of the policy instrument effectively excluded emergent technologies as well as more cost intensive varieties of renewable energy technology. Focus was placed instead on landfill gas and onshore wind developments which provided the lowest risk in investment and financing by comparison and discouraged market entry of emergent technologies (Foxon et al. 2005). Large onshore wind projects proliferated, quickly planned and implemented to comply with short time frames under the NFFO and crowding in wind peak locations. This resulted in a small but strong anti-wind alliance. Forming in the early 1990s, anti-wind protest continued to (adversely) influence wind energy development until the early 2000s (Mitchel and Connor 2004). Planning difficulties were thus identified as a key barrier to deployment besides the risk intensive investment conditions after introduction of the RO (Woodman and Mitchell 2011, p. 3917).

Economic and environmental concerns in energy system design, as well as climate change mitigation targets built pressure on deployment (Winskel 2007, pp. 188–190). The 2008 Climate Change Act set an 80% reduction in CO2 emissions by 2050 based on 1990 levels (HM Government 2008), and the EU commitment to source 20% of its energy renewably in 2020 (European Commission 2010) translated into a national target of 15% renewables in gross final consumption for the UK. In 2008, the share of renewables in the UK amounted to 2.3% of its energy (DECC 2009a). The electricity sector was to deliver the lion's share at 30% renewables by 2020 (ibid.), mainly through offshore wind installations. While offshore energy resources of the UK are considerable, previous deployment rates of offshore planning had been underwhelming (Greenacre et al. 2010). The British government, therefore, acknowledged the role of energy governance over markets in shaping outcomes (DECC 2009b: 36; DECC 2010a, p. 15). Wide-reaching RO reform in 2009 significantly softened its market orientation, reduced risks of investment and acknowledged technology specific requirements for support (Woodman and Mitchell 2011, p. 3920). The introduction of 'guaranteed headroom' countered the risk of prices for RO Certificates crashing in case annual capacity targets were approached or met. Also in 2009, banding replaced technology neutrality. This extended the support granted to emergent technologies. (DECC 2009b) In 2010, small-scale developments up to 5 MW were allowed to opt out of the RO and instead apply for FITs. The FIT is comprised of a generation tariff, specific to technology and scale, and a non-specific export tariff (DECC 2015). This responded to critics pointing to the limitations of the RO to small actors and was interpreted as a lobbying success of small-scale energy actors and environmental NGOs (Wolfe 2016). The reforms implemented lessons learned from the hesitant deployment of renewables under previous policy conditions, yet the mixed messages of market-based and regulatory instruments further increased incoherence of energy governance structures (Kern et al. 2014; Wood & Dow 2011; Woodman and Mitchell 2011).

In 2009, local resistance against wind energy developments resulted in a just 25% approval rate of applications for wind development (British Wind Energy Association 2009, cited House of Commons 2010, pp. Ev 173–174). Planning regulation also did not resolve difficulties of grid access for renewable energy projects, an issue exacerbated by large-scale wind being the dominant project type. To minimize risk, developers would seek the windiest sites, which in turn proved difficult for integration into existing transmission infrastructures. (Woodman & Mitchell 2011: 3917) The strong focus on medium and large-scale wind energy development also stands in direct connection to the limited opening of the energy market to new actors. While associated with relatively low risks, the initial investments and resources required for project realization effectively excluded smaller actors (Gross et al. 2007). Smaller actors faced further impediments through the relatively high transaction costs of the initial policy design, which required generators to secure their demand through negotiations (Woodman & Mitchell 2011, p. 3916). The 2010 reform of the RO introducing feed-in tariffs specific to technology and scale remedied this somewhat, creating certainty for both prices and markets as well as offering a simple policy mechanism accessible to small scale actors (Woodman & Mitchell 2011: 3919). Deployment of renewable energies picked up, with an especially dynamic offshore wind sector (Toke 2011).

A 2011 White Paper announced further policy targets to underscore the government's commitment to European targets for energy and climate change mitigation, including Carbon Price Floors complementary to the European Emissions Trading System, feed-in tariffs for emergent technologies, and emissions performance standards (DECC 2011). In 2013, however, UK ranked last in a European comparison of progress towards the 2020 targets for renewables in gross final consumption, with only 5.1% achieved and 15% targeted (Eurostat 2015). Current implementation still does not see the UK on track of meeting its energy targets (Vaughan 2015), with the RE share of overall energy consumption at 8.9% in 2016, despite the share of RE in electricity generation rising to 24.6% (thus not far from the 30% targeted for 2020 in 2010 government plans). (DUKES 2016, p. 153) The failure to accomplish targets has been attributed to the drastic cutbacks on renewable energy support implemented in by the new Tory government that took office in May 2015. They reduced support for onshore wind, solar and biomass energy generation as well as rolling back investment and tax programs while at the same time reinstating the dominant position of market incumbents and the nuclear power industry (Vaughan and Macalister 2015).

Although the government published a 'Community Energy Strategy' in 2014 (DECC 2014), this did not entail specific targets or measures to foster community involvement in renewable energy projects beyond acknowledging that this should be the case (ibid., pp. 97–98). The 2015 cut backs, however, did not only roll back support for technologies. By considerably reducing the support for onshore wind and solar energy generation, as well as halting programs for home owners in reducing energy consumption and emissions, and cutting subsidies for alternative transportation, small scale and individual citizen's potential for action was once again impaired (Vaughan and Macallister 2015).

Holyrood: Assuming UK leadership in renewable energy and community support
The Scottish national parliament chose to interpret devolution of powers in renewable energy support widely, creating a specific Scottish policy arena for renewable energies (Winskel 2007, p. 186). This was partly due to Scotland's significant renewable energy resources and corresponding opportunities for technology innovation and industry development, and partly an opportunity for the Scottish Government to surpass Westminster (Mitchell and Connor 2004, pp. 1936–1937). The Scottish Government pursued energy policies through competencies in planning and community development, which had been devolved (Winskel 2007, p. 185; Smith 2007a).

In parallel to the 2003 UK White Paper on energy system reform (Department for Business, Energy and Industrial Strategy 2003), the Scottish Executive (2003) published its own ambitious set of targets that surpassed UK plans significantly. The share of renewables in energy generation was to increase to 18% by 2010 (11% in 2003), and 40% in 2020. In addition to its powers in planning and community development, the Scottish executive established institutional support for research, development and community deployment of renewables. This included the expert advisory group 'Forum for Renewable Energy Development Scotland', whose task it was to monitor policy delivery, and the 'Intermediary Technology

Institute for Energy', to financially support innovation in the energy sector. Institutional structures were also established to foster awareness for sustainable energy alternatives and aid project realization on community levels. Community Energy Scotland (CES), an independent charity derived from the government's Highlands and Islands Enterprise (HIE) supports communities in planning and realizing energy projects as well as implementing strategies for reduced consumption, energy efficiency and consumer awareness. Local Energy Scotland is a consortium of government agencies and institutions managing the government's community energy schemes SCHRI (Scottish Community and Householders Renewables Initiative) and later CARES (Community and Renewable Energies Scheme), providing assistance, as well as financial and ideational resources to communities realizing projects of renewable energies, energy efficiency or related topics.

Scottish planning policies, published in National Policy Planning Guidelines after devolution and since 2002 in Scottish Planning Policy (SPP) strategies every four years (Scottish Government 2017a, 2014, 2010), presented the first key instrument to Scottish energy governance. All SPP documents emphasize the RE sector as a key target. Key policy drivers were the economic potential of renewable energy industry development (in 2013, over 11.000 jobs were provided in the sector, Scottish Renewables 2014), as well as consumer relief and community building through community centered energy projects. Project planning was required to reflect community demands regarding both size and siting. This did not only hedge against siting difficulties as encountered in the British 'wind rush' in the early 1990s, but has laid the foundation of what to date remain high levels of public acceptance for renewable generating capacities and support for further RE deployment (Scottish Renewables 2015). Scottish ambitions to surpass UK targets in energy policy making were underlined by the Enterprise Committee of the Scottish executive calling for a comprehensive Scottish energy policy strategy (Scottish Parliament 6/30/2004). In 2007, the position of Junior Minister for Enterprise, Energy and Tourism was created, reporting to the Cabinet Secretary for Finance, Constitution and Economy who assumed overall responsibility for the portfolio (HM Government 1998, sections 47 and 49). In 2018, energy and transportation responsibility were combined, creating a Minister of Energy, Connectivity and the Islands, who reports to the Cabinet Secretary for Transport, Infrastructure and Connectivity. In 2010, the junior minister position on Environment was extended to include Climate Change; in 2014, land reform was added to the portfolio. In 2016, this was turned into a Cabinet position, creating a Cabinet Secretary of the Environment, Climate Change and Land Reform.

The 2008 Climate Challenge Fund extended community support by annually providing £10 million in grants for projects designed to either improve energy supply and/or performance of communities as well as raising public awareness and providing information on sustainable development. Climate change mitigation also provided a strong impulse to Scottish policy making beyond UK targets, resulting in the 2009 Climate Change Act (Scottish Government 2009a). Ambitious by international comparison, this Act sets a target of an 80% reduction in greenhouse gas emissions by 2050 (in respect to 1990 levels) and an interim target of 42% reduction by 2020. In addition, 80% of this must be realized through domestic emissions reductions. The 2009 Scottish Renewables Action Plan (Scottish Government 2009b) set a target of 20% renewables in overall energy use by 2020, with sectoral targets of 50% of gross electricity consumption (31% by 2011), 10% renewable transport, and 11% renewably sourced heat (Scottish Government 2009b, p. 97). Community energy, as a broad term covering projects partially or completely owned by communities and targeted at improving community economies and social structures through energy projects, again features prominently in the document with specific targets quantifying the role of community projects for sectoral development to be delivered via the support schemes (ibid., pp. 48–53). Regarding support for community energy, CARES should replace the previous program SCHRI, which supported over 400 community projects between its launch in 2007 and 2009 (Scottish Government 2009b, p. 52). CARES was officially launched in 2011. Besides offering support through advisory and network structures, this program provides financial aids for high risk projects of a generating capacity of up to 5 MW which require financial investment prior to planning consent but provide significant community benefits (Scottish Parliament 2011).

When in 2011 renewable energy generation had already surpassed the 31% target set for the electricity sector by this deadline, the Scottish government published a reformed '2020 Routemap for Renewable Energy in Scotland' (Scottish Government 2011). The 2020 target for renewables in electricity generation was adjusted to 100% (ibid. Section 1.2.3) in combination with actions on energy efficiency and demand reduction. Local and community ownership by 2020 was to cover 500 MW of electricity (ibid. Section 1.1.5). This ambitious road map was endorsed by sectoral industries, pushing government to set targets that were more stringent and continue sectoral development (ibid. Section 1.2.3). The community focus in Scottish policy strategy remains dominant. A 2015 Community Energy Policy Statement set communities as the "central tenet of future energy systems" (Scottish Government 2015a: 1), emphasizing again the transformative potential for local economies (ibid.). The target of 500 MW of renewable generating capacity in community or local ownership was achieved in 2015, five years early (Scottish Parliament 2017a). Public acceptance of community energy projects is consistently high, increasing with proximity to the developments (Cowell et al. 2011; Warren and McFadyen 2010).

Support for generating projects notably dwindles when project size goes beyond decentralized supply for Scottish communities, instead targeting to support England with its higher energy demand and lower potential generating capacity by large scale generating projects (see for example John Muir Trust 10/30/2015). This is reflected in conflict over expanding transmission capacities which will be necessary to realize overall energy sector targets as devised in both the Scottish and the UK declarations for 2020 and beyond (Tobiasson et al. 2016). If communities plan renewable energy capacities themselves, examples for the UK have shown that communities choose to install capacities beyond their individual demand to accommodate adjacent communities with less favorable natural resource conditions (PlanLoCal 2016). But the decrease in project size in community planned installations in Scotland could point to a tense relationship between communities and strategies for renewable deployment. Some Scottish communities see Scottish community energy leadership as a result of, predominantly, Scottish investments of both the Executive and individual communities, neglecting the role of British infrastructure investments (Bunting 2014). Also, while successful in establishing community energy actors in the sectoral margins, Scottish policies so far were unable to surpass dominant sectoral structures (centralization, large scale investments and utilities, etc.) institutionalized on UK level (Strachan et al. 2015).

Part II
Theory and Methods

The second section focuses on the theoretical and methodical underpinnings of this study. Chapters 2 and 3 discuss the relevant literature and develop the theoretical approach. Chapter 4 will introduce methods. Those readers most interested in the case stories and the lessons to be learned for policy-making and regulation can take a short cut and proceed directly to section II of this study, which dives into the innovation biographies of community energy. For those interested in the underlying theoretical debate of the political sciences, this chapter gives an overview of the literature, and explains how social innovation can improve our understanding of power.

This study focuses on community energy contributions to energy system change through social innovation, changes in power relations and improved capabilities through project activities. Power, in its dynamic (re)distribution and often highly context dependent shapes, is a central element of the analysis. Power references abound in the community energy literature, frequently in the context of marginalization and empowerment of small-scale actors in markets dominated by incumbents. This section begins by reviewing the community energy literature with regard to projects' abilities to alter the energy system and respective power relations (2.1). This discussion is essentially about empowerment. The second part relates this to concepts of power in the political science literature more generally, ultimately focusing on frames and framing as the processes of determining meanings and respective actions (2.2). Chapter 3 turns to social innovation as the underlying theoretical perspective assumed in this study, relating the study of power relations and the "power to do or be what one has reason to value" (Nicholls and Ziegler 2017(2015)) to the study of frames and framing as expressions of power. The chapter introduces resource mobilization theory as one focus point for studying how community energy projects assume, wield, defend or lose material and non-material resources in their attempts of energy system change. The chapter will conclude by summarizing theoretical considerations and key terms to answer the question of how social innovations in community energy projects may explain energy system change in the electricity sector. Chapter 4 introduces situational analysis as the methodical approach for analyzing data in this study.

Community Energy and Power

2

2.1 Literature Review

Academic recognition of community contributions to energy sector change is not new. Early arguments include the exploration of "soft energy paths" (Lovins 1979 (1977)), small-scale development (Schumacher 1993 (1974)), or "appropriate technologies" (Dunn 1978). This indicates two relevant dimensions of energy system change that remain dominant in community energy research to date: Changes to the physical power infrastructure, namely the acceleration and facilitation of technology deployment to increase renewable energy production, and changes to actor and decision making structures, as well as underlying norms and values. Over the past two decades, the literature has significantly expanded and grown in complexity. Research emphasizing the potential of community initiatives to facilitate and accelerate technology deployment and/or its social acceptance (Wüstenhagen et al. 2007), was widened to acknowledge simultaneous changes to technological and social structures (Schreuer 2015; Cowell et al. 2011), as well as implications of community identity, trust and social norms beyond avoiding or overcoming technology protest (Kalkbrenner and Roosen 2016; Goedkoop and Devine-Wright 2016; Devine-Wright et al. 2017). Energy systems are characterized by both a constant "reproduction" and a certain "stickiness" of their technical, cultural and ideological interdependencies (Child and Breyer 2017; Verbong and Geels 2007). The role of community energy actors, both politically and academically, is framed in reference to an agenda of sectoral change (Seyfang et al. 2014; Garud and Karnøe 2003; Ornetzeder and Rohracher 2013). Changes to the physical energy infrastructure are often discussed as processes of energy system

© The Author(s), under exclusive license to Springer Fachmedien Wiesbaden GmbH, part of Springer Nature 2021
A. Colell, *Alternating Current – Social Innovation in Community Energy*, Energiepolitik und Klimaschutz. Energy Policy and Climate Protection, https://doi.org/10.1007/978-3-658-32307-3_2

transformation in the literature (Child and Breyer 2017), while energy system transition, by contrast, refers to a "higher order perspective, one that includes the complexities of societal motivation, facilitation, cost and benefit" (ibid., p. 19). These categories of change and related power shifts will therefore provide a rough structure to the review of the literature.

Academic and non-academic debates on 'community energy' lack an unambiguous definition of the term. This is partly due to the "emergent" (van der Schoor and Scholtens 2015) character of community energy projects and their considerable diversity, but also a result of diverse underlying interests underpinning academic and non-academic publications which then adjust definitions accordingly. Consequently, research includes concepts of "community" (Walker and Devine-Wright 2008; Walker et al. 2007; Hinshelwood 2001), "bottom up" (Leibenath 2013), "niche" (Geels 2004), or "grassroots" (Seyfang et al. 2014; Hargreaves et al. 2013; Seyfang and Haxeltine 2012; Middlemiss and Parrish 2010; Seyfang and Smith 2007) energy projects. While the degree of citizen participation and involvement will vary, definitions typically refer to governance structures, asset ownership, participation in decision-making processes, intensity of (often local) activities, a shared energy agenda, and local effects or benefits as characteristic features (Kalkbrenner and Roosen 2016; van der Schoor and Scholtens 2015; Hoffman and High-Pippert 2010). Community energy projects can involve all aspects of energy related services (generation, procurement, distribution, consumption and conservation, efficiency, or storage) and share a connection to (often local) citizens (Kalkbrenner and Roosen 2016; Romero-Rubio and Andrés Díaz 2015; Boon and Dieperink 2014; Seyfang et al. 2013; Sadownik and Jaccard 2002).

Often, research compares different community energy projects within one national setting (van der Schoor and Scholtens 2015; Seyfang et al. 2014; Bomberg and McEwen 2012; Kunze 2011; Middlemiss and Parrish 2010; Walker and Devine-Wright 2008; Hinshelwood 2001). Michele Betsill and Harriet Bulkeley (2006) provide an early example of comparative research on community actions in climate governance across different national settings, which has since been expanded upon in the energy context (Nolden 2013; Bauwens 2016; Bauwens et al. 2016; Fuchs 2016; Pohlmann 2018). Studies of energy system change in Germany, Denmark and the UK, recognize citizen participation and community energy engagement as fundamental elements of sector transformation and transition (Debor 2018, 2014; Lipp 2007; Seyfang et al. 2013).[1] Energy cooperatives provide an important vehicle for citizens in the German energy transition, connecting environmental, economic, and social aspirations (Yildiz et al. 2015; Yildiz 2014; Eckhard Ott 2014; Klemisch and Maron 2010; Schröder and Walk 2014). In the UK, development trusts, small enterprises, and community groups are key examples of a more diversely organized community energy sector (Seyfang et al. 2014; Haggett et al. 2013; Harnmeijer et al. 2014). Denmark shows an even larger share of citizen ownership than Germany or the UK. Yet, ownership structures are increasingly changing from initially dominant cooperative or individual models as professional investors enter this market segment (Olesen et al. 2002; Bolinger 2001; Oluf Danielsen 1995; Jørgensen and Karnøe 1995).

2.1.1   Community Energy Projects and Energy Transformation

Three categories can organize research on the role of community energy projects in energy sector transformation: changes to the physical power system, e.g. installations and/or accelerated deployment; acceptance of these changes or the avoidance of protest; and, lastly, their economic effects.

Increasing and accelerating renewable energy roll-out
Community energy contributions to the increased and potentially accelerated rollout of generating capacity are a recurring theme in the literature (Holstenkamp and Müller 2013). These studies highlight the installed capacity owned or partially owned by community energy projects, and the ability of community energy actors to invest in installations (trend:research and Leuphana Universität Lüneburg 2013).[2] Germany, Denmark and the UK are well researched on these points and Germany and Denmark frequently feature as best practice examples of strong market positions for community energy actors (Schreuer 2015; Nolden 2013; Toke 2005). The literature identifies key context factors for community energy investments in generating capacity. These are stable and secure conditions of investment and the ability to fund projects based on appropriate loan conditions (Nolden 2013; Enzensberger et al. 2003; Bolinger 2001), cultural factors such as

---

[1] The relevance of community energy engagement for sector transformation and transition is also acknowledged beyond these countries (for example in Romero-Rubio and Andrés Díaz 2015; Sadownik and Jaccard 2002).

[2] Some organizations are also involved in other sectoral services such as energy savings (Kalkbrenner and Roosen 2016: 61).

the existence of an anti-nuclear movement in Germany (Toke 2007; Agterbosch et al. 2004; Enzensberger et al. 2003), socio-demographic factors such as the distribution of financial resources (Enzensberger et al. 2003), and organizational characteristics such as the strong dominance of cooperatives in Germany (Debor 2014; Yildiz 2014; Holstenkamp and Müller 2013; Holstenkamp and Ulbrich 2010).

Various authors have recognized a concentration and commercialization of renewable energy installations and ownership structures in front-runner regions of community energy such as Germany or Denmark. Market analysts pointed to this effect in German citizen-owned PV projects (Bettzieche 2009a, b), whereas Enzensberger and colleagues (2003) found this tendency in wind energy installations. This coincides with the market entry of large commercial actors, especially in wind energy installations, leading to a concentration of market structures and marginalizing previously dominant structures of citizen ownership (Bettzieche 2009a; Agterbosch et al. 2004; Olesen et al. 2002; Enzensberger et al. 2003). In Denmark and Germany, this development was spurred by the introduction of tenders, auctions wherein the most cost effective bidders are chosen to install generating capacity. Although policies included exemption clauses to benefit community energy actors in tendering processes (Lundberg 2019), tenders have been shown to have investor specific adverse consequences for community energy projects (Grashof 2019).

Increasing acceptance or avoiding protest

An early and influential study on acceptance by Wüstenhagen and colleagues (2007) distinguishes "socio-political acceptance" of key stakeholders and policy makers, "market acceptance" of consumers and investors, and "community acceptance" relating to distributional and procedural justice as well as trust (pp. 2684–2686). These dimensions form interrelated and mutually influential facets of acceptance (Sovacool 2009). Early debates on the acceptance of renewable energy technologies centered on socio-political and market acceptance (Huber et al. 2016 (2012); Phadke 2010), but community acceptance has recently taken center stage (Rand and Hoen 2017). Research has found comprehensive participation, for example through local or regional citizen (co-)ownership of installations, to build higher levels of acceptance (Rau et al. 2012; Musall and Kuik 2011; Zoellner et al. 2011), or avoid protest (Musall and Kuik 2011; Martins et al. 2011). Ownership of installations or shares, however, does not solve issues of acceptance or acceptability. Researchers caution that locals must trust the operation and its protagonists, and see conditions of procedural fairness fulfilled (Ceglarz et al. 2017).

The individual's contribution to energy system change is seen either through the lens of the consumer and the relevance of green electricity product design, or the power to accept or reject technological innovation (Batley et al. 2001). Studies of building acceptance or avoiding protest (as well as related policies) often lean on the concept of "communities of the affected" (Batel 2017), wherein local proximity is equated with affectedness and financial benefits are associated with acceptance. Local responses to renewable energy technologies are recognized as such if they are uttered by those 'affected' by energy system decision—presumably not made by those affected who only react—and if these facilitate or impede their realization (Batel 2017; Aitken 2010). Protest should be avoided or overcome, rather than acknowledged as a contribution (re)politicizing the debate of energy system change (Callon et al. 2009; Batel 2017; Devine-Wright and Batel 2017). Wüstenhagen and colleagues (2007) assumed that opposition to renewable energies originated at the local level, whereas socio-political acceptance was associated with public acceptance (pp. 2684–2685). The local and the national, or the community and the public, were studied as distinct spheres. But the utility of this approach is increasingly questioned (Batel 2017). This is a result of the growing recognition of simultaneous processes of technological and social change underpinning energy system change (Batel and Devine-Wright 2017; Demski et al. 2015; Bickerstaff and Agyeman 2009), and communities of intention bound by shared belief systems surpassing communities of the affected in academic assessment (Kalkbrenner and Roosen 2016).

Effects on local economies
Numerous publications highlight the financial benefits of energy projects (co-)owned by (often local) citizens (Fuchs and Hinderer 2014; Moss et al. 2015; Bomberg and McEwen 2012; Kunze 2011; Yildiz 2014; Middlemiss and Parrish 2010). However, some caution that communities may see financial benefits as compensation for adverse impacts, rather than an impulse to shift social attitudes (Cowell et al. 2011; Munday et al. 2011). Beyond financial benefits, the literature points to social benefits such as education and employment opportunities for local residents (Berlo and Wagner 2013). Yet, the dominant demographic of projects shows members often are characterized by over-average levels of education and income (Kalkbrenner and Roosen 2016), and the geographic locations typically exclude less affluent areas (for the UK see Haggett et al. 2013, for Germany see Klemisch 2014). This indicates that community energy projects might benefit a rather specific cohort and qualifies claims of empowerment (Schreuer 2015). Projects' effects may require case-by-case evaluation.

Implications of different ownership structures for local communities, for deployment and for acceptance have also been explored (Yildiz 2014; Enzensberger et al. 2003). Renewable energy projects owned or co-owned by citizens or communities take diverse legal forms, often corresponding to the legislative or institutional support for an organizational model, as well as a shared understanding of its meanings or symbols. In Germany, for example, this has resulted in a proliferation of energy cooperatives (Debor 2014; Holstenkamp and Müller 2013; Holstenkamp & Ulbrich 2010), while a similar dominance of the development trust as the organizational blueprint can be found in Scottish community energy projects (Haggett et al. 2013; Bomberg and McEwen 2012). The dominance of organizational form can be framed as a case of replication (for cooperatives in Germany, see Schreuer 2015) where community initiatives can build on material and non-material (knowledge, shared symbols, etc.) resources made available by the organizational form.

Unsurprisingly, studies of community energy effects frequently focus on instrumental rationales underpinning citizen owned energy projects, most often in reference to policy makers or project developers. Policy makers' dominant instrumental rationales echo, or drive, the importance of acceptance and acceptability on the academic agenda, especially with regard to wind energy (Warren and McFadyen 2010; Agterbosch et al. 2004; Breukers and Wolsink 2007; Walker et al. 2007). Other, often connected, rationales are the raising of awareness and education of the public (Walker et al. 2007), support for the market development of renewable energy technologies (Enzensberger et al. 2003), benefits of distributed generation (Warren and McFadyen 2010; Barry and Chapman 2009), and local or rural development agendas (Barry and Chapman 2009; Walker 2007; Walker et al. 2007). Project developers often share the above-mentioned instrumental rationale of increasing public acceptance. This is augmented by project developers exploiting their knowledge of policy makers' preferences for strong local ties (Enzensberger et al. 2003). Other instrumental rationales the literature identifies relate predominantly to the accelerated and improved deployment of renewable energies through alternative or additional funding (Enzensberger et al. 2003; Warren and McFadyen 2010; Barry and Chapman 2009), benefits from local knowledge for example in siting (Breukers and Wolsink 2007), and operational benefits through locals remarking on unusual events (Enzensberger et al. 2003).[3]

Motivations of citizens were less present in the literature (Schreuer and Weismeier-Sammer 2010, p. 41). Instrumental rationales of citizens engaged in

---

[3]On the importance of early citizen adoption of renewable energy technologies for subsequent technology development in Denmark see Jørgensen and Karnøe 1995.

community energy are strongly tied to local development or regeneration agendas (Breukers and Wolsink 2007; Devine-Wright 2011), and the opportunity of a socially and environmentally sound investment (Maruyama et al. 2007). More recently, related research in the water sector indicates that citizen and community engagement is underpinned by strong normative rationales of community self-help and environmental stewardship (Ziegler 2017b). While financial benefits and business partnerships feature in self-declared statements of the projects' utility for members, the creation of new bonds of friendship and knowledge gain outrank these factors (ibid. p. 93).

2.1.2   Community Energy Projects and Energy Transition

Community engagement in energy transformation predominantly features as an implementation of technological innovation. In the context of energy transition, a growing literature addresses the simultaneous development of technological and social innovations in community settings (Mattes et al. 2015; John Barry 2011; Frantzeskaki et al. 2013; Hoffman and High-Pippert 2010). Community energy projects are seen to provide a playground of "social innovation" (Mulugetta et al. 2010, p. 7545). Community energy actors are not implementing or adapting technological innovations; rather innovations, which include social parameters, originate in community or grassroots actors (Schreuer 2015; Seyfang et al. 2014; Hargreaves et al. 2014; Hielscher et al. 2013).

Much of the research on citizens and communities making use of technical opportunities of small scale renewable energy generation implicitly or explicitly assesses the mobilization of resources by these actors (Gubbins 2007, p. 82; see also; Schreuer 2015; Nolden 2013; Seyfang et al. 2013; Bomberg and McEwen 2012; Middlemiss and Parrish 2010). Local public authorities and the resources they provide to activities of climate change mitigation and renewable energies are one of the more dominant comparative units in the research (Betsill and Bulkeley 2006).[4] Research often focuses on the mobilization of resources based on available local and national financial support structures (Seyfang et al. 2013; Brickmann et al. 2012; Boomsma et al. 2012; Hinshelwood 2001), as well as policy support more generally (Nolden 2013; Brickmann et al. 2012; Middlemiss and Parrish 2010; Gubbins 2007). As new forms of cooperation between local public authorities and citizen or community-owned projects proliferate, so too does related

---

[4]On the role of local governance see also Goldthau 2014; Kunze 2011; Pohlmann 2011; Peters et al. 2010.

research (Debor 2018, pp. 118–126; Becker et al. 2013, p. 24; Kunze 2011). Trust or social capital as well as shared social norms and networks find increasing recognition as "symbolic resources" mobilized by community energy projects (Seyfang et al. 2013, p. 985; Goedkoop and Devine-Wright 2016; Hielscher et al. 2013; Bomberg & McEwen 2012; Brickmann et al. 2012; Walker et al. 2010a; Hinshelwood 2001; Devine-Wright 2001). The literature discusses the explicit mobilization, and the implicit effects, of these symbolic resources at the level of the individual and the community.

The Individual: (Energy) Citizenship and Responsibility
The transition towards renewable energies is widely associated with new roles for the public (Walker and Cass 2007). Yet, the proliferation of community energy projects and their ascribed beneficial effects for sector diversification and community building (Wüstenhagen et al. 2007) stood against largely unreflective practices of energy consumption and passivity with respect to sector transition among the larger public (Ricci et al. 2010; Walker and Cass 2007). This has been characterized as a function of the strong institutional 'stickiness' of the energy sector, wherein conventional systems of generation, distribution and consumption are institutionalized to seem beyond consumer influence (Heiskanen et al. 2010b; Heiskanen et al. 2010a). Ideas for individuals as newly empowered citizens coincided with realities of rather powerless consumers. Power generation in conventional energy systems is physically remote from consumers, which creates psychological remoteness (Shackley and Green 2007). This kind of structural disempowerment of consumers in infrastructure design is not unique to the energy sector but also manifests in other "networked infrastructures" (Neuman 2006, pp. 3–4) such as water (Ziegler 2017b, pp. 8–9). Community engagement in energy systems, on the other hand, may foster emotional or psychological engagement with renewable energies (Rogers et al. 2012, p. 243) and raise awareness (Romero-Rubio and Andrés Díaz 2015). This is connected to the physical manifestation of infrastructure (in this case energy) services in close proximity—seeing might really be believing (Rogers et al. 2012, p. 243; Ziegler 2017b). This underscores the recognition of material participation as public participation in transition processes (Marres 2015 (2012), 3–4, 11; Ryghaug et al. 2018).

Research on individual aspects of community engagement also investigates connections to concepts of citizenship and identity. Engagement in community energy projects has been found to promote energy responsibility (Frantzeskaki et al. 2013, p. 102). Beyond potentially bridging the gap between attitude and behavior in energy consumption (Kalkbrenner and Roosen 2016), this also points to concepts of "ecological citizenship" (Dobson 2000, p. 43, 2006, pp. 447–449) or "energy citizenship" (Devine-Wright 2007, p. 63) which include concepts of political obligation and civic responsibility. This line of research integrates insights from the socio-cultural embeddedness of energy system transition and applied social and environmental psychology. Beyond "situat[ing the] individual human experience and action within processes of socio-cultural communication and contestation" (Devine-Wright 2007, p. 63; Alisat and Riemer 2015), this may re-politicize the study of psychological engagement to ask questions of governance (Ockwell et al. 2009).

The Community: Trust, Social Norms, and Community Identity
Public participation in energy projects has been found to increase social capital within a community (Walker et al. 2010b), where interpersonal networks and relationships foster trust and reciprocity and secure societal functions (Robert D. Putnam 1993, pp. 170–171). Pre-existing social capital or cohesion seems to predicate and ground these projects (Kalkbrenner and Roosen 2016; Haggett et al. 2013). Strong neighborhood ties in general promote engagement in community energy (Hoffman and High-Pippert 2010). Trust, the "mutual confidence that no party to an exchange will exploit the other's vulnerability" (Sabel 1993, p. 1133), has been found to be a key factor in the development of community energy projects as "both a necessary characteristic and a potential outcome" (Walker et al. 2010a, p. 2657; Wiersma and Devine-Wright 2014). Trust and civil engagement are connected by positive correlation (Putnam 2000, p. 137), but trust is also positively connected to volunteering and overcoming group boundaries (Delhey and Newton 2003). Trust also facilitates financial investments (Ding et al. 2015). General levels of trust have been found to interact positively with willingness to participate in community energy, with trust and strong social norms mediating the effect of community identity on willingness to participate completely (Kalkbrenner and Roosen 2016, pp. 65–67). Yet, trust remains "under-appreciated" in the study of energy systems (Greenberg 2014) and analysis of trust specifically in the context of community energy has so far often been neglected.[5]

Social norms, or a person's belief of social (dis)approval of a specific behavior (Ajzen 1991, p. 195), are also studied for their influence on participation and community energy engagement (Owens and Driffill 2008; Bamberg 2003). Social norms have been studied in respect to pro-environmental behavior and consumption in the energy sector generally (Dwyer et al. 2015; McDonald and

---

[5]Prominent and important exceptions are Walker et al. 2010a and Yildiz et al. 2015.

Crandall 2015; Gadenne et al. 2011; Thøgersen and Grønhøj 2010). Recent studies demand more specific research on the interconnections of social norms and their behavioral influence and energy policy design (Rathi and Chunekar 2015), as well as acknowledgement of the more complex interactions of individual and social factors influencing pro-environmental behavior (Gifford and Nilsson 2014). Social norms are expected to demand respect, specifically, in combination with personal or institutional leadership (Mulugetta et al. 2010, p. 7544). If respected community members actually observe or even appear to follow specific norms, this lends power to the subjective norm (Kalkbrenner and Roosen 2016). This is underscored by research in community engagement in the agricultural sector, wherein legitimacy is identified as a linchpin resource to community projects enabling the mobilization of other important resources (van Oers et al. 2018). While energy related social norms have been found to positively influence willingness to participate in community energy projects in a general context (Kalkbrenner and Roosen 2016, pp. 65–67), their specific implications within structures of community energy engagement warrant further academic inquiry (Batel 2017).

Trust and social norms underscore the importance of a person's social ties to their community when assessing their enthusiasm to contribute to or comply with community actions (Hoffman and High-Pippert 2010; Sydney Verba 1995; van Vugt and Cremer 1999). Both are positively influenced by a strong sense of community (Bomberg and McEwen 2012). Community identity, as "feelings of attachment to the community, taking pride in the community, and having friends within the community" (van Vugt 2002, p. 797), may shift self-oriented behavior towards community-oriented actions (van Vugt 2001). Much as with levels of trust in the community, research has found solidarity and community identity to be both a facilitator of community energy engagement and its outcome (Bomberg and McEwen 2012; van der Horst 2008). Hoffmann and High-Pippert have found a shared spirit of making the community a "better place" to be an important success factor of community energy projects (2010, p. 7570), and Hagget and Aitken (2015) highlight the importance of community identity in collective action for energy system change. But what 'community identity' means as well as the explication of its effects or effectiveness often remains vague (Bomberg and McEwen 2012). Also, research on the willingness to participate among the general public indicates that trust and social norms may mediate the influence of community identity completely (Kalkbrenner and Roosen 2016, pp. 65–67).

Community engagement, not only in the energy sector, relies on proactive participation in relevant processes and structures of decision making (Sovacool and Brown 2015; Hoffman and High-Pippert 2010; Walker and Devine-Wright

2008; Florin and Wandersman 1990). With respect to community energy, long-term engagement and participation of the locality, as well as financial resources, expertise and government support structures are identified as key factors (Bomberg and McEwen 2012; Mulugetta et al. 2010). Yet, while the relevance of participation for sustained collective action is widely acknowledged (Foster-Fishman et al. 2013), academic understanding of "effective involvement and collective pro-environmental action" is still flawed (Kalkbrenner and Roosen 2016, p. 61; Alisat and Riemer 2015; Broman Toft et al. 2014). Community benefits seem to outweigh personal benefits in motivating long-term participation (Hoffman and High-Pippert 2010). Leadership is discussed controversially as a driver of participation and a potentially misleading focus on agency over structure.[6] Political process approaches overemphasized structure (Goodwin and Jasper 1999). Resource mobilization theory overemphasized agency (McCarthy and Zald 2001). Morris and Staggenborg (2004) in an attempt to integrate structure and agency in analyses of leadership in social movements argue in favor of reciprocal frameworks, acknowledging leadership types as well as stages of movement development. They also point to the dynamic interactions of social movements and their surroundings, wherein political opportunity structure is not necessarily a context factor but is also itself potentially influenced by social movements (ibid. p. 191).

Leadership in community projects has been studied in a range of issue areas,[7] but systematic analysis of leadership for community energy is rather recent and underrepresented (Martiskainen 2017). Analyses of leadership for community energy emphasize the reciprocity of agency and structure both within the project, as the ability to engage and sustain meaningful participation of members, and vis-à-vis its surroundings in their ability to mobilize material and non-material resources for their projects (Martiskainen 2017, p. 86; Onyx and Leonard 2011). They underscore the "diverse and non-linear manner" (Martiskainen 2017, p. 86) in which community energy engagement unfolds (Seyfang et al. 2014; White and Stirling 2013).

Normative rationales underpinning community energy engagement include an agenda of environmentally sound modes of energy generation, distribution and consumption (Flieger and Klemisch 2008; Maruyama et al. 2007; Agterbosch et al. 2004); improvement of community cohesion or the support of

---

[6]Morris and Staggenborg 2004 provide a detailed overview of the study of leadership in the social movement literature.

[7]See Bukoski et al. 2015 for examples from the educational sector, Trapence et al. 2012 for health, Sullivan 2007 for local government, Bénit-Gbaffou and Katsaura 2014 for local politics.

local/community identities (Walker et al. 2010a; Rogers et al. 2008); democratization of the energy sector and increased legitimacy of transition processes (Flieger 2011; Klemisch and Maron 2010; Breukers and Wolsink 2007); and empowerment of the individual (Flieger and Klemisch 2008). Flieger and Klemisch (2008) show how organizational form and agenda connect, highlighting that energy cooperatives in Germany most often explicitly pursue political goals of democratization and empowerment, while limited partnerships frequently pursue a predominantly economic agenda (Enzensberger et al. 2003). For the British case, by contrast, Walker and colleagues (Walker et al. 2007) show how a less closely determined understanding of 'community energy' and respective rationales coincide with more experimentation in ownership models. This could vary across the UK, as the proliferation of development trusts with distinct agendas of local development and clear rules of collectivizing benefits in Scotland indicates (Pohlmann 2018).

2.1.3   Research gap: Stories of Social Innovation

The literature review focused on how community energy projects connect to changes of physical power systems and power relations in the energy sector, organizing the literature into two strands. One emphasizes the ability of community energy projects to foster and accelerate (physical) power system changes that result from technological changes in the energy system. A project's value is instrumental, and the ability to change power relations derives from technological opportunity, and the capacity to benefit from political or financial incentives. Beneficial effects of community engagement beyond physical changes to the power system further build the case for political support or financial incentives.

The power of community energy projects can be summarized in their ability to create a frame of physical power system change that finds public approval (Walker et al. 2014). The second strand emphasizes the ability of community energy projects to create alternatives to physical power systems that incorporate technological and organizational innovation. Project value is not tied to instrumental or normative rationales, although it is acknowledged that both may manifest in empirics, and technological as well as political or financial opportunity structures are included as potential sources of (or impediments) to community capabilities, alongside other socio-organizational factors. The power of community energy projects can be summarized in their ability to create frames of physical power system change immediately including changes to structures of decision-making and agenda setting, and the values or meanings attached to them. Both strands share the assessment that community energy projects can alter the image of energy system change, in its technological and/or organizational implementation, while providing different answers to the question underpinning all processes of conciliating diverse interests: "Who gets what, when, how?" (Lasswell 1936). This question can also structure the summary of gaps in the literature.

Who Gets What?

The community energy literature is strongly shaped by ideas of strategic niche management (SNM) and the concept of "strategic niches" (Schreuer 2015, p. 28), which provide protected spaces for radical innovation to develop beyond the pressure of the market (Kemp et al. 1998, p. 186). Niche development is facilitated by respective "managers", and occurs through learning, the establishment of actor networks, and shared expectations (Schot and Geels 2008, p. 540).[8] While not always explicit, the view of community energy actors as managers of strategic niches wherein the deployment and acceptance of renewable energy technologies can be grown is particularly influential in the study of community energy projects in energy sector transformation (Batel and Devine-Wright 2017; Matthew Cotton and Devine-Wright 2011). This perspective offers important insights into the robustness of niche innovations, their challenges and opportunities. But it falls short of understanding how community energy projects create frames wherein implementation of technology innovations may be nurtured because it neglects projects' socio-organizational dimensions.[9]. Authors that study civil society or grassroots actors as origins of innovation highlight the mismatch of SNM theory and empirics, indicating that SNM perspectives prove unfit to analyze "grassroots innovations [...] *neither strategic nor managed*" (Seyfang et al. 2014, p. 41, original emphasis). Development occurs in the absence of learning (Seyfang and Longhurst 2013a, pp. 888–889) or converging expectations (Seyfang and Longhurst 2013a, pp. 887–888; Smith 2007b, pp. 441–443).

Strategic Niche Management (SNM), developed as a theory of socio-technical change and a policy approach, aligns well with a research agenda following the instrumental rationales of policy makers and project developers. But it assumes an outside, managerial perspective which is of limited help in understanding bottom-up innovation in civil society context and its often fragmented and highly varied, contingent configuration (White and Stirling 2013, pp. 839–840). Strategic Niche

---

[8] See Kemp et al. 2001 for a study of the Danish wind energy sector through the SNM lens.

[9] Smith offers a critical appraisal of the utility of SNM in studying innovation in civil society settings in the organic food sector (2007, pp. 441–443), Seyfang and Longhurst 2013b on community currencies (see also: Seyfang and Longhurst 2013a).

Management also does not capture variability in meanings and desires underpinning community energy settings (for diversity of meanings and interpretations
of 'community energy' in the UK, see Walker and Devine-Wright 2008). While
the literature of innovation studies acknowledges that different actor groups may
hold diverse perspectives regarding the functions of a specific innovation (Pinch
and Bijker 1984, pp. 421–424), SNM seeks convergence, shared expectations and
alliances. This fosters an understanding of strengths and weaknesses, but discards
variability and "interpretative flexibility" in community energy groups (Schreuer
2015, pp. 28–30). This neglect of social embeddedness of technology and respective sector change is not unusual in research on the deployment of technological
innovation. A literature review on vehicle-to-grid services for electric driving
and renewable energies indicates that here, even research in public acceptance or
awareness is lacking, with larger socio-technical system considerations neglected
almost entirely (Sovacool et al. 2018). This indicates an under-complex academic representation of sector transformation and transition (Sovacool et al. 2018,
pp. 12–15; Child and Breyer 2017), which is especially unfortunate in sustainability sciences given the number and urgency of transition processes underway.
This is all the more surprising, as in innovation studies socio-technical embeddedness and the relevance of narratives, expectations and visions is well established
(Benjamin K Sovacool et al. 2018, p. 13; Geels and Verhees 2011; Schot and
Geels 2008; Borup et al. 2006). The study of the niche has since been extended
to conceptualize different mechanisms of protective spaces. Smith and Raven suggest "nurturing" and "shielding" as key protective functions of the niche, while
"empowerment" is suggested to capture analyses of agency and power previously
neglected in SNM (2012, pp. 1025–1026).

   The framing of community energy projects as 'niche' actors resonates with
challenger-incumbent representations of energy sector change, wherein community energy projects feature as marginalized actors facing incumbent market actors
in a setting of unevenly distributed power resources, namely money and decision-
making or agenda setting power of vested interests (Schreuer 2015, p. 161).
Challengers can either be blocked by incumbents, suffer further marginalization
by the incumbent's uptake of niche innovations while the original innovator is
pushed aside, or be successful through his or her empowerment in a platform of
collective action (Seyfang and Smith 2007), and the replication of challengers'
socio-organizational approaches (Debor 2018). Similar to studying community
energy actors as niche actors, challenger-incumbent settings suggest a distinct
position within the energy sector and a corresponding narrative. Empirically, this
is called into question by cases wherein neither the organizational setting nor the

narrative chosen correspond to the image of challenger and incumbent (Pohl-mann forthcoming, also olegeno in this case study). Projects themselves risk inadvertently creating a streamlined version of events neither historically true nor necessarily helpful for understanding or fostering sectoral change (Papazu 2018).

Ziegler (2017b) offers an alternative understanding of the niche. It is innova-tive in its departure from both the managerial multi-level perspective prevailing in SNM studies, as well as the concept of niches as protective spaces (Smith and Raven 2012; Smith et al. 2016). Ziegler considers community engagement in "ecological niches" (2017b, p. 21). Actors within the niche do not attempt to establish new practices beyond contextual factors but instead exist precisely in their interaction with contextual factors. The specific configuration of these inter-actions may present a comparative advantage, or protection, for the actors of the niche. But its development is contingent, rather than managed, and directed at maintaining or adjusting its niche position, rather than creating a dominant model for practices beyond the niche (ibid.). This interpretation of the niche is supported by empirical studies of leadership in community energy, underscoring the recipro-cal relations of community project leaders with their socio-organizational context (Martiskainen 2017).

An alternative suggestion for understanding community actors in energy sys-tem change are so-called communities of relevance (Batel 2017, p. 357). Studies of social acceptance, explicitly or implicitly employing the frame of the niche, frequently focus on communities of the affected. This concentrates attention on the micro level of people's interactions with renewable energy technologies, dis-regarding larger socio-technical relations (Batel and Devine-Wright 2017). The concept of communities of relevance, instead, acknowledges the 'public' to have direct interests and be directly affected (e.g. by renewable energy technology development) but also as an unaffected community as it cannot participate in respective debates or deliberative processes which are often macro-level (Batel 2017, p. 357; Marres 2015 (2012)). This concept of community, and the public, counters the tendency to focus on the local (e.g. protest is local, so its remedy must be local also) and to compartmentalize participation (e.g. what are domestic settings of every-day practices). Instead, it acknowledges the multi-faceted enga-gement of communities and citizens with energy system change (Chilvers and Longhurst 2016, p. 587). Studying energy transition processes within communi-ties of relevance adopts a relational ontology (Shose Kessi and Carolina Howarth 2015, 2015, 2015; Batel et al. 2013), in line with the simultaneous study of socio-organizational and technological innovations in processes of system transition (Batel and Devine-Wright 2017; Cotton and Devine-Wright 2011).

The research gap of 'who gets what' in energy system change therefore refers not to the ambiguity of definitions of community energy, but instead to a lack of ambiguity in conceptual assumptions in the literature. Narrowing the organizational position of community energy actors by placing them in 'niches', as well as suggesting dominant frames of their actions in challenger-incumbent settings clouds academic assessment of the diverse contributions of community engagement to system change. Instead, analytical frameworks should consider heterogeneous factors of the research situation (individual and collective actors, formal and informal institutions, frames, technologies and installations, etc.) and their interactions in negotiations for technological and organizational changes underpinning the *Energiewende*.

...When, How?

Numerous studies explain the ability of community energy projects to exercise power in energy system change—to mobilize support or protest in reference to renewable energy technologies, influence policy decisions or agenda setting, or to realize their own projects—by analyzing the mobilization of resources. Attention focuses primarily on the mobilization of material resources such as money or equipment, often in combination with policy-based financial incentives, and the mobilization of organizational resources in reference to the legal form community organizers choose. This echoes a strong focus on an organization's ability to access resources in combination with the political opportunity structure in the resource mobilization literature (Edwards and McCarthy 2004, p. 125; McCarthy and Zald 2001, pp. 544–545; Jenkins 1983, p. 530; McCarthy and Zald 1977), as well as a focus on different types of material resources (Jenkins 1983, p. 533). In the literature on resource mobilization, the study of different resource types has since broadened to include land or other natural resources (Tilly 1978), as well as non-material resources such as shared interpretive frames (referring to shared meanings guiding the interpretation of events and actions, McCarthy and Zald 2001), and knowledge resources (Schreuer 2015).

Shared symbols or values such as a sense of community, trust, or a democratic pretense in project actions are gaining attention in some contributions to the community energy literature (Batel 2017; Kalkbrenner and Roosen 2016; Goedkoop and Devine-Wright 2016; Schreuer 2015; Greenberg 2014; Walker et al. 2010a). However, explanatory power with regard to how trust matters to community energy projects remains limited. Kalkbrenner and Roosen (2016), for example, analyze the relevance of trust, social norms and community identity for willingness to participate in community energy projects (finding that general trust and favorable social norms matter, while community identity does not).

Yet, the authors' questions are unspecific to community energy settings and were posed to the general public, thus providing limited insights into the motivational structures of community energy projects (a shortcoming the authors acknowledge themselves, p. 67). Bomberg and McEwen (2012) in a study involving over 100 community energy projects in Scotland seek to understand the role of symbolic resources in emergence and sustenance of community energy engagement, in an effort to counterbalance the dominant focus on structural and material resources typical of the study of opportunity structure (pp. 436–37, 440–442). Yet, their study is exemplary for highlighting the importance of non-material resources while not explaining how projects define and subsequently mobilize these resources. The authors point to 'community identity' and 'autonomy' as key symbolic resources, without explaining how respective projects conceptualize or wield either (ibid. 440–442). Similarly, while Schreuer (2015)in a study of citizen owned wind energy projects in Germany and Austria considers various resource types, neither the definition and mobilization of non-material resources nor their relation to material resources in project emergence or establishment is explained (p. 161). Moreover, non-material resources such as trust, social capital or community appear both as conditions and outcomes of community energy projects in the literature (Walker et al. 2010b, p. 2657).

The analysis of material and to some extent organizational resources has garnered important insights into the ability of community energy projects to mobilize especially financial support for energy system change. But the community energy literature stands to gain from an adaptation of resource mobilization frameworks enabling on-par analysis of non-material resources with respect to their definition, mobilization, and maintenance or alteration by community energy actors. Moreover, comparing the practices of resource mobilization indicates similarities and differences across projects rather than across countries, as in most previous analyses. This provides new insights into the effects of policy support or impediments, and projects' abilities to act upon them.

The research gap this study aims to bridge lies in the understanding of the social and organizational power community energy projects hold by providing alternative frames of energy system design and mobilizing material and non-material resources accordingly. Social change, particularly when attempted by civil society based actors (Seyfang and Smith 2007, 597, 599–600; Hess 2013, p. 848), almost always implies a change of power relations. The literature review has shown that two interpretations of power in reference to community energy actors underpin most analyses: concepts of empowerment of civil society or community actors, emphasizing the possibilities community energy projects hold, and concepts of powerlessness, emphasizing limitations of community energy actors within the established structures and actor constellations of the energy sector (Schreuer 2015, p. 161). However, while implicitly or explicitly referencing concepts of power frequently, the theoretical considerations underpinning the study of power in the community energy literature often remain limited (ibid., Shove and Walker 2007, 2008; Avelino 2011). Before turning to theories of social innovation as the suggested explanatory approach within this study, the following section briefly refers the study of community driven energy system change to concepts of power in political science.

2.2   Concepts of Power

Studies of power, very basically, refer to the struggle over diverse and typically scarce resources—hence the laconic caption of Lasswell: 'who gets what, when, how?' (1936). Power can be understood as a productive, enabling or obstructive, preventative force. It can be analyzed in reference to the different forms it may take, such as military, political or economic power. It may be understood in its distribution, concentrated among actors or diffused within institutional structures. And it may be understood as a static image of power distribution at a given point in time, or as dynamic processes of (re)assuming power over time.

This study focuses on social innovations within community energy projects, based on a productive definition of power as capabilities "to do or be what one has reason to value" (Nicholls and Ziegler 2017(2015), p. 13), as well as a relational understanding wherein power is dynamically, and typically unequally, (re)distributed among actors and institutions (ibid.). Social innovations focus on changes to these power relations. The academic goal is therefore to assess the explanatory potential of these productive processes of power assumption and redistribution for energy system change. This refers to material changes of the energy sector (such as energy technologies, installations and services, or ownership and investment structures) in line with the literature discussing material participation as a form of public participation (Batel 2017). And it refers to changes in the social structures embedding technologies, installations, and services: policies and respective decision making, agenda setting, and the frames of reference these are subjected to. Rather than presenting an exhaustive discussion of the study of power in the literature, this section will therefore briefly relate the study of social innovations to concepts of capabilities, decision-making, and framing.

2.2.1   Capabilities, Coercion and Conciliation

Power is typically interpreted in terms of relations and the, often unequal, distribution of power. Dahl describes his "intuitive idea of power [as] something like this: A has power over B to the extent that he can get B to do something that B would not otherwise do" (1957, pp. 202–203). Power can be interpreted through the sources of power, structuring the kinds of abilities attainable. This immediately connects to the relative powers assigned to actors and structures in theoretical concepts. Research on community energy groups can be framed in an actor-oriented definition of power, connecting actors to power sources and their ability to influence outcomes in their interest (Giddens 1984). However, outcomes of energy system design can be influenced by multiple factors such as climatic or topographic characteristics, technology specifics, or price developments in energy resource markets that affect outcomes and qualify the relative power of actors (and structure). Returning to Dahl's concept of power, power is not (only) connected to actors but enacted in relations. This includes powers diffused in structure, which feature not as "power in and of itself, but rather a power source for an actor" (Krott et al. 2014, p. 37; Scharpf 2000).

Procedural qualities of power refer to how power is exerted, and have come to be discussed as the 'three faces' of power in the literature. Dahl's (1957) relational perspective of power as the ability to achieve behavioral change in others is the first face. Bachrach and Barratz (1962) suggest "two faces of power" (p. 947). Power is not only the ability to make decisions and achieve compliance in others, but also the ability to control or prevent decisions by manipulating the context of decision-making (ibid. p. 948). Lukes (2005(1974)) adds a third face: the ability to influence what others think or feel with respect to an issue (p. 24–25). This complements an actor-oriented approach to power by acknowledging structural and discursive powers, strengthening the relational understanding of power. Boulding (1990 (1989)) assessed these three faces of power with respect to how they are exerted, describing destructive powers or means of threat, productive powers of exchange, and integrative power or love. He concludes that integrative powers prove most important: "The stick, the carrot, and the hug may all be necessary, but the greatest of these is the hug" (Boulding 1990 (1989), p. 250).

In reference to power sources and processes, community energy projects can indeed appear as relatively powerless actors. Civil society actors engaging in energy system change can often draw on a limited keyboard of political and economic power sources. To exert negotiating powers of 'stick, carrot, or hug' would also require the counterpart to recognize community energy projects as actors sufficiently powerful to wield these tools. Relative power, and/or potential for empowerment, of community actors may lay in community energy projects' ability to either define contextual dimensions of decision-making, or influence what others think or feel about an issue up for decision.

2.2.2 Decision-Making and Agenda Setting

Power can be studied based on the questions of "who participates, who gains and loses, and who prevails in decision-making" (Polsby 1963, p. 55). Academically, the decision-making process was initially considered as an open process wherein any interest could find representation (Dahl 1961, pp. 91–93): "the independence, penetrability, and heterogeneity of the various segments of the political stratum all but guarantee that any dissatisfied group will find a spokesman" (ibid., p. 93). The fragmentation of political government secured that interests of minorities were "usually attended to" (Polsby 1963, p. 118). The idea that conscious actions of interest groups or their spokespersons could influence decision-making processes introduced the concept of actor preferences to the study of decision-making.

The concept of agenda setting, by contrast, refers to the ability to prevent decisions, or non-decision-making, and the control of the political agenda (Bachrach and Baratz 1962; Schattschneider 1975 (1960)). Agenda setting research acknowledges the influence of issue definition, analyzing campaigning or the ability of promoting an issue on the decision making agenda and lobbying or the ability to prevent certain decisions. Agenda setting research, moreover, assigned power to public opinion by relating the setting of the political agenda to what was prioritized in public discourse and news coverage (Kosicki 1993, p. 116; McCombs and Shaw 1972). Various approaches discuss different ways of public discourse influencing agenda setting (Strömberg 2004; Hawkins 2002; Rogers and Dearing 1988; Manheim 1987). News coverage may influence public perception of policy making, in turn creating policy expectations, which then influence agenda setting (Strömberg 2004; Baumgartner and Jones 1993). Or it may influence policy makers' perceptions of the public opinion (Rogers and Dearing 1988, p. 579), acting as a "proxy for public opinion" (Bakir 2006, p. 69; Hawkins 2002).

The concept of social innovation combines an actor-oriented understanding of power as the capabilities to do or be what one values, with a structural setting of power embedded in relations. Social innovation refers to such changes of power relations through improvement of related capabilities. While concepts of decision-making would require community energy actors to influence the energy policy process or utility decisions directly, agenda setting research focuses on their ability to influence public opinion and discourse and determine the structure wherein decisions may be made. Examples of this in the literature are the analysis of how community ownership influences acceptance of renewable energy projects (Walker et al. 2014), or the shaping of public opinions of energy system change (van der Schoor and Scholtens 2015; Baumgartner 1989). While this can indicate empowerment of community energy actors vis-à-vis political and economic decision makers of the energy sector, it does not sufficiently capture interactions of community energy projects with their immediate communities. These form an integral part of their activities and should therefore be included in analyses of capabilities attained or lost within them (Pohlmann forthcoming). This refers to Lukes' (2005(1974)) concept of the third face of power, the ability to influence what people think or feel. This will be the focus of the final section contextualizing the study of social innovation in concepts of power in the literature.

2.2.3   Frames and Framing

The study of frames and framing connect to an understanding of power as the ability to shape thoughts, needs or requirements of individual and collective actors (Lukes 2005(1974)). Frames are "schemata of interpretation", enabling individuals to "locate, perceive, identify, and label" the instances of their everyday lives (Goffman 1974, p. 21). Frames provide organizational principles guiding actions of the everyday life, which in turn "implies agency and contention at the level of reality construction" (Benford and Snow 2000, p. 614). Frames result from, often collective, negotiations over shared meanings—the process of framing (Gamson 1992, p. 113). Contention refers to both the struggle over meaning within a frame and the realization that new frames may challenge existing frames. New frames must prove themselves in competitive settings of established meanings and belief systems (Finnemore and Sikkink 1998, p. 897).

Successful establishment of a frame as a shared object of meaning is attributed to its ability to "be effective or 'resonate'" (Benford and Snow 2000, p. 619; Snow and Benford 1988). Frame resonance refers to the ability of a frame to effectively mobilize support of a larger community. Benford and Snow (2000) attribute this to a frame's credibility and relative salience (p. 619). Credibility hereby refers to the degree of consistency between a social movement organization's frames and actions, the empirical credibility of a frame measured in the ability to connect empirical evidence to the frame's claims,[10] and the credibility of those articulating and promoting the frame (Benford and Snow 2000: pp. 619–620; Snow and Benford 1988). Salience of a frame is assessed in reference to the targets of mobilization, referring to how important the values underlying a frame's claims are to its intended audience, how well the frame resonates with the daily experiences of said audience, and the embedding of the frame in established and culturally shared narratives (Benford and Snow 2000: pp. 620–22). While traditional interpretations of power differentiate between the capability to assert power over others and the ability to influence or persuade (Heywood 2002, pp. 10–11), frames and framing blur the lines between coercive and persuasive powers (Finnemore 1996, p. 341; Finnemore and Sikkink 1998, p. 914; Payne 2001; Lynch 1999).

Concepts of frames and framing were established in sociological research on social movements quite early. The early political science literature adopted these concepts primarily as a means of explaining mismatch between the intent of a policy and its subsequent implementation (Schön and Rein 1995 (1994)). They have since grown in relevance in the political science social movement literature, establishing themselves alongside studies of resource mobilization and political opportunity structure (van Hulst and Yanow 2016, pp. 100–110). Frames and framing are employed both as methodologies of explaining problem settings, and as a concept for storytelling, wherein frames provide an interpretive context to bind features of the research situation in coherent patterns (Rein and Schön 1996). Frames and framing feature in the study of social movements, in reference to the development of shared belief systems (interpretive frames, McCarthy and Zald 2001) as well as the analysis of social movement organizations. Benford and Snow's (2000) concept of collective action frames builds on underlying shared meanings to organize everyday experiences or individual actions, and phrases "action-oriented sets of beliefs and meanings that inspire and legitimate the activities and campaigns of a social movement organization" (p. 614).

Agency features importantly in this respect. The ability to "connect new normative ideas to established ideas" (Payne 2001: 39) creates a non-material basis of power relatively independent of material power sources (Lynch 1999: 265). Actors capable of forging these connections are therefore attributed important mobilizing powers (Finnemore and Sikkink 1998, pp. 906–907; Nadelmann 1990, p. 482). This underscores the importance of considering the processes of defining and mobilizing non-material resources in community energy contexts, as well as

---

[10]This does not refer to the factual credibility of the frame as such, but to the ability to empirically verify the claim in the eyes of the intended constituency (Benford and Snow 2000: 620, Snow and Benford 1988, Gamson 1992). Empirical credibility "is in the eyes of the beholder" (Jasper and Poulsen 1995, p. 496).

the study of individual and community leadership (Walker and McCarthy 2010). The study of interpretive frames as non-material resources hereby refers not to the, arguably questionable (Paynes 2001), ability of frames and framing processes to alter norms or belief systems. It refers to the capabilities these non-material resources provide to community energy projects, and the relative importance of material and non-material resources.

Frames and framing offer an analytical approach to community energy projects that focuses not on powerlessness measured in actors' lack of resources but on empowerment understood as the ability to alter power relations. By creating new frames for energy services, or connecting energy-related activities to established frames, community energy projects can alter power relations of the energy system. This refers not only to their potential to influence policy agenda setting or decision making processes, but also to processes of dynamic (re)distribution of power within projects, as shared meanings of energy system change are negotiated. This dynamic understanding of power underpins this study of social innovations in community energy projects. The explanatory power of concepts of social innovation for energy system change thereby connects to the power of community projects to generate alternative references structuring the experience of and the structure for energy system services.

Biographies of Social Innovation and Resource Mobilization

3

3.1 Social Innovation

Social innovations, while including material power resources such as money, technologies and installations, consider non-material power resources, such as decision-making processes, ownership structures, or narrative as equally important for system design and change. The community energy literature is increasingly referencing social innovations to recognize the agency of civil society or grassroots actors ('grassroots innovations', Smith et al. 2013; Hargreaves et al. 2013; Seyfang & Smith 2007). Avelino and colleagues (2015) connect the study of social innovations to frames of sustainability transitions and community engagement, studying social innovations not as "a dimension of technological innovation [but] an object of innovation in itself" (p. 3). But the study of social innovation so far focuses primarily on organizational innovations and sectoral upscaling (Smith et al. 2016). This academic focus is in line with instrumental rationales dominating policy debates on social innovation, which are similar to those on community energy introduced above. The European Commission (2019) defines social innovation as "new ideas that meet social needs, create social relationships and form new collaborations", which could be "products, services or models". Yet, the only objective mentioned in this respect refers to their market uptake (ibid.). Previous studies of the social embedding of technological innovation predominantly focus on their organizational integration (Schreuer 2015, 28, 185; Walker and Cass 2007). Drawing on Nicholls and Ziegler (2017 [2015]: pp. 8–9), this study considers social innovations as intentional changes of power relations which occur through interventions to the relations between social networks, institutions and

© The Author(s), under exclusive license to Springer Fachmedien Wiesbaden GmbH, part of Springer Nature 2021
A. Colell, *Alternating Current – Social Innovation in Community Energy*, Energiepolitik und Klimaschutz. Energy Policy and Climate Protection, https://doi.org/10.1007/978-3-658-32307-3_3

interpretive frames. The theoretical and analytical concepts underpinning this are summarized in the following.

Social innovation is "the development and delivery of new ideas and solutions (products, services, models, markets, processes) at different socio-structural levels that intentionally seek to change power relations and improve human capabilities as well as the processes via which these solutions are carried out" (Nicholls & Ziegler 2017 [2015], p. 2; see also: Chiappero-Martinetti et al. 2017; Ziegler 2017a). The focus of this concept of innovation is predominantly social, as new ideas and solutions are intended to improve human capabilities and alter power relations to this end. These innovations primarily address "societal purposes […] rather than market opportunities, and […] and do not avoid political conflict" (Ziegler 2017b, p. 18). This definition structures the study of community energy projects at hand. Key terms of this concept are capabilities, power relations, and new ideas and solutions.

Capabilities

This definition of social innovation draws on the capabilities approach of Sen (1999) and Nussbaum (2006) to analyze human development and empowerment of marginalized communities. Capabilities refer to one's ability—what one is capable of—and reflect the freedom to achieve valuable (often non-economic) "functionings" such as being happy or having self-respect (Sen 1992, p. 5). Nicholls and Ziegler (2017 [2015]) summarize this by defining capabilities as "a form of power: the real opportunity to do and to be what one has reason to value" (p. 13). The study of capabilities is set in a relational and procedural understanding of power (Nicholls and Ziegler 2017(2015), pp. 5–6).

Power relations

Concepts of power as introduced above address predominantly its performance qualities and refer to how power is exerted over another. The discussion of power by Nicholls and Ziegler (2017(2015)) refers to its relational dynamics. The authors also distinguish different sources of power based on the literature. Mann (1986, p. 22) identifies four ideal-typical sources of power, namely political, economic, ideological and military power. Political power refers to regulatory authority exercised primarily through government and administration (Mann 2012, p. 11). Economic power resides predominantly in capitalist market structures (Mann 1993, p. 7). Ideological (or cultural, Heiskala 2016, pp. 13–14) power derives from shared norms and values (Mann 2012, p. 7). Military powers appear as the capacity for organized violence (Mann 1993, p. 7; 2012, p. 11, 2013, pp. 1–2). Mann's power sources are derived from historical analysis rather than theoretical considerations (Heiskala 2016, 2014), and are conceptualized as analytical tools (Mann 1986, p. 4). Heiskala (2016, pp. 31–34) adds artifactual power, as the often knowledge-based power of technological alterations to the world, and natural power, represented as the advocates of natural or environmental structures in their interactions with human society. But the assessment of power sources returns to underlying relational and procedural qualities of power, as the choice of power sources and their relative prowess is set in interactions of diffused power structures and actors attempting to assume power within them (Mann 1993, p. 10).

The relational dynamics of power are captured by Beckert's (2010) concept of the social grid.[1] Social innovations occur in a triangular dynamic shaped by social networks, institutions, and cognitive frames, which Beckert defines as follows. Social networks are structures and patterns of social relations between individual and collective actors within society, creating unequal distributions of power between them. Institutions are rules and norms of the context wherein social networks are established, constraining or facilitating actions within the social network. Cognitive frames (akin to the above-introduced concept of interpretive frames) are shared meanings structuring and guiding the interpretation of the social network and its institutions. Beckert (2010, p. 606) focuses analysis of power on the dynamic and changing relations of these three constituting elements of the social grid. Nicholls and Ziegler (2017 [2015]) extend Beckert's model of the social grid by introducing six ideal-typical dimensions of power based on the concepts of Mann and Heiskala introduced above: (1) political; (2) economic; (3) cultural; (4) security; (5) artifactual; and (6) natural. The choice of power sources and their relative prowess is set in interactions of diffused power structures and actors attempting to assume power within them (Mann 1993: 10). This further characterizes power relations in the social grid, explicitly extending the study of social innovations beyond economic powers (Nicholls & Ziegler 2017 [2015]:15). However, rather than providing categories of power (and indeed of social innovation) existing alongside or even separately of one another, the authors aim for a "promiscuous" understanding of social innovations to "impact on and involve social power in different, interrelated ways" (Nicholls & Ziegler 2017 [2015]: 15). Broad categories for power dimensions therefore are material and non-material.

Political and economic power, as well as security, artifactual, and natural power feature as material power dimensions. Political power as it is enacted by state

---

[1] Beckert's model was originally developed for the study of markets. Its application to community contexts beyond the market is supported by the theoretical underpinnings of social networks, institutions and cognitive frames which are not limited to the study of market dynamics, as well as the (related) acknowledgement that the alteration of power relations is not confined to market structures (Beckert 2010: 606; Nicholls & Ziegler 2017 [2015]: 9).

and administration is material in its manifestations through regulation, executive, legislative and judiciary powers. Economic power is material in its reference to revenues and the ability to invest, and create material structures of investment, development and employment. Artifactual power is material in its manifestations as technologies and installations. Natural power is material when considering its ecosystem dimensions, such as the provision (or collapse) of ecosystem services, as well as the natural conditions or events such as climatic conditions or wea-ther events. While this aspect of natural power poses a challenge in the study of social innovations as it arguably cannot be held responsible for its actions or be wielded as a social power source, the context of community energy merits con-sideration of the material dimension of natural power as climatic conditions. The availability of land may importantly determine the ability to establish renewable energy system services. Cultural power is a non-material dimension of power. This refers to interpretive frames and institutions, and immediately connects the study of social innovation to concepts of framing as social power. In community energy contexts which rely importantly on non-paid labor, this has an organiza-tional dimension in that frames not only affect how people think or feel about energy services, but also how they act upon these beliefs. In addition, natural, artifactual and political power dimensions each have non-material aspects. For natural power, this refers to environmental advocacy and activism attempting to implement alternative human-nature relations (Nussbaum 2006: p. 77; Nicholls and Ziegler 2017[2015]: p. 12). Artifactual power has a non-material dimension in its grounding in scientific knowledge. Especially in the context of sustainability studies, scientific knowledge and counseling are highly relevant for the develop-ment of artifactual solutions as well as political decision-making (Mann 2013: p. 363; Heiskala 2016). Political power, lastly, has a non-material dimension in processes of agenda setting and decision-making, as well as framing. It closely connects to the cultural power dimension.

New ideas and solutions
Innovation per definition seeks the new. This connotation can be challenging, since many citizens and communities engaging in sustainability dislike fra-mes wherein their activities are seen as a departure from established structures. Seeking the new is associated with the radical, while these individuals and com-munities will often understand their work as a conservative or restorative practice (Ziegler 2017b: pp. 17–18). But the concept of social innovation is not pre-dominantly concerned with the unprecedented quality of processes or solutions under study, but rather the changes to power relations and capabilities. Innova-tion might also be a new combination of previously existing processes, solutions

or practices. A new combination of ownership models, technology and decision making processes building on established patterns of cooperative ownership in the German Black Forest to gain control of local energy systems with a view of independence from nuclear power may be an example of social innovation (Colell and Neumann-Cosel 2016). The fight for the restoration of local wells and the protection of respective watersheds in rural Bavaria to prevent centralized water treatment is another example (Ziegler 2017b). Innovation may, therefore, also be found in the reconfiguration of existing socio-technical systems, in their protection, adaptation or renovation (Ziegler 2017b: pp. 94–95; Smith and Raven 2012). Focus lies on the agency of (often marginalized) civil society actors, and intentional alterations of power relations connected to their actions.

This concept of social innovation reconnects the study of innovation in community energy settings to the study of power as frames and framing while acknowledging projects' agency. This refers to the understanding of Benford and Snow (2000), wherein 'collective action frames' provide guiding lines of interpreting actions and events, and serve to legitimate actions (p. 614). By conceptualizing social innovation as an intervention to the relations of social networks, institutions and interpretive frames drawing on a combination of material and non-material sources of power, this focuses academic attention on energy system change as processes of social change (in organizational networks, institutions, interpretive frames etc.), and the agency of civil society actors. At the same time, social innovations in community energy projects combine interpretive, or non-material, interventions to power relations with material interventions. Shared meanings of energy system services are combined with respective material changes to the energy system. The concept of capabilities refers to an active understanding of empowerment. Citizens and communities engaged in energy system alternatives are not passively empowered by the establishment of social innovations, such as community energy projects, but shape these social innovations themselves (Jacobi et al. 2017).

Social innovations "are neither good nor bad" (Nicholls & Ziegler 2017[2015]: p. 18). In community energy studies and in the study of social innovation in civil society more broadly, there is a tendency to sympathize with citizens in an energy system dominated by strong organized private and/or public interest. This shows in the initiatives themselves telling stories of a 'small Gaulish village' mentality (Ziegler 2017b: p. 20). It also speaks from an, often implicit, assumption of decentralization and actor diversification through community energy engagement as a 'good thing' in academic story lines (Haggett and Aitken 2015; Seyfang et al. 2014; Hielscher et al. 2013).

3.2   Innovation Biographies

While changes in power relations imply a temporal dimension, this is not explicated in the adapted social grid setting suggested for social innovations by Nicholls and Ziegler (2017[2015], pp. 7–9). Innovation biographies address chronological aspects of innovations while not assuming linearity in their development (Rammert 2000). The biography of an innovation characterizes its emergence as an "interactive and recursive" process within reflexive networks of actors, institutions, and socio-economic, technological and natural structures (Bruns et al. 2011, pp. 11–12; Rammert 2000). This refers to forces driving or impeding the development of a shared identity, as well as corresponding roles of actors and institutions. It is a chronological assessment of the development of project identities and actions within respective contexts (Lenzen 2007, pp. 11–12).

Various publications have researched the development of energy system change over time (Köppel 2016, pp. 303–306). Concepts of path dependency have been applied to energy system change, referring to the historical dependencies of processes impeding innovation (Meyer and Schubert 2007; Arthur 1994), or lock-in effects driving innovation through contingency and reinforcement (Sydow et al. 2009). In the Danish wind energy sector, related studies indicate the importance of political windows of opportunity, technology learning, and incremental developments driving larger processes of innovation (Garud et al. 2010; Garud and Karnøe 2001). Analyses of actor constellations (Bruns et al. 2011) and coalitions (Gründinger 2017) show the relative power of alliances and coalitions in processes of energy system change. Exogenous shocks may prove important drivers of system change in some cases. The German anti-nuclear movement grew exponentially after the devastating nuclear accident of Chernobyl in 1986, importantly driving the political decision to phase-out nuclear power in Germany in the late 1990s, and regained strength after the nuclear incident of Fukushima in 2011, forcing the government that had subsequently reversed phase-out plans to return to initial goals (Schreurs 2014).[2]

The concept of innovation biographies has been applied to the study of energy system change in Constellation Analysis, a methodology of studying the contributions of social, technological and natural structures to innovation processes (Bruns 2011; Bruns et al. 2008; see also: Ohlhorst and Schön 2015). Bruns and

---

[2]Wakiyama and colleagues (2014) suggest that the same was not true for energy system change in Japan in the aftermath of Fukushima, illustrating the varied effects of exogenous shocks.

colleagues (2011) studied the development of various renewable energy technologies in Germany over time, drawing on actors, technologies, symbols and natural factors as equally important elements of analysis. The resulting configurations of elements represent "phase constellations" (Ohlhorst and Schön 2015: p. 259). Over time, these constellations of different phases of development construct the biography of (in this case) a technology, assessing "its own specific, individual character" (Ohlhorst and Schön 2015: p. 267). This concept implies that beyond contextual factors or lock-in mechanisms, as captured in the study of path dependencies, or actors and alliances, as focused in the study of networks or coalitions, the interactions of these social, technological and natural or environmental factors in the research situation create distinct patterns of innovation development over time. This study, by contrast, applies the concept of innovation biographies to an organizational form: community energy projects. Innovation biographies of community energy are characterized by social innovations, changes in power relations and related capabilities, over time.

3.3   Mobilizing Material and Non-material Resources

The study of capabilities or 'powers to' connects to the study of power as a struggle over resources. A resource is "any social, political, economic asset or capacity that can contribute to collective action" (Jenkins 2001, p. 14368). Resource mobilization theory (RMT) is a strand of social movement research (McCarthy and Zald 2001; 1977). Social movements are "collective attempts to bring about change in social institutions" (Jenkins 2001, p. 14368). They are often oriented towards protest, and typically employ non-institutionalized means (Jenkins 2001, p. 14371). Community energy projects may "not necessarily [be] a social movement in themselves" (Schreuer 2015, p. 61). Still, previous studies have argued the applicability of social movement theory focusing on its procedural setting: community energy projects face similar challenges to social movements in their establishment, and their methods can be analyzed with similar means (Schreuer 2015, p. 61). The applicability of social movement studies can also be argued based on the framing of community energy projects as social movement organizations (SMO) (McCarthy and Zald 1977). An SMO is a "complex, or formal, organization which identifies its goals with the preferences of a social movement or counter-movement and attempts to implement those goals" (McCarthy and Zald 1977, p. 1218). Community energy projects in their organization and actions are not only influenced by the institutional context of the respective energy sector, but also by the social movements underpinning civil society struggles for energy system change (Bauwens et al. 2016; Oteman et al. 2014).[3]

Resource Mobilization Theory (RMT) rejected the previously dominant focus on deprivation and/ or grievances as key explanatory factors in movement formation. Grievances, it was argued, are trivial—they are essentially too frequent and pervasive to have strong explanatory power (Jenkins 1983, p. 530). The availability of resources instead took center stage in the explanation of individual participation but also agenda setting of social movement organizations (McCarthy and Zald 2001). Leaders of social movements were conceptualized as entrepreneurs having to mobilize, i.e. acquire and employ resources and develop strategies of action for their social movement (McCarthy and Zald 1977, 2001; see also Jenkins 1983, 2001; Edwards and McCarthy 2004). Early applications of RMT focused more on the ways resources were acquired and employed, rather than their typology or respective relevance (McCarthy and Zald 2001, pp. 544–5; Edwards and McCarthy 2004, p. 125). When considering types, key resources identified were material resources such as money and physical equipment (Jenkins 1983, p. 533), as well as the organizational structure of movements (Jenkins 2001: p. 14369). Recent categorization has regrouped this into 'social-organizational resources', meaning organizational structures suitable for recruitment, the distribution of information or any other organizational task (Edwards and McCarthy 2004, p. 127), including labor as the resource critical for execution of tasks within the organizational structure (Schreuer 2015, p. 63). Material and organizational resources remain dominant resource categories to date (Schreuer 2015, pp. 61–65). Tilly (1978, p. 69) introduced land as a resource, which was however rarely researched in social movement contexts as these often seek discursive action.

Since the 1990 s, RMT research has seen the establishment of more elaborate typologies as well as more attention to abstract or intangible resources (Cress and Snow 1996; Edwards and McCarthy 2004). Cress and Snow introduce informational resources (1996, pp. 1094–96). In line with more recent examples of SMT research seeking to incorporate the construction of issue frames and identities (Benford and Snow 2000; Polletta and Jasper 2001), cultural or symbolic resources are also now included in RMT research more explicitly. Edwards and McCarthy speak of cultural resources referring to cultural products including literature, music or videos to aid with recruiting as well as ongoing mobilization (2004, p. 126). Symbolic resources, the "collective understandings

---

[3]This is frequently argued for Germany, where community energy projects are set in the context of strong nuclear power protest (e.g. Bauwens et al. 2016; Schreuer 2015). While many German projects reference anti-nuclear protests, however, analysis will show that community energy projects are not necessarily able to mobilize this argumentative proximity as a resource.

and interpretations that render something meaningful and desirable" (Schreuer 2015, p. 64), refer to interpretative frames that may be derived from cultural environments. McCarthy and Zald discuss symbolic resources (2001, p. 558), cautioning that these may be relevant for social movement work but might also evade categorization and measurement.

These extended resource types have found their way into the community energy literature. Community energy projects frequently access not only land but also other natural resources such as specific climatic conditions, warranting the consideration of natural resources (Schreuer 2015, p. 63). Knowledge resources are included to account for both explicit forms of knowledge such as information and tacit forms (Schreuer 2015, p. 63). Shared symbols or values such as a sense of community, trust, or a democratic pretense in project actions are gaining attention in recent contributions to the community energy literature (Batel 2017; Goedkoop and Devine-Wright 2016; Schreuer 2015; Greenberg 2014). Trust, for example, appears frequently in studies of symbolic resources in community energy settings (Kalkbrenner and Roosen 2016; Goedkoop and Devine-Wright 2016; Walker et al. 2010). Bomberg and McEwen (2012) in a study involving over 100 community energy projects in Scotland seek to understand the role of symbolic resources in emergence and sustenance of community energy engagement, in an effort to counterbalance the dominant focus on structural and material resources typical of the study of opportunity structure (pp. 436–37, 440–442). Schreuer (2015) in a study of citizen owned wind energy projects in Germany and Austria considers five resource types (material, socio-organizational, natural, knowledge and symbolic), as well as structural resources as an expression of political opportunity structure (pp. 61–65).

To study social innovations as interventions to power relations drawing on diverse material and non-material sources of power, this study broadly categorizes resources into material and non-material types (Table 3.1). Material resources include those established in the literature, namely money, equipment and office or assembly space. Material resources also include artifacts such as technology installations, or installed infrastructure. Lastly, they include natural resources such as land, climatic conditions or energy crops. Non-material resources refer to organizational aspects of a project including labor, as well as knowledge and symbols. Knowledge resources include information, skills, and expertise—all forms of explicit or tacit forms of knowledge. Symbolic resources refer to shared interpretive frames rendering actions and/or initiatives meaningful and desirable, both within the organization and in its cultural context. Certain traits of a community energy project can feature in both categories. For example, the organizational form of a cooperative in Germany or a development trust in Scotland provides legal and administrative opportunities, such as eligibility for funding or tax exempts, which can be mobilized as material resources by such projects. However, the establishment of both organizational forms as well-known and trusted means of organization beyond the energy context also provides non-material resources to projects as an interpretive frame of accountability and predictability.

Table 3.1 Typology of resources

Type of Resource	Specification
Material	Money, equipment, office/assembly space Technological installations Land, climatic conditions, energy crops
Non-material	Knowledge: Information, skills, expertise Forms of social organizations (such as institutionalization, networks), procedural resources, labor Shared understandings and interpretations of meaningful/desirable actions and/or initiatives both within the organization and its cultural context

The resource types as introduced present analytical rather than ontological categories. For example, the relevance of natural resources is derived from the specific challenges and agendas of community energy projects rather than assumptions on the intrinsic properties of social movement organizations. The resource typology allows for a certain degree of "interpretative flexibility" in reference to what actors may understand and mobilize as resources (Pinch and Bijker 1984, pp. 421–24). This applies specifically, although not exclusively, to symbolic resources. Differences in the interpretation of a technology may result in corresponding variation of its design (Pinch and Bijker 1984, p. 423). Pohlmann (2018) specifically highlights this in reference to the emergent and diverse character of community energy practices, wherein energy system design follows community values independent of energy or technology considerations. In the study of resource mobilization, this refers not only to the definition of what projects may (or may not) consider as a resource but also to interactions in the mobilization of resources. Interpretive frames shared by a project (non-material resources), for example, may matter not only as a shared understanding of meaningful actions, but also as determinants of the mobilization of other resources. This implies a dynamic understanding of resource mobilization as processes of accessing, and releasing, power in its various dimensions. What can be mobilized as a resource varies not only by project and context, but potentially also over time in reference to the development of both.

The RMT literature finds different approaches to treating political opportunity structure or context variables more generally. This refers mostly to questions of whether context variables may feature as resources (Avelino 2011), and whether the ability of projects to access opportunity structure can be assessed (Schreuer 2015). Bomberg and McEwen (2012) in their research on community energy projects in Scotland introduce structural resources (p. 438). Structural resources include all aspects of the larger political setting, contextualizing, enabling or impeding opportunities for mobilization. Their focus lies on movements' (in)abilities to capitalize on existing structural advantages or overcome barriers (Bomberg & McEwen 2012, p. 438). However, political support structures evade specific ownership or mobilization (Avelino 2011, p. 70). Also, social movement organizations' abilities to mobilize on the basis of these structural variables may vary considerably. Schreuer suggests including structural resources as a type of 'meta-resource' referring to "formalized resource allocation mechanisms" (Schreuer 2015, p. 64). She points out that this resonates with RMT researchers' interest in the way resources are made available to movements. However, the difficulties of ownership and mobilization prevail. This study therefore considers context variables if projects mobilize resources on their behalf. For example, financial support mechanisms appear as material resources if projects make use of them. Similarly, support for community development in the agenda setting process of national governments may find its expression in the mobilization of interpretive frames linking energy system change and community support (non-material resources). Focus lies on how projects interact with context variables, and whether they define and mobilize resources accordingly.

3.4 Summary

Community energy projects contribute to physical transformation of the energy system towards a larger share of renewable energy sources, and to transition processes addressing their social and organizational structures. Approaching community energy projects as social innovations focuses on the social and organizational powers these projects wield by providing alternative frames of collective action for energy system design. Social innovations are intentional changes to power relations and include the improvement of human capabilities, as the powers to do or be what one has reason to value. These power relations develop dynamically between individual and collective social actors, institutions, and interpretive frames of a given issue area, drawing on diverse combinations of material and non-material sources of social power relating to political, economic, security, artifactual, cultural, and natural power dimensions. The social actors, institutions, and interpretive frames, and the dynamic power relations between them are the constituting factors of the energy system context. Social innovations create interventions to these relations, also drawing on these power dimensions. Power dimensions do not categorize social innovations. Rather, the various power dimensions indicate that interventions to power relations typically rely on a variety of different power sources at the same time.

Social innovations are studied through collective action frames developed by community energy projects. Collective action frames provide shared beliefs initiating and legitimating actions. Community energy projects mobilize material and non-material resources, namely money, equipment, organizational, artifactual, natural, knowledge and symbolic resources, to act upon these shared beliefs. Processes of resource mobilization are the capabilities or powers to act upon shared beliefs of community energy projects. As a result, community energy projects present alternative interpretations of energy system design referring to both the social and organizational structures and the subsequent choices of technologies and services. These are interventions to existing power relations of the energy system, improving human capabilities in the process. The analysis of social innovations in community energy projects over time creates innovation biographies of community energy projects, which through comparison may indicate patterns of shared conditions for innovation.

Resource mobilization depends on collective action frames set by the project. Resources can originate within the project, networks or partnerships, or within context variables, such as the political or cultural opportunity structure. Resource employment, however, depends not on the availability of a resource but on the ability of projects to mobilize the type of resource on their behalf. The agency of community energy projects in constructing energy system alternatives is acknowledged through the concepts of resource mobilization and framing as opportunities to create interventions to existing power relations. The processes of negotiating meaning and defining and mobilizing resources on its behalf are shaped by individual and collective actions within the projects. Processes of resource mobilization are dynamic and may change over time. Social innovations, accordingly, develop dynamically and reciprocally in reference to the respective energy system context.

Research Design and Methods

4

This study is a qualitative analysis of five cases in three countries. It involved two rounds of field research. In the first round, interviews and document analysis provided data on the individual community energy projects, as well as on respective contexts embedding project activities including actors, institutions and political and technological structures. In the second round, additional interviews, focus groups and workshops offered reflections on the analysis of data collected in the first round. Methods of data collection and analysis were chosen based on research interest and research setting; research question and methods developed reciprocally (Silverman 2013, p. 11; Flick 2009, p. 15). Methods were developed based on the methodology and tools of situational analysis (Clarke 2003).

This chapter explains the research design and method in detail for the interested reader. Again, readers may take a short cut. Those seeking to learn more about the individual cases, their innovation biographies and the patterns of resource mobilization defining community energy may continue directly to section III of this study. For those interested in methodological detail, this chapter provides a detailed account of the approach. The first section of this chapter explains the qualitative comparative case study design and case selection. Section 4.2 presents methods of data collection and discusses the combination of qualitative and participatory research methods in focus groups and workshops. Section 3 details the approach to data analysis.

Electronic supplementary material The online version of this chapter (https://doi.org/10.1007/978-3-658-32307-3_4) contains supplementary material, which is available to authorized users.

4.1   Research Design and Case Selection

This study compares processes of social innovation in community energy settings, focusing on the definition and mobilization of material and non-material resources over the course of project development. Two methodological considerations follow from this theoretical approach. For one, social innovations develop in interactive and reciprocal networks comprising individual and collective actors, institutions such as rules, norms and values, as well as socio-economic, technological, natural and political structures (Schön et al. 2007, p. 18), and their relationships. Social, technological and natural structures are equally influential to the development of innovations (Ohlhorst and Schön 2015, 259, 264–65). Secondly, development processes can be characterized as innovation biographies (Rammert 2000, p. 185); a chronological assessment of the development of project identities and actions within respective contexts (Lenzen 2007, pp. 11–12). Innovation biographies are characterized by distinct, individual configurations of identities and activities in reference to their respective contexts. A comparison of these biographies can indicate similarities in patterns of development, in key factors supporting or impeding development, and in circumstances (Ohlhorst and Schön 2015, p. 267). Regarding the research question and the explanatory power of social innovations for energy system change, this offers new insights into the types of resources driving or impeding community energy projects, as well as corresponding actor constellations, changes in power relations between these actors, and the potential and limitations of policy intervention.

4.1.1   Qualitative Case Study Comparison

Qualitative research, in its methodological and methodical "way of thinking about and studying social reality… [and] procedures and techniques for gathering and analyzing data" (Strauss and Corbin 1990, p. 3), acknowledges the subjectivity of social realities and those engaged in their (re)production (Hitzler 2014, pp. 64–65; Silverman 2013, pp. 6–7; Hollstein and Ullrich 2003, p. 35). The principles of contextuality, second-order interpretation and interactiveness underpin qualitative inquiry, resulting in three methodical requirements. The principle of contextuality requires participants of the research project to be enabled to describe and explain their social realities based on their own categories of pertinence (Denzin and Lincoln 2003, p. 3; Hollstein and Ullrich 2003, p. 36; Przyborski and Wohlrab-Sahr 2014, p. 17). The second principle refers to academic analysis as a second-order interpretation of first-order constructions of reality by participants

(Przyborski and Wohlrab-Sahr 2014, p. 13). Academic interpretation of participants' knowledge enables inference beyond the immediate setting (Haraway 1988, pp. 587–590; Rose 1997, pp. 314–315). Theoretical assumptions on the research object are challenged by interpretations of participants. Pulla refers to this as the challenge of keeping an open, while not empty mind (2014, p. 19). Transparency and accountability in presenting research methods show how 'methodically controlled understanding of others' ("methodisch krontrolliertes Fremdverstehen", Przyborski and Wohlrab-Sahr 2014, p. 14) is achieved.[1] Qualitative results provide an understanding of a specific social phenomenon at a given point in time. Interactiveness, thirdly, refers to the process of acquiring and analyzing data between researcher(s) and participant(s). Research settings and narratives are jointly constructed by interviewer and interviewee (Silverman 2013, p. 45). These principles and requirements resonate with the call of community energy researchers to acknowledge diversity and ambiguity of community energy projects, rather than seeking generalization and scalability in their understanding (Pohlmann 2018, pp. 268–269; Seyfang and Longhurst 2013a, pp. 887–889; Smith 2007b, pp. 441–443; Pinch and Bijker 1984, pp. 421–424). Qualitative inquiry enables an in-depth understanding of what may be mobilized as material (money, technology, land, etc.) or non-material (labor, knowledge, symbols, etc.) resources in community energy projects in diverse context settings, and how. The concept of innovation biographies is hereby applied not to technologies or sectors (such as in the analysis of renewable energy innovation biographies by Bruns et al. 2011), but rather to an organizational form: community energy projects. Each community energy project was studied from the time of its initiation up until December 2017, when the period of inquiry was closed for this project.

Case studies "emphasize detailed contextual analysis of a limited number of events or conditions and their relationships" (Dooley 2002, p. 335; Vandenbroucke 2001, p. 331), and have therefore been called the "implicit companion" of qualitative inquiry (Yin 1981, p. 58; see also: Yin 2014, p. 62). New contingent

---

[1] The scientific nature of qualitative methods has been at the core of longstanding debate on the ability of qualitative research to meet standards of validity, reliability and objectivity as developed for quantitative methods. One response lies in the adjustment of scientific standards to fit the paradigm of qualitative interpretation, rather than bending qualitative methods to quantitative requirements. This highlights replicability and transparency as core principles of the qualitative research process, as well as the appropriateness of methods in respect to research interest and question. On methodically controlled understanding as a strategy of qualitative research see Przyborski and Wohlrab-Sahr 2014, p. 21; Meyer and Meier zu Verl 2014.

insights on a phenomenon and its context are central (Yin 1981, p. 59). Community energy research predominantly compares community energy projects within single countries (among many see Hinshelwood 2001; Walker and Devine-Wright 2008; Middlemiss and Parish 2010; Kunze 2011; Bomberg and McEwen 2012; Seyfang et al. 2014; Radtke 2016; Ruggiero et al. 2018). In recent years, the number of publications studies considering different national settings is growing (Pohlmann 2018; Bauwens et al. 2016; Fuchs 2016; Schreuer 2015; Nolden 2013). International comparison frequently focuses on aspects of the political or technical opportunity structures for community energy projects, such as supporting policies or the ability to access technologies and related services (Bauwens et al. 2016; Schreuer 2015). These publications often make use of country rather than project level case studies (Schreuer 2015; Soutain 2015). The comparison of energy policies and sector development has informed our understanding of different policy strategies for managing sector change (e.g. Gullberg et al. 2014; Lewis and Wiser 2007; Mitchell et al. 2006). The definition of community energy, however, often simply considers whether there is some connection between an energy installation or service and an adjacent community (Nolden 2013, p. 546). This neglects how communities interact with technological innovation(s) and the often complex embedding of technological innovation in socio-organizational structures. Respective research struggles to explain the different forms energy system change assumes in community settings. The diverse configurations of community energy projects are increasingly acknowledged in national comparisons (Goedkoop and Devine-Wright 2016; Walker et al. 2010; Walker et al. 2007), but remain an exception in international comparison (Pohlmann 2018). Studying "commonalities in the conditions for innovation [...] central driving and restricting [...] factors, and [...] identify[ing] typical development patterns" (Ohlhorst and Schön 2015, p. 267) across national contexts, however, strengthens the explanatory power of social innovations for understanding the role of community actors in (diverse settings of) energy system change. It focuses attention on the specific features that community energy actors share in this respect.

Unlike generalization through representative sampling (Lamnek 1995, p. 21), qualitative case sampling seeks variation and even ambiguity in the information on the phenomenon under study (Lamnek 1995, p. 22; Glaser 1992, p. 102; Glaser and Strauss 1970). A case speaks not for a larger population (Gomm et al. 2000, pp. 99–103) but to a conceptual level positing applicability in alternative settings (Yin 2014, p. 68; Mitchell 1983, p. 207). Significance of the case is "extended" beyond its immediate situation (Burawoy 2009(1991), pp. 271–280, 1998, p. 5). The goal is "analytical generalization" (Yin 2014, p. 40), lessons to be learned empirically as well as with respect to the development of theory.

The comparison is set up as a 'most different system' design (Przeworski 1987, pp. 38–41; Przeworski and Teune 1970), referring to differences in energy system contexts between cases. The energy system is understood as the "rule-set or grammar" (Rip and Kemp 1998, p. 338) governing a particular combination of technologies, social practices and institutions (Geels 2002, p. 1260). This includes governmental and non-governmental actors and administrative organizations, as well as institutions such as rules, norms and values. It also encompasses fossil and renewable energy technologies employed (or rejected) in an energy system and their organizational embedding and cultural connotations, and natural conditions for energy system services. The most different system classification of this comparison refers to the context of community energy projects, assessing relevant factors at national and sub-national levels in Germany, Denmark and Scotland over the course of project development.[2] The countries share an overall favorable approach to community energy projects in energy policy and regulation, as well as a supranational context supporting energy system change towards renewable resources and sustainable practices at the level of the European Union (see above at section 1.3). Differences prevail regarding the respective country strategies for energy system change, the political and regulatory opportunity structures for community energy engagement, and ownership and investment structure of the energy sector. Differences also show in energy system configuration at sub-national and local levels, acknowledging that factors determining individual decisions of the socio-technical configuration are often highly specific (Pohlmann 2018). What is relevant in the context must be understood from the perspective of the 'niche' (Pohlmann forthcoming). Moreover, projects were initiated at different points in time resulting in different political and socio-economic conditions for community energy actors. For example, the case selection includes two cases from Germany but their different location, services, organizational background, etc. results in different roles in the local energy context. And although located in the same country, policy contexts varied considerably as projects were initiated and established at different times. For example, EWS in Schönau was successfully established as a community energy project prior to the introduction of feed-in tariffs supporting small market actors, and by the time that olegeno in Oldenburg was attempting market entry the German feed-in tariff had been abolished.

Similarities are expected with respect to the innovation biographies of community energy projects. Innovation biographies, in this context, refer to the definition and mobilization of material (money, land, etc.) and non-material (knowledge, symbols, etc.) resources by community energy actors and resulting changes of

---

[2]EU regulation was considered when applied in national contexts, see section 1.3.

community capabilities and power relations over time, respective actor constellations, as well as technological, institutional, and structural contexts. Technological context refers to technologies available or rejected in each case setting, and respective actor constellations, belief systems and structural mechanisms. Institutional context refers to shared norms and belief systems relevant to energy system services for each case. Structural context refers to government or administrative actors and policy instruments. For example, communities attempting wind energy installations in Denmark could benefit not only from a number of local or national providers offering installations and maintenance, but also from a national discourse favorably portraying windmills as symbols of national development and positive change and policy instruments tailored to increased wind energy capacities and community involvement.

4.1.2   Community Energy Projects in Germany, Denmark & Scotland

Community energy projects, the unit of analysis, are defined here through their concern with renewable energy (RE) related services, ownership and majority management by (often local) citizens, and the collectivizing of benefits.[3] RE services include direct enabling or own renewable energy generation, distribution, retail, or related services of facilitating and/or educating on efficiency improvements, reduced consumption, or the like (similar definitions in Boon and Dieperink 2014, p. 298). Services must be co-produced. This means the community project is directly involved in the provision of services, not only in ownership and organization (Becker et al. 2016, p. 64). To avoid focusing on rather well researched energy generation projects, a project's activities have to cover at least two of the above listed service areas. Only projects were chosen that had been active for more than two years at the outset of this research project in 2015. This age limit enriches data in reference to established and/or altered processes of social innovation over time and avoids selecting cases reacting to policy changes underway in all countries under study and reducing policy support for community energy actors at the time. Lastly, locational characteristics (urban, rural, remote, island, etc.) were considered.

---

[3]Citizen management refers not to honorary management but rather to the control of assets and procedures through community structures, as opposed to utility management of projects that enable financial involvement or utility-controlled modes of participation in decision-making (Bonn and Dieperink 2014, p. 298).

Country locations of cases satisfy criteria of basic comparability. All three countries were members of the European Union while under study and had declared their commitment to a shift of the electricity sector towards renewable resources, efficiency improvements and reduced consumption, as well as providing (different yet) overall beneficial conditions to community engagement in the energy sector (Lipp 2007; Lewis and Wiser 2007; Mitchell and Connor 2004).[4] Community energy engagement for renewable energy proliferation, measured in the involvement in installed capacity, the overall number of projects and the overall financial investment of community actors, is relatively high across countries. This coincides with a leading position regarding the penetration of renewables in the energy sector (Soutar 2015, p. 74).

Five community energy projects were chosen located in Schönau and Oldenburg, Germany, Samsø, Denmark, and Fintry and the Isle of Mull, Scotland. Projects were identified through personal contacts as well as online research. A co-founder and board member of an energy cooperative in Berlin (Germany) since 2011, the author had established national and international contacts to community energy projects. This network enabled direct contacts to cases in Germany and Denmark. Cases in Scotland were identified with no prior relations. Pragmatic criteria for case selection (Yin 2014, p. 95), included accessibility of projects (projects were visited multiple times) as well as shared language (interaction occurred in German or English without assistance of translators). Table 4.1 summarizes project characteristics in relation to the definition of community energy projects as projects majority owned and managed by citizens and communities wherein benefits are collectivized, as well as the added selection criteria of diverse renewable energy services, age and diversity of location. The first column lists projects by name and location. The second column indicates the year wherein initiatives were founded, which in 2015 corresponded to an age range of 28 years in the oldest case to four years in the youngest case.

The third column refers to ownership, management and collectivized benefits. Both German projects are organized as citizen owned cooperatives that also manage and operate energy assets and services. While EWS was based on non-paid engagement in the first years of its campaign, establishment of DSO services in 1996 coincided with establishment of paid staff. Returns are reinvested in

---

[4]The choice to treat Scotland as a country in this project was made based on the difference in opportunity structure of community energy projects through policy and regulation in Scotland vis-à-vis other parts of the United Kingdom. It is worth noting that renewable energy installations in Scotland are also influenced by UK planning for deployment, which favors Scotland in installations due to topography and wind energy potential (Vaughan 2015; Toke 2011).

projects of the energy transition, with a limited share of revenues being distributed among members. Investments towards sustainable energy services are not bound by membership. In Oldenburg, olegeno is organized in un-paid structures.[5] So far, no financial revenues were redistributed. Educational and advisory work is free of charge and not limited to members. The Danish projects on Samsø shows diverse forms of ownership ranging from individual citizen ownership, over collective forms of ownership, to municipal ownership and partnerships with utilities. Ownership of energy assets is direct. The majority of projects is owned by citizens, although this does not necessarily mean cooperative organization but includes individual ownership. Management structures are diverse. The Energy Academy, as a project moderator, relies on paid staff, while cooperatively organized, and also some individualized energy projects rely on unpaid staff structures. Commercially and municipally operated structures include paid staff. Financial revenues are redistributed according to ownership structures, and therefore include individualized benefits, cooperative benefits and municipal benefits. Educational, awareness raising and advocacy work of the Energy Academy is provided to islanders free of charge, as are facilitating formats for community development. The Scottish projects are organized as a Community Development Trust in the case of Fintry, and a Community Benefit Company in the case of Mull, which is, however, closely connected to the Mull and Iona Community Trust. Returns in the Scottish cases are reinvested in the communities holding the Trust or Benefit Company respectively, independent of membership.

The fourth column refers to renewable energy services. All projects engage in educational and awareness-raising work for energy related issues within their community, and in the facilitation of renewable energy related projects. Most are engaged in renewable energy production, while technologies of energy production vary. Renewable energy production is fully grid integrated in all cases. Two projects are engaged in electricity retail (one of which also offers retail of natural gas), two projects include distribution heating retail services. Projects engaged in generation do not necessarily include retail. Only one project (EWS in Schönau) is active as a distribution system operator (DSO). Two projects, in Fintry and on Samsø, include district heating networks. One project (GEM on Mull) engages in off-grid trials to increase renewable energy capacity beyond national network capacity. All projects operate fully grid integrated energy related services, although one project (GEM on Mull) is currently simulating operations integrated

---

[5]The terms 'paid' and 'non-paid' were chosen to indicate that while all projects required professionalization as increased management and operations expertise, not all (can) pay their staff.

in the local grid but disintegrated from the regional and national network. The fifth and final column specifies characteristics of the location in reference to the number of inhabitants and proximity to small, medium or large urban centers. Location ranges from rural, including varying degrees of remoteness from city centers, to small urban.

4.2 Data Collection

Data collection occurred in two rounds, including document analysis and interviews in the first, and focus groups and workshops in the second. Both rounds involved site visits and observation of project meetings or events hosted by projects by the researcher, as well as joint planning sessions for workshops or focus groups. Interviews, workshops and meetings were held face to face. In some cases, subsequent personal communication by telephone or email supplemented meetings. Participation in interviews or workshops was possible in all stages of the research process. Projects could also extend own invitations to meetings and workshops, which resulted in public or partially public workshop settings. In one case, the local press covered a workshop. Participation in the research process varied across projects, with three projects more actively engaged, and two projects engaging when asked to. The intensity of participation correlated with the intensity of own advocacy in projects (projects engaging in political advocacy work beyond immediate energy services participated more actively, involved more members and shared content within their organization), but also with resource availability (projects with full-time employees typically sustained more prolonged participation).

The first round of field research yielded comparable interview numbers and access points across projects. All projects agreed to site or office visits and openly shared information and opinions. In the second round, focus groups and workshops were held with two out of five projects; several additional in-depth interviews were conducted with members of a third project when a workshop date proved unfeasible. Workshops involved preparatory meetings and personal communication, as well as additional interviews to brief and debrief one another prior to and after the event. The limited number of workshops was mostly the result of logistical constraints of non-paid project structures and the researcher herself, but did not adversely affect the explanatory power of the data, since the workshops held also proved a saturation of data with regard to the research question.

Table 4.1 Overview of community energy project characteristics

PROJECT NAME & LOCATION	FOUNDING YEAR	OWNERSHIP, MANAGEMENT & COLLECTIVIZED BENEFITS	RENEWABLE ENERGY SERVICES PROVIDED	LOCATION CHARACTERISTICS
Elektrizitaetswerke Schönau (EWS), Schönau, Germany	1986	Citizen owned cooperative direct ownership of energy assets paid management returns directed towards projects of the energy transition, limited revenues to members educational and advisory work free of charge, advocacy	Distributions systems operator for electricity (low voltage) and natural gas, RE generation (wind, solar, hydro), retail education and facilitation for energy efficiency, reduced consumption, sustainable energy sourcing	Rural community (2.400 inhabitants), Black Forest in Baden Wuerttemberg ca. 30 km to small urban center (230.000 inhabitants)
olegeno, Oldenburg, Germany	2011	Citizen owned cooperative non-paid management returns to members Educational and advisory services to community free of charge	Electricity retail, RE generation pending (solar) Application for DSO, education and facilitation for energy efficiency, reduced consumption, urban applications of RE	Small urban center (165.000 inhabitants) in Lower Saxony

(continued)

Table 4.1 (continued)

PROJECT NAME & LOCATION	FOUNDING YEAR	OWNERSHIP, MANAGEMENT & COLLECTIVIZED BENEFITS	RENEWABLE ENERGY SERVICES PROVIDED	LOCATION CHARACTERISTICS
Energy Academy (EA), Samsø, Denmark	1997/8	Diverse forms of citizen ownership, municipal ownership, partly co-ownership with utilities; direct ownership of energy assets paid management (non-paid management in some ownership forms) financial returns individualized or collectivized in municipal household educational and advisory services to the community free of charge, advocacy	RE generation (wind, solar, biomass), district heating, alternative fuels for transportation (on land and water) facilitation and education community development	Island in the Kattegat (Baltic Sea), rural community (3.700 inhabitants) 60 km to medium sized urban center (370.000 inhabitants)

(continued)

Table 4.1 (continued)

PROJECT NAME & LOCATION	FOUNDING YEAR	OWNERSHIP, MANAGEMENT & COLLECTIVIZED BENEFITS	RENEWABLE ENERGY SERVICES PROVIDED	LOCATION CHARACTERISTICS
Fintry Development Trust (FDT) Fintry, Scotland	2003	Community Trust, citizen owned, no direct ownership of energy assets non-paid management, paid operations financial returns collectivized in community independent of membership	RE generation (wind), district heating, insulation facilitation and education community development	Rural community, (700 inhabitants), Western Lowlands 35 km to large urban center (600.000 inhabitants)
Green Energy Mull (GEM), Isle of Mull, Scotland	2010 (1997)	Community Benefit Company, citizen owned, direct ownership of energy assets non-paid management, operations support through paid management and staff of Mull & Iona Community Trust (MICT) returns to comm. through benefit company independent of membership	RE generation (hydro); subsequently supply and demand management for RE grid integration; facilitation and education	Island in the Inner Hebrides, rural community (2.400 inhabitants) ca. 190 km to large urban center (600.000 inhabitants)

4.2.1   Interviews: Talking to Experts

Interview partners included members of community energy projects, members of associations working in (financial, administrative, institutional or content-related) support of community energy, and academics working on community energy within the region. All interviews were treated as expert interviews, in the sense that the interviewee is chosen for their knowledge of the research situation and asked to share insights into specifically related actions and framings (Pickel and Pickel 2009, pp. 452–454). Characterizing members of community energy projects as experts is based on their involvement and actively influential role in the field of study and corresponding advantages of the expert in accessing information and knowledge, rather than an observant role or their professional status.[6] In all projects, interviews included different hierarchy levels (founders, management, team members, etc.). Participants were not chosen with regard to their status in their respective organizations and hierarchy effects were disregarded because of their multiple meanings and varied constructs (Marmot 2007, p. 1153). The analysis does consider access to material and non-material resources based on capabilities acquired through project engagement, as they manifest in the data.

Three exploratory interviews confirmed case selection and refined the research question. Interview partners included a community energy practitioner, also involved in advocacy and support associations, a senior employee of an association supporting community energy and an academic professional with relevant sectoral and regional knowledge. In addition to their direct involvement in the field of study, these interview partners were able to assess larger developments throughout the sector because of their long-term involvement in community energy and/or their senior positions within respective organizations. Conceptually, this referred to the variation a case adds while allowing for replication (Yin 2014, p. 65). Empirically, practitioners' views on the relevance of a community project for its peers guided the selection.

The first round of field research included 31 interviews with 28 practitioners of community energy projects and members of associations working in support of community energy projects were subsequently conducted (Table 4.2 gives an overview of interviews and participation per country). Interviews with members

---

[6]As a community energy professional whose personal experience was known to participants, the researcher held the role of a knowledgeable, friendly outsider (Minkler 2004, pp. 688–690). The invitation to participate in additional rounds of interview or workshops was phrased as an opportunity to share knowledge between projects, as well as between the researcher and participants, a common frame for interactions with research participants (Minkler 2004, p. 689). This was confirmed in how projects themselves framed their motivation to participate.

of supporting associations were conducted exclusively in Scotland because of direct and close ties between the associations and the projects, either through consultancy or through actual project cooperation. Contact with potential interview partners was made personally during events if possible, or by email. All interview partners received introductory information as well as details of the two rounds of field research planned. Verbal consent was given in all cases. In addition, interviewees that chose to engage in follow up interviews and/or workshops signed a form (included in Annex 2) explaining methods of participation, data analysis, communication and storage in accordance with the principle of confidentiality (Legard et al. 2003, p. 146). This included the option of making all personal data anonymous (which none of the participants selected), and the affirmation that all audio recordings would be deleted following transcription and personal information would be separated from data used in the scientific process. Interviews were conducted face to face during site visits, with the exception of follow-up interviews, which often occurred via telephone, or email. One interview with a Danish expert was conducted via email without prior personal contact. In one project, participation in an internal meeting was possible, while another project offered participation in a public meeting. Both meetings were documented through notes and not recorded. This was followed up by interviews via telephone with some attendants. Personal communication via telephone or email was helpful when prior personal contact had been established but yielded poor results when this was not the case (see Christmann 2009 on telephone interviews; on email communication see Gläser and Laudel 2010, p. 154). All interviews were individual and typically took between 45 minutes and one hour. The author recorded and subsequently transcribed interviews.[7] German or Danish sources were translated to English for direct citation by the author. This followed the spoken word and content directly while transferring idioms to the according language.[8]

Interviews were conducted in a semi-structured fashion (see Pickel and Pickel 2009 on methods of expert interviews). The interview guide was based on the

---

[7]Interviews were conducted in German and English, transcription occurred in the respective languages. Transcription followed the interview word for word, with the exception of repetitions of the same word as well as "ehm" sounds, which were deleted. Breaks were marked as (…), non-spoken noises such as laughter, coughing etc. as ((…)). Normal font indicates normal spoken word, **fat** indicates words spoken with special emphasis.

[8]For example the German expression "wie das Kaninchen vor der Schlange", indicating a person frozen in fright, was translated to the English idiom "like a deer in the headlights", rather than translating word for word as "like a rabbit in front of a snake". This was done to maintain legibility while leaving the content of the quote unchanged.

Table 4.2 Overview of interviews by country and affiliation, 1st round of field research

Country	Organizational Affiliation of interviewee	# of interviews	# agreeing to long-term participation
Germany	Community energy project	14	11
Denmark	Community energy project	5	4
Scotland	Community energy project	3	3
	Supporting association	9	9
Total #		**31**	**27**

research question and theoretical frame of social innovation and resource mobilization (see Figure 4.1 for an overview of topics). Prior desktop research included academic literature on community energy in the respective regions and the technological, institutional, and structural context of projects, as well as information made available by projects themselves or associated organizations and policy documents. Aspects raised in previous interviews or made available through newly published documents could be added (Strauss and Corbin 1990, p. 419). This interview guide was typically shared with respondents by email beforehand. The sequence of questions as well as the time assigned to individual issues could be varied in reference to individual preference, knowledge or interests of the interviewee (Legard et al. 2003, p. 141). Based on whether the participant was a practitioner of community energy projects or a member of supporting associations, interviewees would be asked to elaborate on details of the project and services, underlying motivations and rationales, their legal form and institutionalization, and the processes and institutions of the project themselves, or on measures of project support and the details of cooperation. Both groups were asked about the meaning of community energy projects and related material and non-material values. Interviewees were also asked to specify their view of the community energy sector more generally within their country.

Follow up interviews to prepare workshops were conducted as necessary; three additional interviews were conducted Samsø and one in Germany. Also, two interviews were conducted with academics specialized in the field. The total number of interviews conducted was 47. A list of interviews is included in Annex 1.

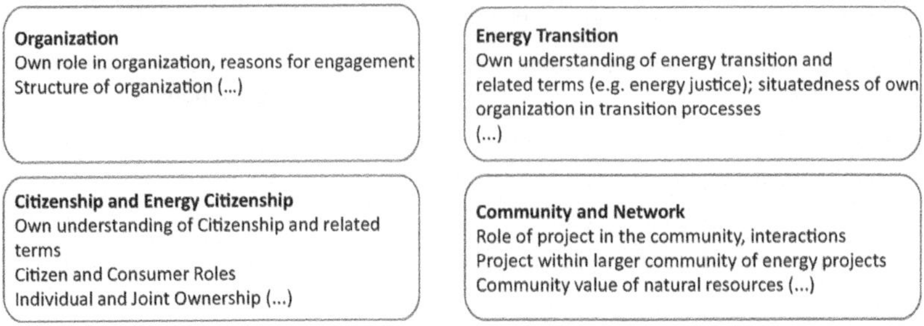

Organization	Energy Transition
Own role in organization, reasons for engagement Structure of organization (...)	Own understanding of energy transition and related terms (e.g. energy justice); situatedness of own organization in transition processes (...)

Citizenship and Energy Citizenship	Community and Network
Own understanding of Citizenship and related terms Citizen and Consumer Roles Individual and Joint Ownership (...)	Role of project in the community, interactions Project within larger community of energy projects Community value of natural resources (...)

Figure 4.1   Topic guide of interviews. *Source* AC

4.2.2   Workshops: Connecting Expert Knowledge

Following analysis of the data obtained in the first round of interviews, focus groups (internal project meetings focusing on certain aspects) or workshops (public events discussing the overall analysis) were held. The addition of focus groups or workshops was planned early on to jointly reflect on results of analysis and create a reflux of knowledge into participating projects (Minkler 2004, p. 689). This indicates a combination of qualitative and participatory research methods in this project. Participatory approaches are advocated in sustainability sciences more generally (Ziegler and Ott 2011), and in parts of the literature on community energy engagement specifically (Catney et al. 2013). In the sustainability science literature, the discussion of participatory methods refers especially to the normative and/or political underpinnings of research, the potential (and limits) of engendering change through research, the academic validation (and legitimation) of practitioners' arguments, and the delineation of research and consultancy (Kates et al. 2001; Funtowicz and Ravetz 1992 (1991)). The difference between participatory and qualitative research is not so much epistemological, as it is political (Ziegler and Ott 2011, p. 36). This study has a qualitative research design, but community energy research can be charged with inherent political and normative underpinnings.

The study of resource mobilization in community energy projects as social innovation is political in two ways. First, it is an academic echo of a political agenda of energy system change supporting community and/or citizen involvement. Much of the research in community energy engagement is in this sense inherently political (Smith et al. 2016; Spreng 2014; Stirling 2014). In addition, many of these publications portray community energy engagement as an inherently 'good' thing without questioning the normative underpinnings of this understanding (see van Veelen 2018 for a discussion of 'energy democracy' in this respect). Secondly, this research project positions itself against a larger narrative supporting a streamlined understanding of civil society engagement (Stirling 2014, pp. 88–89; see also Chilvers and Longhurst 2016). Framing community energy engagement as social innovation counters a discourse on energy system change dominated by innovation in technology and economics (Stirling 2014, p. 88). Politically, an academic argument of community energy engagement as social innovation can therefore provide leverage to community actors. Academic considerations determined the decision to frame community energy projects referring to social innovation (Seyfang et al. 2014, p. 41; Ziegler et al. 2017; Ziegler 2017a). To academic assessment, innovation in itself is neither 'good' nor 'bad' (Ziegler et al. 2017, p. 295). But the term carries diverse connotations in the public discourse making this study implicitly political in interviews and workshop settings. In market oriented arguments, innovation is understood as a positive driver of development (Okereke 2008). In the context of community organizing, innovation is predominantly connected to technological or economic paradigms and carries adverse connotations: "Social innovators don't like to talk about innovations" (Ziegler 2017b, p. 112).

Focus groups and workshops shared analytical results, but did not constitute processes of co-inquiry as participatory research would require (Reason 2011(1994)). Instead, inquiry of researcher and non-academic participants coincided with diverging interests and ends. Lastly, proximity to non-academic participants may blur the lines between research and consultancy (Stoecker 1999). Participatory researchers often openly advocate for this (Jordan and Kapoor 2016). Consultancy through academic publication is supported in cases of marginalized communities (Lundy and McGovern 2006, p. 49), or in reference to the urgency of an issue (Ziegler and Ott 2011, p. 37). Sharing knowledge and networks indicates an indirect form of consultancy involved in this research project. However, this study was neither commissioned nor financed by non-academic actors involved, nor does it provide personalized services to projects. Methods, analyses and conclusions of the author are made transparent according to academic standards.

Upon completing interview analyses, participants were contacted to confirm and specify their interest in a second meeting. Follow up interviews were conducted to this end. Projects could specify the preferred setting for exchange and extend the invitation to third parties. One project included the workshop in a larger event on renewable energies in the region and invited local press. Participation ranged between 5 and 10 in number. In preparation of meetings, eight interviews were conducted with project members previously involved to specify content planning (see Figure 4.2 for a topic guide of focus groups and workshops). Focus group and workshop participants were briefed that data presented were part of unpublished research. Country and project information of other participants as well as characteristics of locations were shared among the projects, however all participants' quotes were made anonymous. The author did not ask questions but rather initiated discussion among participants based on analysis across all cases. In non-public settings, participants were asked to treat others' contributions confidentially. Workshops were documented through photos of the room and materials, and through notes during the discussion. No recordings were made.

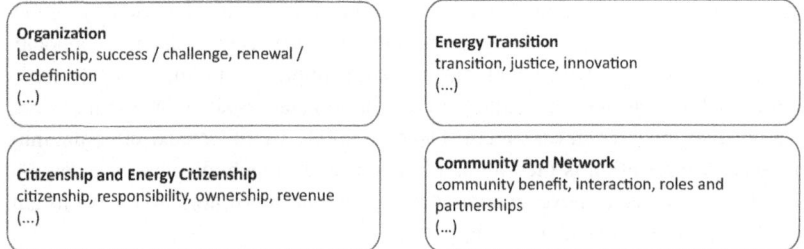

Figure 4.2   Topic Guide of Workshops. *Source* AC

Policy developments by country and at EU level as well as project developments were shared with project participants on a timeline (figure 4.3), beginning in 1973 with development of wind energy in Denmark and ending in 2017. Illustrations included project developments, as well as policy developments to provide participants with a general background across cases.

In all workshops, three kinds of maps were shared based on the mapping tools developed in document analysis: (1) maps of the research situation of each project, which recorded actors, technologies and installations, and institutions, and the relationships between them, as well as characteristic resources for each relationship, (2) maps of relevant arenas of negotiation and the resources mobilized within them, and (3) maps of argumentative positions and their connection to material and non-material resources. How data were analyzed with these tools is detailed in section three of this chapter. Maps of argumentative positions within projects assembled anonymous quotations across projects and countries, reflecting similarities and differences. Arguments were grouped in the categories of ownership, community, citizenship, challenge and renewal, as well as ambiguities and dissent in the underlying data (figure 4.4). Argumentative positions were

Figure 4.3   Workshop materials. *Source* AC

also mapped according to their affiliation with different material or non-material resources in each project, indicating how an argument connected to a certain (type of) resource. These maps illustrated what could be a resource within a given project, and a certain point in project development or over time, and whether or not this corresponded to other projects.

Maps of the research situation included relevant human (individual and collective actors) and non-human (technologies, institutions, policies, etc.) elements of each case and relationships between them.[9] Different kinds of material and non-material resources were then discussed in reference to these relationships (see figure 4.5 in section 4.3.2 for an exemplary illustration). The recognition of what could (or could not) be mobilized as a resource based on argumentative maps, was complemented by discussion of who could (or could not) wield this resource. Partially permanent maps of the research situation were created on blackboards to be used as an interactive tool of process reflection for projects themselves, wherein

---

[9]Clarke refers to human and non-human actors and actants (2003, 561–563.).

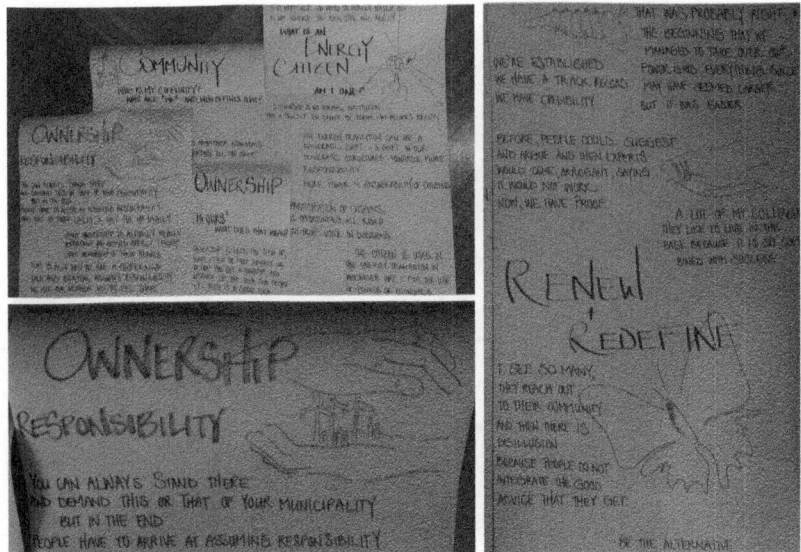

Figure 4.4 Positional maps, examples from workshops. *Source* AC

elements of the research situation were marked permanently but resource relations suggested by the researcher could be erased. Lastly, maps of arenas of negotiation were discussed to understand and recognize the coalitions and issue areas structuring negotiations. Resource relationships could be analyzed in reference to the different kinds of material and non-material resources, as well as their origin (e.g. within the project or within connecting arenas).

Maps proved helpful for visualizing the analytical process and its results for participants. At the same time, the author's maps took on a life of their own in the individual projects as participants developed representations to cater to their immediate organizational interests. Mapping in this sense provided a way "into the data" (Clarke 2003, p. 570), yet in the workshops also offered a way 'out of' the academic analysis of data.

Throughout the analysis, interviews are referenced [I/Wn:123] by indicating the type of interaction (I = interview or W = workshop), the interaction's reference number (n) and the line of transcription (123). While organizations are not made anonymous in this research project, individual interview partners or workshop participants are. In referencing direct quotations, the pronoun 'they' (their etc.) avoids identification by gender. In some accounts, identification of members would have been inevitable based on their comments. In these cases, the date of the interview serves as a reference rather than the numerical identifier to protect other sensitive quotes from direct identification.

4.3   Data Analysis: Mapping With 'Situational Analysis'

Qualitative research often produces maps of context, process and meaning of the research situation (Ritchie and Lewis 2003, p. 4). Methods of Grounded Theory are popular due to their good epistemological fit and their more systematic processing of data (Robson 2002, p. 191), and are employed in diverse settings (Fram 2013). In socio-technical contexts, mapping tools must reflect both dimensions appropriately (Ohlhorst and Schön 2015: 259, 261–62). Frameworks including social and technical factors of the research situation and offering mapping tools of some sort include Actor-Network-Theory (Latour 1987, 1991, 1996), Strategic Niche Management (Kemp et al. 1998; Schot and Geels 2008), Multi-Level Perspective (Kemp 1994; Schot et al. 1994; Geels 2006), Constellation Analysis (Schön et al. 2007; Ohlhorst and Schön 2015), and Situational Analysis (Clarke 2003).

Actor-Network-Theory was developed as a tool to include non-human elements of the research situation in social science analysis. However, the central role of actors in networks within this model results in a strong focus on enforcement (Latour 1996). Also, actor networks conceptually are expected to either grow or die (ibid.; see also Schön et al. 2007, p. 53). Strategic Niche Management focuses on transformations of socio-technical systems through the analysis of protective niches wherein technological innovations, social practices and regulatory conditions can develop in experimental social settings (Kemp et al. 1998, p. 186; Kemp et al. 2001). Developed as a tool for innovation and policy management, however, the approach centers on development mechanisms of experimentation and upscaling with clear expectations for processes of learning and consolidation within the niche (Schot and Geels 2008, p. 540), neglecting empirical evidence to the contrary (Seyfang and Longhurst 2013a). Multi-Level Perspective analyses, similarly, set the context of innovations in micro, meso and macro levels of interpretation, resulting in similar conceptual difficulties (Smith 2007b, pp. 441–443; Seyfang and Longhurst 2013a, pp. 887–888). Constellation Analysis and Situational Analysis, by contrast, offer both more inclusive and more neutral mapping tools for assessing research situations and developing analyses (and theory) from the bottom up (Ohlhorst and Schön 2015, pp. 258–9). Both approaches include human and non-human factors in their understanding of the research situation without ascribing organizational or hierarchical relations, as well as structural factors and conditions of context (Clarke 2003, p. 559; Clarke and Keller 2012, p. 24). Constellation Analysis was developed as an application-oriented tool of inter- and transdisciplinary research (Schön et al. 2007). It maps relationships of actors, technical, symbolic and natural elements to study material and non-material influences to the development of innovations on equal footing (Ohlhorst and Schön 2015, pp. 258–9; Schön et al. 2007, p. 18). Unlike Actor-Network Theory, Constellation Analysis assumes a more neutral position in understanding the role of actors in respect to other factors of the constellation and constellation development (Ohlhorst and Schön, p. 275). Its neutrality also distinguishes Constellation Analysis from Strategic Niche Management and Multi-Level Perspective approaches, assuming neither different levels for the elements of a constellation nor pathways for their developments, while at the same time focusing more on the relationships between elements within the constellation (Ohlhorst and Schön 2015, pp. 275–76). Developed to "build bridges" between the social, natural and engineering sciences (Ohlhorst and Schön 2015, p. 273), Constellation Analysis is a 'pragmatic' tool to create inclusive representations of socio-technical settings that speak to diverse academic audiences (ibid. p. 276). Situational Analysis distinctly tailors its mapping tools to "methodologically sound analytics for social sciences" (Ohlhorst and Schön 2015, p. 276).

Situational Analysis is a post-modern renovation of Grounded Theory, and suggests three kinds of maps (Clarke 2003). The first includes human and non-human elements (such as artifacts, symbols or discourse) of the research situation and their relationships, the second type of map focuses on actor groups respective and arenas of negotiation, and the third type refers to argumentative positions (Clarke 2003, p. 554). Situational Analysis supports the interactionist understanding of the research situation in Grounded Theory (Strübing 2018(2004), p. 8), while integrating instability and power (Clarke and Keller 2012, 26, 37). Instability is understood as the dynamic reconstruction of situations through negotiations of (groups of) actors in reference to their respective technical, institutional and natural surroundings. This refers to pragmatist underpinnings of Grounded Theory; 'truth' is procedural (Strübing 2014, p. 101). It is enacted within the social situation (Strauss and Corbin 1990, p. 279), which is in turn constituted by the interactive (re)production of common meanings of symbols (Blumer 1973, p. 84). According to Clarke, the negotiations of symbols, the underlying features of the situation, occurs within groups of actors that are bound by shared commitments and resources. Interactions occur within and between arenas of negotiation (Clarke 2003, p. 554). Membership in these actor groups is dynamic and non-exclusive and dynamic. Actor groups may overlap regarding individual features. Arenas are issue areas wherein actors negotiate activities and symbols with respect to specific topics (Clarke and Keller 2012, p. 31; Clarke 2003, p. 553).

To address power, Clarke integrates a relational understanding of the term in mapping drawing on Foucault: "who is authorized and not authorized to make what kinds of knowledge claims about whom/what, and under what conditions?" (Clarke 2005, xxv). Inequality and power differences are reflected in relational analyses (Clarke 2005, p. 59). This inclusive yet relatively neutral understanding of the research situation, enabling an assessment of social and technical factors based on the situation itself, in combination with a relational understanding of power motivated the selection of Situational Analysis methods for this study.

This study is based on maps of the single case as well as comparative maps drawing on the analytical tools suggested by Situational Analysis. In Clarke's approach, both mapping and creating memos complement coding, the labeling process organizing data for comparison (Charmaz 2014, 4, 190), and theoretical integration (Holton 2010, p. 21). Mapping in this study alternated between initial and theoretical coding. Initial coding closely follows the data in ascribing codes to expressions, segments, or sentences (Charmaz 2014, pp. 109–112). It is often descriptive or summarizing (Charmaz 2014, 112, 115). Theoretical coding, the integration of initial codes, was based on the theoretical framework of social innovation and resource mobilization. This is in line with later developments of grounded theory applications, acknowledging theoretical assumptions as 'sensitizing concepts' (Charmaz 2014, p. 117; Clarke 2003, p. 559).

Combining situational analysis of community energy projects with resource mobilization in a comparative research design expands upon recent research explaining community energy projects based on their situational analysis (Pohlmann 2018). Mapping enabled both a nuanced understanding of each community energy project as well as insights into overlapping issues and similarities among the different cases. The following sections explain how Clarke's mapping tools were adjusted pragmatically to the research focus of social innovation and resource mobilization while keeping the epistemological and ontological underpinnings of situational analysis.

4.3.1   Mapping the Single Case: Innovation Biographies of Community Energy

The first steps of analysis were taken by mapping the single case based on document analysis and interview data. Transcription, mapping and memo-ing proceeded at the same time. Analytical progress was recorded in memos (Clarke 2003, 560, 563, 570–571). Three kinds of maps were created for each case.

The first type of map records all relevant elements of the research situation. This includes individual and collective actors of the research situation, and factors of the technical, institutional and structural context: installations and technologies, strategies and planning documents, norms and values pertinent to energy services and/or community engagement in the projects' settings, related public and/or policy discourse, policy instruments, or relevant external events (Clarke 2003, p. 554). If document analysis and interview data suggested an actor (group), document, installation, discourse etc. to be relevant to the research situation, they or it were included in the map. Initial mapping considered without hierarchical or other organizational categories. The purpose of this mapping type is precisely to "intentionally work *against* the usual simplifications of scientific work" (Clarke 2003, p. 559, original emphasis). Then, relations of each element on the map with every other factor on the map were indicated and characterized (Clarke 2003, p. 569). This focuses on relationships, showing how elements on the map feature in dynamics of the project and which symbols or other instruments are used in this respect (Clarke 2003, p. 554).

The second type of map groups all elements of the research situation into "arena(s) of commitment within which they engage in ongoing negotiations" (Clarke 2003, p. 559). This kind of mapping focuses on symbols, objects and activities characterizing relationships within the research situation. Actors that share a commitment to joint activities, as well as resources and ideologies are grouped together (Clarke 1991, p. 131). Within a community energy project, this might mean a group of farmers joined by investments in wind energy installations on lands individually owned to ensure the economic well-being of their agricultural business, while complying with larger community plans for sustainable local development. Interactions between different groups on a contested issue or activity can be organized in arenas. In this example, an arena of negotiation might center on wind energy installations. The group of farmers might interact with regulation on wind energy siting, grid connection, and financing, including the respective administrative and government entities, other commercial wind energy developers, citizen groups supportive of wind energy and those rejecting installations. Ideologies converging or conflicting within the arena could in this case be local economic development, sustainable energy production, environmental protection through emissions reduction, or dangers of turbine installations, but also belief systems underpinning regulation such as who should have the right to determine types of energy installations and their siting, or financial responsibilities for energy system change.

Arenas are not manifest or institutionalized, but instead provide an analytic understanding of on-going negotiations and the actors, activities, objects and symbols involved (Clarke 2003, p. 559). Membership is gradual, meaning it may manifest in different levels of commitment; it is temporary, meaning it may fade as commitment fades; and it is non-exclusive, meaning it may may coincide, overlap and interact with membership in other arenas (Clarke 2003, p. 556). This analytic perspective focuses on relational dynamics in the community energy project and its immediate surroundings. It points to differences between actor groups within the project, and resulting differences in the resources or ideologies these actor groups employ. And it indicates that the technological, institutional and structural context of projects may look very different from the perspective of the project, even when projects are located in the same regulatory context, because of the differences in the actors, activities, objects and symbols involved. For example, while the Scottish projects FDT and GEM are both located in areas with geographically beneficial conditions for wind energy installations and could both benefit from feed-in tariff (FIT) regulation for renewable energy generation at the time of their initiation, choices of technology and organizational form differed in reference to negotiations on the local level, as well as differences in the interactions with regulatory agencies governing FIT implementation.

The third type of map records all argumentative positions assumed in the data, without ascribing these to individual or collective actors (Clarke 2003, p. 554). Argumentative positions and relationships between them can be analyzed independent of respective actor(s). Conflicting positions of a single actor can be analyzed without resolving this conflict. This keeps ambivalent or ambiguous positions in the data (Clarke 2003, pp. 556–560).

These different kinds of maps form the analytical basis of case analyses as presented in section three. To return to the language of the theoretical framework, mapping offers different perspectives on the innovation biographies of community energy projects, the development of project identities and actions within respective contexts over time. It concentrates on the diverse elements constituting the research situation, their relationships, negotiations, and argumentative positions. This forms the basis of assessing whether community energy projects can be understood as instances of social innovation, as well as comparing projects for patterns and similarities in their development and their conditions.

4.3.2   Comparative Mapping: Resource Mobilization

Maps were characterized by relating them to material (money, equipment, technology, land, etc.) and non-material resources (knowledge, voluntary labor, symbols, etc.). Relationships between different elements of the research situation were characterized according to the types of resources that were mobilized. Maps of arenas focused on the origins of resources, and the direction of their mobilization. Maps of argumentative positions provided insights into what constituted a resource within certain relations and conditions of negotiation. Figure 4.5 shows an example of how the author characterized maps of relationships by material and non-material resources using different colors. Argumentative maps of the project were consulted to understand the definitions and interpretations assigned to the relationship as well as the kind of resource.

For example, the map of the research situation for Samsø includes the discursive construct of 'Viking leadership' as an aspect of local community identity and a building element of its energy project. Argumentative maps connect this symbol to different definitions and interpretations in interviews. They show how the reference to a local tradition of individual leadership carrying a positive image of island traditions ('Viking leadership') was linked to a similarly established local image of 'good housekeeping' as an understanding of responsibility for one's affairs. These images were actively connected to the development of renewable resources on the island, creating a shared discursive backdrop for new activities. Land, as another example, could feature as material (natural) or non-material (symbolic) resource within the same project depending on the mobilization process. The community energy project on Mull could make use of publicly owned land for renewable energy generating installations, a material resource, under certain regulatory conditions. Using of the land involved navigating its discursive meanings on the island, which related to land as a symbol of community and self-reliance given the confined conditions of living on limited land—an island— together, as well as images of wilderness and nature. Land also featured as a tourist attraction, creating an influx of material resources through other economic activities of islanders.

Maps tracing arenas of negotiation show how a (type of) resource was mobilized within the project: where did it originate, who contributed to its definition, who employed it and to what end, and how does it feature in relationships between different (groups of) actors within the projects, as well as the arenas of negotiation the project engages in. These maps also show, whether a resource originated within the project itself or was part of the technological, institutional or

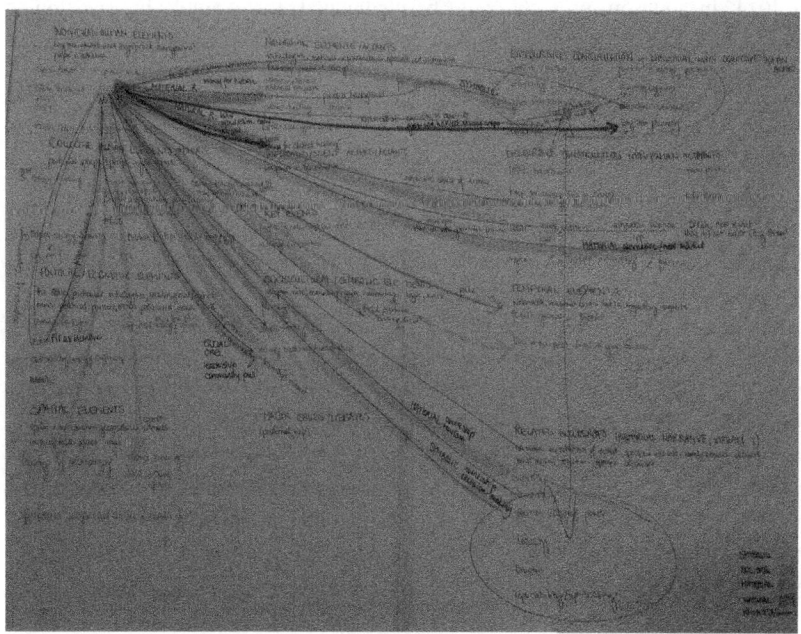

Figure 4.5   Resources/Situational map. *Source* AC

structural context of the project, such as a pre-existing narrative of good house-keeping or a government funding mechanism which could be tapped into. This offered an understanding of the path of mobilization.

Analysis of resource mobilization also indicated that the types of resources relevant to the project varied over time. Three distinct stages of project development occurred. These were project emergence, referring to project initiation and early mobilization, establishment, referring to the completion of the project's self-proclaimed first goal, and maintenance, referring to the upkeep of project activities after establishment. Establishment does not necessarily refer to success, but to the completion of this first phase of project mobilization which could occur either through successful resolution of a goal or through acknowledging that this goal would not be achieved and moving towards other activities. EWS, for example, emerged in 1986 and establishment was complete once grid operations commenced in 1996, whereas olegeno emerged in 2011 and establishment was complete once the city council had assigned grid operations to a competitor in 2014. In addition, projects could be challenged, for example by questioning the continued existence of the project or by failure of project goals. Challenges also showed a distinct pattern of resource mobilization, independent of the stage of project development they occurred in. Challenges in the case of EWS, for example, were the initiation of a referendum by its competitor to fight the concessionary decision during project emergence, as well as the withdrawal of political support for community energy actors in the 2016 REL (EEG) reform in the phase of project maintenance. Definition and mobilization of resources indicated similarities and differences across projects, both with respect to the development stages of project work and regarding the types of resource mobilized respectively.

Positional maps were aggregated across cases to create an understanding of similarities and differences in symbols across cases. These aggregated positional maps of a term indicated the material and non-material resources connected to a concept by project participants. Relational aspects of the term were explored to understand interconnections between resource types, such as the mobilization of organizational or knowledge resources based on symbolic resources. For example, while material aspects of ownership manifest in ownership of energy assets or project shares, non-material definitions of ownership may refer to symbols of responsibility and commitment.

Part III
Innovation Biographies of Community Energy

Innovation biographies trace social innovations—changes in power relations and improved individual and collective capacities—over time. The analysis is led by the three types of maps introduced above: elements of the research situation and their relationships, arenas of negotiation, and argumentative positions. For each case, this creates an overview of the research situation including actors, installations, and technologies, institutions and argumentative positions, as well as their respective interdependencies. Social innovations are interventions to the elements of the research situation and their interdependencies. Innovation biographies trace who and what mattered to the project and how these factors interrelated, how this informed negotiations, and what the dominant arguments and narratives were over time. Each case was mapped multiple times as the project developed. This provides the basis of comparison in the following section, where patterns of resource mobilization will be studied to discern similarities and differences in the quality and conditions of social innovations in community energy.

As the innovation biographies of each project unfolded, distinct features emerged characterizing community engagement in the respective settings. This was captured in five images. The project of Schönau was called electricity rebels, taking up the self-declared label of citizens as critical incidents to the nuclear energy system. The project on Samsø was nicknamed the moderator, referring to its strong procedural focus. Community energy organizers on Mull were called reluctant subversives, paying reference to their origins as a rather conventional energy generating project which turned into a groundbreaking co-operation between community energy actors and incumbents of the infrastructure system concerning grid access. The project in Fintry is called the pioneer, acknowledging its organizational and technological leadership in community energy solutions. Finally, the project in Oldenburg is called the catalyst, for its contributions to many local energy projects ultimately realized without the cooperative.

Energy Rebellion in the Black Forest

5

Starting as a parent's initiative against nuclear power, EWS today is one of Germany's largest independent suppliers of renewably sourced electricity and natural gas, and the only citizen-owned distribution systems operator. Members call themselves 'electricity rebels' to explain their vision for a nuclear free, climate friendly and citizen-owned energy system. Campaigning with the image of citizens as "accidents" (Störfall) to the nuclear power industry, the cooperative overcame resistance of the local incumbent and convinced political leadership to endorse citizen-led operations. The project's innovation biography is strongly characterized by founders' sustained commitment to its founding principles and their engagement over time. These characteristics have become important symbols of the German community energy sector and its hybrid character between social movement organization and entrepreneurship.

5.1 "Just Cut Out Nuclear Power"[1]

The nuclear devastation of Chernobyl in 1986 initiated EWS, shaping energy system choices and narratives of the cooperative to date. Upon project formation, nuclear energy technologies were deeply embedded in utility services of energy incumbents and a political commitment of the German national government as well as most regional and municipal administrations (Schreurs 2014). Schönau was no exception. KWR supported nuclear technologies and failed to respond to local opposition. The municipality backed the local incumbent's business model.

---

[1] "Atomkraft einfach wegsparen", slogan of early mobilization for EWS (Janzing 2016a).

© The Author(s), under exclusive license to Springer Fachmedien Wiesbaden GmbH, part of Springer Nature 2021
A. Colell, *Alternating Current – Social Innovation in Community Energy*, Energiepolitik und Klimaschutz. Energy Policy and Climate Protection, https://doi.org/10.1007/978-3-658-32307-3_5

Throughout Germany, civil society resistance to nuclear energy installations had been mounting since the 1970s. Although extensive, anti-nuclear protest had so far rarely swayed government support of nuclear installations and the utilities holding them (Schreurs 2014, pp. 11–14). The national energy sector and the anti-nuclear movement formed two distinct arenas of negotiation in the German energy system (figure 5.1). It was against this backdrop that the Schönau community energy project linked energy system services and anti-nuclear mobilization.

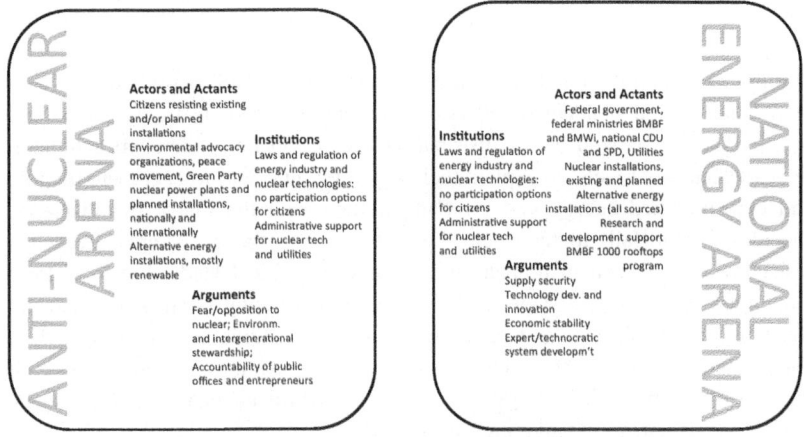

Figure 5.1   National energy and Anti-nuclear arena, Germany 1986. *Source* AC

Schönauers had not been part of anti-nuclear protests prior to 1986. Chernobyl formed a critical moment mobilizing local opposition among villagers and connecting the community to a growing national movement opposing the technology as well as its dominant supporting coalition of utilities and political decision-makers. Central elements to the mobilization of Schönauers included fear of nuclear technologies, fear for the safety and health of family, and solidarity with affected families. Collective organization in the form of direct action towards the energy system became a means to overcome a sense of helplessness. The initial ad in the local newspaper read: "Who is afraid for the future of their children and grandchildren after Chernobyl? Who wants to do something, too, and does not know, how? We are seeking comrades-in-arms, who like us no longer want to stand by as the environment is endangered by radio activity and chemistry!" (EWS 2018b). This call for neighborhood action was without definite ideas for the direction said action should take. It nevertheless assembles key elements of the cooperative strategy to date: Organizational and technology choices were and are dominated *ex ante* by the project's fundamental rejection of nuclear power and related industries. Citizen empowerment vis-à-vis incumbent energy system actors and institutions is a central goal (FUSS 2007). The founding group established the association EfaZ, "Eltern für eine atomfreie Zukunft" ("Parents for a nuclear-free Future"), organized relief and care for families affected in the Ukraine, and began organizing local resistance to nuclear power.

In the beginning, resistance referred predominantly to educating consumers and enabling them to take control of their consumption under the slogan of "just cut[ting] out nuclear energy" (Janzing 2016a). For example, EfaZ members installed functional electricity meters which had been discarded by the local utility and could therefore be obtained free of charge in every household to monitor consumption on the level of individual appliances. While consumption had previously been assessed only at household level, this enabled villagers to learn about consumption patterns of individual activities and appliances and alter their individual consumption as part of an organized group. Schönauers were not particularly energy-savvy, nor were they overly aware of or active in the sustainability movement. As one member points out: "I knew nothing of corporate theory and actions, nothing of the energy industry. It was all Greek to me" (I1: 328–329).[2]

Taking energy-related actions changed how project members engaged with energy service providers. This also included alternative ways of local energy procurement, with project members assuming energy system services without participation of the incumbent utility. Technologically, this included education on consumption patterns, empowerment and incentives to change behavior, and growing interest in alternative energy procurement. Organizationally, this resulted in the establishment of a parent's association, communication and mobilization throughout the village, the integration of local businesses through energy efficiency competitions, and ultimately the foundation of two companies to produce energy from renewable sources and assume grid ownership. Citizen actions forged new coalitions within the village.

---

[2] All German or Danish terms were translated to English by the author. Verbatim terms are included for campaign slogans or specific regional terms. For reasons of legibility and to protect individuals' anonymity in this publication, verbatim quotes are not included in the text. Speaking or being reproduced in a language other than the mother tongue conceals personal details which might make the individual recognizable. Special care was taken to avoid recognition of Scottish interview partners that are reproduced in their native language here. Verbatim quotes are available with the author upon request. For similar reasons, the pronoun 'they' (their etc.) is used in reference to individuals to obscur gender.

These initial activities coined relations between villagers and the local utility, KWR. EfaZ approached KWR, assuming it would want to support more sustainable consumption patterns. Instead, utility representatives threatened to sue the initiative for potentially damaging business (Janzing 2008). A founding member recalls the opposition of KWR as a key driver of subsequent citizen engagement, stating that "had KWR behaved in a cooperative manner back then, there would be no EWS today" (Horst Radny in Janzing 2016a). Spurred by the utility's dismissal, village households reduced consumption by an average of 20%, some even cutting their demand by half (ibid.). Local businesses donated prizes when the initiative started a tradition of energy saving competitions. Project members also initiated small scale installations of renewable energy generation (photovoltaics (PV), combined heat and power (CHP), and reactivated hydro power) in a newly established limited liability company named Gedea.[3] In a non-liberalized, vertically integrated energy market, independence of nuclear power required cooperation of the local utility. Energy generation, distribution and retail were handled by the same provider, in this case KWR, majority owned by Swiss shareholders. KWR opposed citizens' actions and attempts, seeking instead to reinforce its close ties to municipal government by prematurely prolonging its contract for grid operations. In 1990, citizens therefore founded the Netzkauf GbR (Grid buy limited) to gain control of local distribution grid operations and energy provision (Janzing 2016a).

The research situation upon project initiation included four arenas of negotiation (Figure 5.2). The project arena, indicated by the central circle shape, included EfaZ, Gedea mbH and Netzkauf, who mobilized villagers and local businesses in favor of alternative energy system services and against nuclear energy technologies, and found support of the oppositional party "Free Voters Association". The second relevant arena was that of local energy system negotiations. It included KWR and the municipal administration, led by the local Christian Democrats (CDU), as well as villagers in favor of established energy system structures. Locals could organize direct action refering to the energy system in the project arena. But changes to the technological, organizational or institutional parameters underlying the local energy system required negotiations in the local energy arena.

The national energy arena set institutional conditions such as the Energy Industry Law (EnWG) and respective regulation governing a non-liberalized market. According to these conditions, a non-responsive utility could only be replaced by

---

[3]Gedea was short for Gesellschaft für dezentrale Energieanlagen mit beschränkter Haftung (limited liability company for decentralized energy installations).

revoking the concession and installing a new grid operator that would then also procure and provide energy for local demand. The national arena connected the project and local energy system arenas to the general discourse and organizational setting of nuclear energy in the German energy system. It involved coalitions of economic and political decision makers in support of nuclear technologies, including the federal government led by the national CDU, which also controlled energy legislation. Support of alternative technologies was organized by the Ministry of Education and Research (BMBF), which offered financial incentives for small wind and solar energy installations. Conditions of ownership and grid management were not questioned by federal governments. Similarly, the anti-nuclear arena connected villagers to a growing and increasingly well-organized movement.

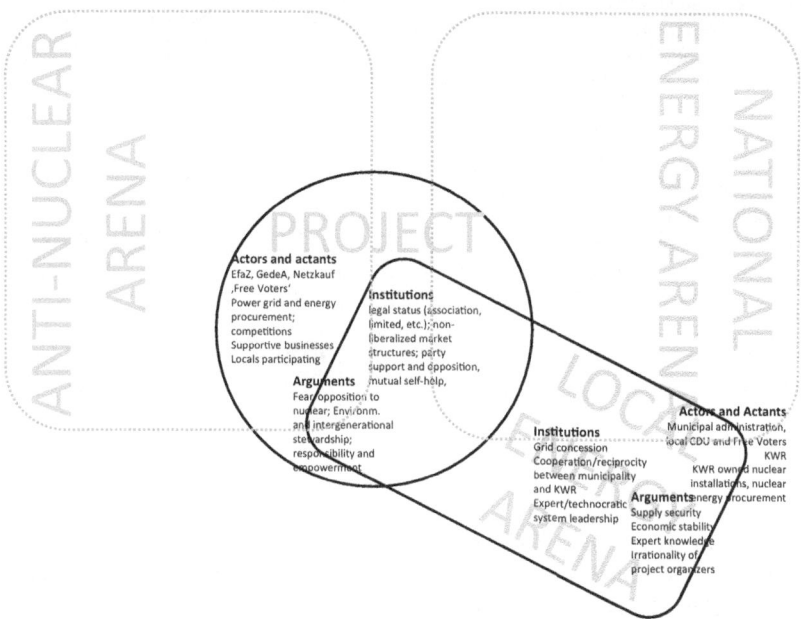

Figure 5.2 Overview of arenas in the research situation of EWS, 1986–1990. *Source* AC

Project members referred to nuclear technologies with frames of fear, opposition and active civil society resistance, as well as environmental and intergenerational stewardship. These positions were flanked by arguments of awareness, education and empowerment, self-reliance and responsibility, and collective organizing. Citizens' arguments challenged the unwavering support for nuclear technology, and dominant frames of security of supply and expert-driven system development shared by utilities and administration. The local utility sought to strengthen its argumentative alliance with the municipality using arguments about the importance of continuity and stability. The positions characterizing this stage of mobilization were predominantly influenced by organizational arguments, such as associative and entrepreneurial alliances or collective action, and symbolic arguments, for example "comrades-in-arms", fear and powerlessness, empowerment and defiance (Figure 5.2).

Figure 5.3 Legend, arenas & interactions (Figure 5.3 provides a legend to all illustrations of arenas and interactions within this section. Different geometric forms for project arena and arenas of negotiation were chosen to improve visual orientation. They are not indicative of analytical differences between different kinds of arenas.)

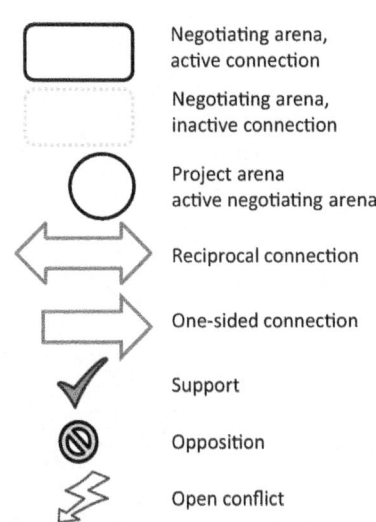

Negotiating arena, active connection

Negotiating arena, inactive connection

Project arena active negotiating arena

Reciprocal connection

One-sided connection

Support

Opposition

Open conflict

Arenas of negotiation are dynamic spaces wherein human and non-human elements of the research situation interact, for example EWS co-founders and villagers experimented with technology and behavioral changes to reach a shared goal of reduced consumption within the local project arena. Arenas of negotiation may also overlap, creating different connections between arenas, Figure 5.3 provides a legend of arenas and dynamics. Figure 5.4 summarizes dominant connections between arenas within the research situation of EWS prior to 1990. The local energy system and the project arena are embedded in the national energy system, for example through legal and regulatory frameworks. This shows that actors and actants of the research situation can be members of multiple arenas at

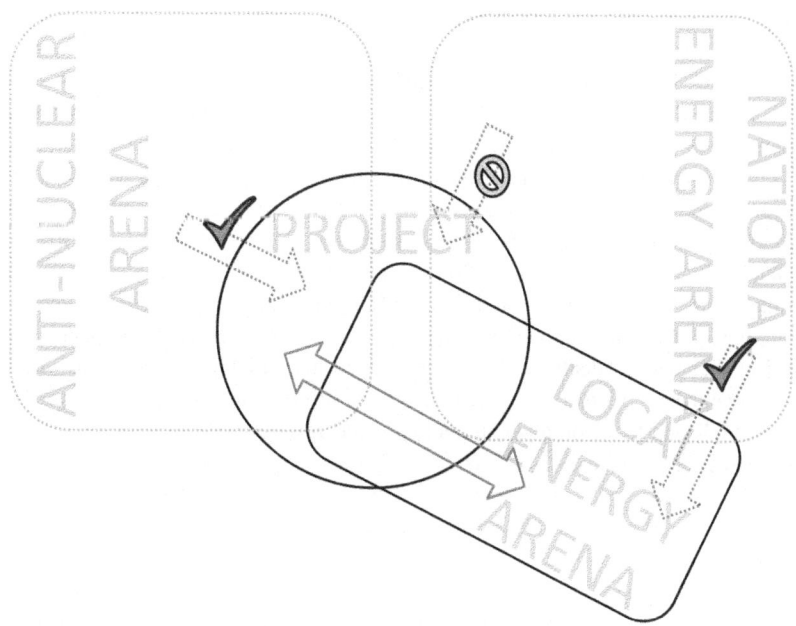

Figure 5.4   Arenas and connections, EWS prior to 1990. *Source* AC

the same time. This can be an expression of an inactive or 'back-drop' function of an arena. Prior to 1990, EWS held no active role in the anti-nuclear arena, but still found its argumentative positions embedded in the arguments shared by the arena. Similarly, the national and local energy arenas shared argumentative positions, as well as organizational similarities in close alliances between utilities and political decision-makers. The national energy arena opposed arguments of the local project arena. Dotted lines and one-sided arrows illustrate the backdrop function of these arenas, indicating opposition and support respectively. The project arena and the local energy arena are the only arenas actively connected in negotiations wherein each side presents own demands, indicated by continuous lines and a double-tailed arrow.

5.2   "Say 'Yes' to Schönau"[4]

After 1990, KWR began noticing a shift in local awareness about energy system services. It feared a political change. As the concessionary contract for grid operations, energy procurement and retail would expire in 1994, the utility offered to pay an increased concessionary fee (the money paid to the municipality by grid operators in return for permission to run service lines on public land) of 23.300 DEM annually over the course of the next four years if the municipality extended the contract by another 20 years (Janzing 2016b). EfaZ approached the utility, suggesting Schönau could be a real-life-laboratory for sustainable energy solutions, but KWR declined (Janzing 2016a). A utility representative dismissed project organizers "idealists": "We were treated like pitiable lunatics," one villager remembered the dismissive attitude of the utility (Janzing 2016b). Netzkauf was established to accomplish an alternative energy system "not with KWR, but against her. The gloves are off." (Janzing 2016b).

Negotiations on the local energy system did not change the setting of four arenas of negotiation, but altered their configurations, adding or subtracting elements of the research situation and changing the way these elements interacted. In 1990, KWR added money as an actant in the local energy arena and changing interactions between actors within the local energy arena, as well as between the local energy and project arena. The utility offered financial incentives to the municipality to prolong established contractual relations, which it could mobilize based on its business operations. The municipality had limited financial resources and struggled with unemployment rates of 10% (SPIEGEL 1996, p. 135). The community energy project so far had not required financial resources, even cutting household spending by reducing electricity consumption and related bills. To buy time to assemble a competitive offer to apply for grid operations against KWR, Netzkauf sought 250 investors who would pay 100 DEM annually to the municipality over the course of four years. Should the municipality award the concessionary contract to the community energy project, the money paid would be transformed into company shares including interest. Should KWR prevail in the concessionary competition, the money would be lost. Within four weeks, 282 shareholders of Netzkauf signed financial commitments (Janzing 2016b).

In January 1991, the municipal administration ruled to delay contracting decisions for three months, demanding a feasibility study including a financial,

---

[4]"A heart for Schönau – say YES" (Ein Herz für Schönau—sag JA), EWS slogan in the campaign for the 1991 referendum countering premature contracting between the municipality and KWR (EWS 2016b).

technical and economic concept for grid operations from the community energy project, as well as an indication of its ecological benefits (Vorholz 1991). The cooperative enlisted its first independent consultant, an electrical engineer. This provided a competitive advantage to the project, being "one of the few companies with access to the specialized knowledge of energy providers while simultaneously being independent of corporations" (consultant Wolfgang Zander, quoted in Janzing 2016b). The regional authority reviewing the feasibility study for the municipality was convinced by its concept, and advised the municipality against premature contracting with KWR and in favor of Netzkauf (Janzing 2016b). The local energy arena now included new actors: the national, independent energy experts the community energy project consulted became involved in local energy decision making, and regional authorities began monitoring municipal actions. The national press grew interested (women's magazine 'Brigitte' 1989 in Janzing 2016b; national weekly newspaper DIE ZEIT, Vorholz 1991). The project's position changed discursively. The evaluation of the feasibility study proved that citizens could present a competitive energy system alternative; proof "that you can build different structures… that energy operations are not rocket science, like large utilities like to make out" (I5: 35–40).

The alliance between the municipal administration and the local utility and their opposition to EWS' suggestions remained unchanged. Schönau's mayor called for the immediate acceptance of KWR's offer and an end of negotiations with Netzkauf. The utility and political decision makers presented similar arguments in the local energy arena. A declaration of the CDU disqualified the community energy project's application based on the "considerable economic risks in the case of Netzkauf operations" (Janzing 2016b), echoing the utility's classification of community energy organizers as 'idealists'. In July 1991, the municipal council ruled in favor of KWR's offer. This had been anticipated by Netzkauf supporters, who requested a municipal referendum annulling the council decision in the same meeting (ibid.). In an otherwise mostly unchanged setting of relationships between arenas, the dominant paradigm of interactions between the project arena and the local energy arena was conflict, with opposing positions on the technological qualities of energy system services and their organizational embedding between EWS members on one side, and the utility and municipality on the other (figure 5.5).

The community energy project mobilized the required support of 15% of the electorate for the referendum within four weeks, and the council was obligated to take a formal vote. The campaign of the community energy project again drew on organizational and symbolic frames to mobilize support. Various rally formats were chosen to engage different groups within the village electorate,

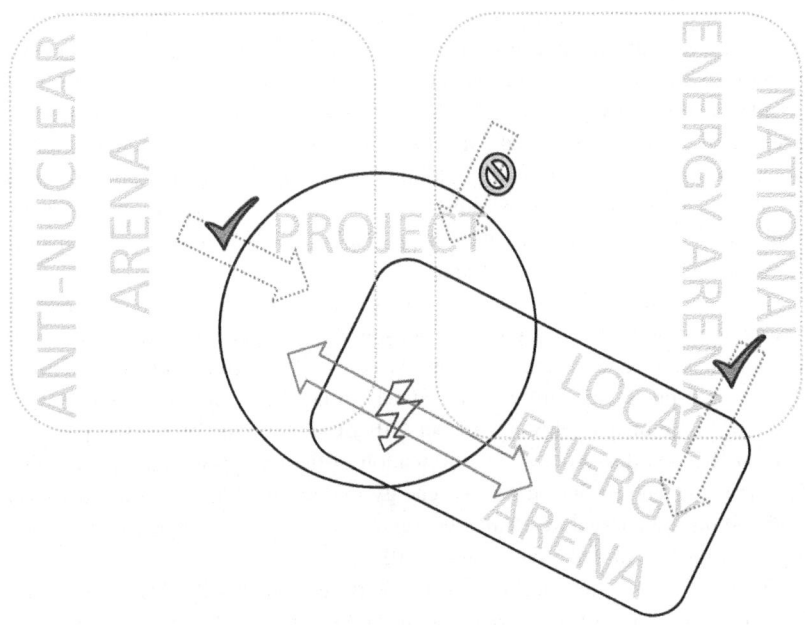

Figure 5.5   Arenas & interactions, EWS 1991–1993 (AC)

including the elderly and children, as well as outdoor activities. The utility, on the other hand, campaigned through brochures and information booths on markets, basing its communications on the assumption of a lack of technical and economic experience within the community energy project (Janzing 2008). On the day of the election, project campaigners baked 1.000 heart-shaped cookies reading "A heart for Schönau – Say YES" (EWS 2016b). "These hearts lay on every breakfast table," one co-founder remembered, "everybody knew what the right answer was" (EWS 2018b). The project had presented a feasibility study underscoring its position with technical, economic and environmental arguments to municipal and regional authorities. But the local campaign strongly relied on symbols of self-reliance and localism and suggested a shared understanding of what was 'right' in the village. With a voter turnout of 75%, 56% of the Schönau electorate voted to revoke the decision of the municipal council and accept the offer of Netzkauf to take over grid operations (Janzing 2016b). In February 1992, Netzkauf opened for business.

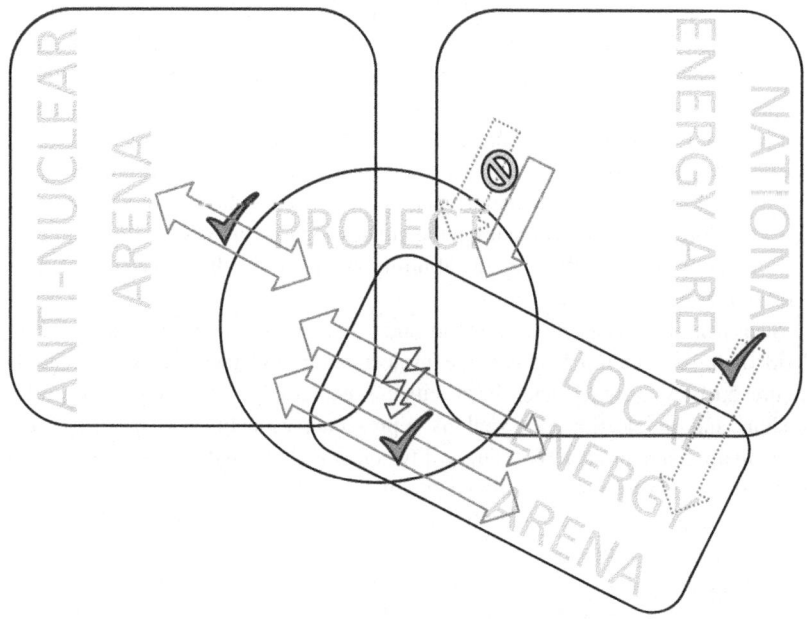

Figure 5.6 Arenas & interactions, EWS 1994–1999 (AC)

The argumentative positions of the community energy project did not change during this second stage of mobilization. Its activities in the project arena concentrated predominantly on becoming a grid operator. Two dominant narratives emerged: the community energy project itself was a protest against nuclear energy technologies and its ideas for community self-reliance and responsibility were to be forwarded by arguments in favor of the technical, economic and environmental feasibility of their claims. The alliance between the municipality and utility, on the other hand, extended its argumentation of reliability and expertise. In the process, they also sought to belittle the competition and reject citizens' ideas as utopian idealism. The successful referendum shifted majorities in the local energy arena. The political support of the electorate now outnumbered the opposition in the municipal administration. Its endorsement, in turn, meant that the project had to engage with the regulatory conditions of grid operations. This included compliance with the energy industry law (EnWG, Federal Republic of Germany 7/7/2005), which had to be approved satisfactory by the Ministry of Economics in order to obtain a permit for grid operations. This connection to the national energy arenas shows as a one-sided arrow in figure 4.6, which is neutral in that a requirement of compliance with regulation was neither framed as support or opposition. Again, the project chose a strategy of networking with external partners, cooperating with the local utility of a nearby municipality to prove its technical and economic capacities (Janzing 2017a). Municipal elections in 1994 changed the political make-up of the council; 'Independent Voters' came to hold a majority of seats. This secured a majority in favor of the community energy project when the council had to decide upon the concessionary contract. In 1995, Elektrizitätswerke Schönau (EWS), the local utility newly established by Netzkauf, was chosen as the concessionaire.

Seen from the perspective of arenas and interactions, this development openly divided the connection between the project arena and the local energy arena (figure 5.6). A growing actor base, citizens expressing their view in the referendum and in elections as well as representatives acting on behalf of the municipality, supported EWS. The utility remained opposed. Both parties could draw on the backdrop of the national energy arena. It was the right of the municipality to choose a concessionaire based on legal requirements and political favor. The regional authorities and the BMWi had cleared the technical capacity of EWS to perform as a grid operator, and had endorsed its concept. At the same time, the utility could draw on still-established coalitions of utilities and administrations on the national level, opposing citizen-owned grid operations. A coalition of the Christian Democrats and the local utility therefore began fighting the concessionary decision with a second referendum, asking to revoke the choice of Netzkauf and reinstate KWR. The answer in support of the community project was now 'No'. This was a challenge for campaigning, EWS running the danger of accidentally losing supporters as they unwittingly repeated the answer of the first referendum (Janzing 2017a). The utility mobilized heavily, opening campaign offices in the village, holding information booths and public debates, printing brochures, and running large ads in local and regional newspapers. Exhibition of expensive appliances and installations including price tags were hosted, the utility questioning how a community initiative should be able to mobilize such resources (Janzing 2017a). The village population was deeply divided: "Local restaurants are no longer chosen by their menu, but for their energy policy beliefs" (SPIEGEL 1996, p. 136). Campaign rhetoric of project opponents was hurtful and personal, for example accusing the village physician, co-founder of the community energy project and campaign leader, of manipulating elderly patients (ibid.). Voter turnout reached 85%, with EWS supporters scoring 52% of the vote (Janzing 2017a). The concessionary decision of the municipal council was confirmed, prompting project co-founders to call EWS the "world's most democratically legitimized utility" (EWS 2018b).

The amplification of concessionary dispute also activated connections between all four arenas of the research situation (figure 5.6). According to regulation, the municipality determines the political right to own and operate energy infrastructure through the concession. The new concessionaire, however, must then buy infrastructure assets from the previous owner. KWR demanded 8.7 million DEM, roughly twice the amount energy experts had calculated (SPIEGEL 1996, p. 136). The utility could exploit an ambiguity of the EnWG: principles of pricing are not clearly determined by the law (§ 46 Energiewirtschaftsgesetz; Berlo and Wagner 2013, p. 22). Price calculations are indicative of a fundamental discrepancy in the reasoning of the old and new concessionaires. While previous owners calculated fees based on the replacement values of all infrastructure components, new concessionaires typically based calculations on the performance value of installations. While EWS could have sued immediately for exaggerated pricing and this would have risked political support in a prolonged legal investigation. The project decided to pay the price calculated by their consultants and pay the additional sum with reservations, announcing to sue for exaggerated pricing (Janzing 2017b).

Mobilization of the exaggerated sum required donations. Company shares could not be raised this high. The Ministry of Economics would have withdrawn its consent for operations arguing that refinancing such a large sum from grid operations would threaten economic performance of the distribution systems operator. A co-operation with a German bank committed to sustainable investments enabled the project to commission a trust fund to raise donations. A campaign was drafted, coining the project slogan "Ich bin ein Störfall" ("I am a critical accident")—the phrase playing off of the words used to describe a major accident in a nuclear power plant (Janzing 2017b). Interactions with the anti-nuclear movement in Germany were intensified and mobilization drew heavily on its networks (see figure 5.4). Citizens were presented as critical accidents to the nuclear energy industry. The agency commissioned a nation-wide campaign free of charge, cinemas showed its video clip, radio programs broadcast the ads and national print media was engaged. Major environmental NGOs including Greenpeace and the World Wide Fund for Nature called for donations. Mounting pressure on KWR forced the utility to drop its price calculations to 6.5 million DEM, and then 5.7 million DEM. In July 1997, EWS paid up while at the same time announcing it would be suing the company (EWS 2018b). In 1999, KWR and EWS agreed in court to commission an external consultant to calculate prices, which the court ruling would then be based upon. The calculated value amounted to 3.5 million

DEM undercutting even EWS' calculations (Janzing 2017b). The parties agreed to settle, KWR repaying the excess money including interest. Beyond the financial significance for the energy project, this law suit initiated the now established German practice of performance-based calculations supported by court orders (Berlo and Wagner 2013, p. 59). The negotiating position of the new concessionaire in application processes was strengthened (so-called Kaufering ruling by the Federal Court of Germany 1999, see Berlo and Wagner 2013, p. 59; Becker and Templin 2013). This successfully concluded Germany's first take-over of grid operations by a community energy project.

The changes to relations, argumentative positions and negotiating arenas were extensive. For the first time that citizens entered the energy arena not to push for change based on discursive positions, which for example the anti-nuclear alliance had previously successfully done by squatting construction sites and building argumentative pressure on energy incumbents (Schreurs 2014). Instead, citizens formed a commercial organization and took over energy services (Figure 5.7). A narrative of anti-nuclear activism ('Atomkraft einfach wegsparen'), self-reliance and empowerment was created. This narrative was immediately connected to material participation, first by altering consumption patterns, then by changing production modes, and ultimately by assuming grid control. Direct action for local energy system change could draw on discursive frames of anti-nuclear protest employed on the national level. This created additional support. One direct form of support was the notion of a larger community of shared values beyond the local. As EWS' campaign progressed and the conflict with local decisionmakers and the utility intensified, the national movement also mobilized support beyond the local level, for example the endorsement of environmental advocacy organizations on the national level. By aligning political beliefs of the community energy project with the shared beliefs of the anti-nuclear movement and national environmental organizations, the Schönau project could therefore draw on nationwide networks to raise awareness, create public debates, and mobilize discursive and ultimately financial support. The community energy project also combined emotional narratives of 'grandkid-suitability', community empowerment and self-reliance with technical expertise in energy systems. The successful establishment of this agenda through the choice of EWS as concessionaire countered the predominance of technocratic arguments and dismissal of participatory or community oriented values in energy system services previously established in the energy arena on both local and national levels. The community energy project created new argumentative and organizational parameters of experiencing the local energy system. Energy system design was tied to shared values of project founders, which were debated in the larger community and ultimately established and legitimized through public votes.

The community project's relative position of powerlessness in the local energy arena changed. Upon initiation, the Schönau project was faced with a local utility unresponsive to its energy system demands. The utility was supported by the municipal authorities with whom they had long-term contracts. This coalition echoed parameters of the national energy system, where energy utilities were invested in nuclear technologies and had the support of the national government. With no free choice of supplier, potential actions of the community project were limited. Yet, limited options both gave direction to activities and provided a sense of self-efficacy in pursuing them. This created immediate experiences of community and success for those involved, for example through the competitions for reduced consumption. Community support for project positions grew, lending confidence to project initiators still campaigning against a coalition of utility and municipal decision-makers. The change of municipal leadership was also an expression of frustration with the former mayor and the leadership of the conservative majority, which although not uniquely caused by energy system conflict in the village was nevertheless tied to it. Two referenda, one prior to the change of municipal leadership and one afterwards, proved that the majority of the electorate supported the project's agenda, which was embedded in a national-level belief system of anti-nuclear protest, community participation and empowerment.

The community energy project did not make use of regulatory support provided for energy system change for taking over grid operations. Policy support for alternative energy systems concentrated on research and development for renewable energy generation provided by the Ministry of Education and Research, as well as a small FIT scheme after 1989. After 1991, support for renewable energy generation also included non-discriminatory grid connection and compensation of small scale renewables (Ohlhorst 2009, 114, 132; Mautz 2008, p. 54; Jacobsson and Lauber 2006, pp. 261–264). Regulation provided financial support for increased renewable generating capacities. They did not address alternative configurations of ownership, procurement and consumption. EWS could, however, make use of regulation enabling civil society participation beyond energy-specific contexts, namely the option of a municipal referendum secured by the regional constitution of Baden-Wuerttemberg (Baden-Wuerttemberg 11/19/1953). The referendum could legitimize project's actions and lent support to its narrative.

5.3   National Operations, Membership and Community of Intention

The project's nation-wide campaign for donations substantiated a tendency immi-
nent in the EfaZ campaign from the start: The creation of a community of shared
intentions (Kalkbrenner and Roosen 2016). EfaZ formed in response to a threat
which did not immediately affect the local community, the category previously
used to understand community energy engagement (Batel 2017; Chilvers and
Longhurst 2016). Rather, Chernobyl created a shared sense of urgency and responsibility driving villagers to act. Similarly, financial donors and supporters from
outside the municipality would not be affected by the change of grid operator
in Schönau. Yet, the struggle of citizens to establish sustainable energy system
services and responsiveness to local demands in service operations against an
incumbent utility playing dirty mobilized a community of intention, creating ideational support and raising millions. In the innovation biography of EWS as a
community energy project, the formation of a nation-wide community of intention based on sustainable energy system services and citizen participation was
reinforced by two developments: The uptake of nation-wide retail services following electricity market liberalization in 1998, and the legal reformation of the
project as a cooperative in 2009.

In 1999, EWS entered the national energy market, providing renewable sourced electricity (EWS today also offers natural gas, serving approximately 14.000
natural gas customers nationwide in 2017, EWS 2018a). Electricity retail is immediately connected to extending renewable capacity via a fixed investment in new
generating capacity, the so-called sun cent ("Sonnencent", see section 1.2.1). In
addition, the company requires its electricity to be sourced renewably from providers without connections to nuclear or fossil fuel generation, and 70% must
be sourced from installations younger than six years (EWS 2017, pp. 69–70).
The cooperative actively encourages reduced consumption and increased energy
efficiency. At 2.330 kWh p.a., its household customers undercut national average
annual consumption of ca. 3.500 kWh per household by roughly 30% (EWS 2017,
p. 61). Other independent renewable energy providers also showed reduced consumption rates. German provider Naturstrom reported an average consumption of
around 2.700 kWh in 2013 (Brieden 2013). Still, the savings rate was particularly
significant in the case of EWS, underscoring the hypothesis that EWS customers
formed a community of intention in support of company values. While retail
forges connections to customers, revenues are limited. In Germany, distribution
fees account for 25.6% of the retail price per kWh, regulatory fees and taxes for
another 55,1%, which leaves only 19,3% influenced by energy procurement and

services (Stromreport 2017). Profit margins in energy retail are very low. EWS income is more strongly determined by regulated service fees for distribution services and income from energy generation. However, analysis of market roles in community energy projects also indicates that, overall, profit expectations rank lower than in commercial energy developments (Agora Energiewende 2017).

Figure 5.7 Campaigning for effiency in competions for reduced consumption (above, 1992) and the PV "window of creation" (below). (EWS 2018; 8/28/2018)

EWS' initial organizational form, a partnership under the civil code (GbR), created shared but individual liability depending on members' private capacity to be liable. A GbR does not operate by representation. Return is similarly shared among all principals. Upon becoming the concessionary of the municipality, a limited liability company was established as a full subsidiary to professionalize operations of generation, retail, and distribution. A change in legal status to a cooperative in 2009 reorganized and further limited the individual responsibility of members. Still, it extended the opportunity to assume responsibility. More people could join the cooperative and support the cause of EWS in more than a simple customer relationship or political commitment, while establishing more

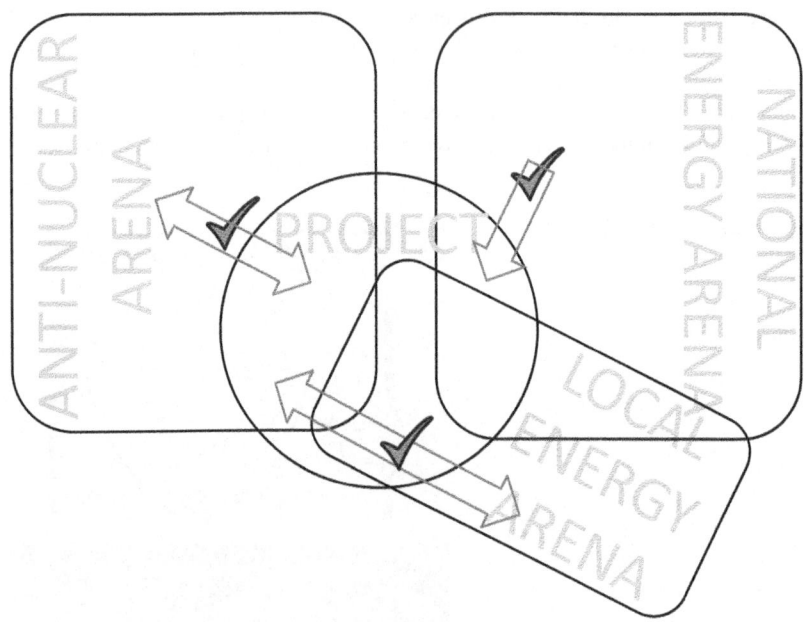

Figure 5.8   Arenas and interactions, EWS expanding energy services in 1999. *Source* AC

suitable legal conditions of organizational governance, liability and representation. The General Assembly elects a Board of Directors that selects and monitors
company management (GenG § 9, Federal Republic of Germany 2006).

   This shifted the company base from its strong local ties (the GbR had around
650 predominantly local members). Membership was purposefully opened to non-
locals in support of the cause and numbers rose significantly (to over 6.000 in
2017, EWS 2018a, p. 3), turning EWS itself into the locality of engagement.
EWS as an ideational community, in its specific configuration of technologies,
community based actors and partnerships and their rules and principles, replaced
the geographic community of Schönau as the rallying point of system change it
had been upon establishment in 1991. This underscores the important differentia-
tion between communities of affectedness based on locality and communities of
interest introduced above (Becker et al. 2017). Another example of EWS seeking
to forge such bonds is the prize "Electricity Rebel of the Year"[5] awarded annually

---

[5]A gallery of previous laureates is available online (EWS 2019).

by the cooperative since 2001 to people with an outstanding contribution to civil society based renewable energy systems. Remembering the spirit of rebellion of the 'nuclear incident' campaign in 1996, one interviewee said how "unholy motivations... are just as important. That I know I can kick one of those large energy utilities where it hurts, to put it bluntly, that they have to change... – I love that. To prick that dinosaur with a pin and make it squeal" (I3a: 285–292). Speaking in terms of arenas and interactions, this can be interpreted as project actors connecting actively to all three arenas and successfully creating a support base in each of them (figure 5.8).

Rising membership numbers changed the negotiations in the project arena. The project's actions were no longer exclusively determined by locals. Each member now had one vote in the General Assembly and could affect strategic decisions. Management and operations of the cooperative, however, remained local. The cooperative has roughly 100 employees, advancing to one of the principal employers and corporate tax payers locally. Said one on the cooperative's relationship to the municipality, "We [are] the largest corporate tax payer in the village. Of course, that pleases a municipality and implies a certain reciprocal behavior" (I2a: 163–64). At the same time, this person continued to underscore members' commitment to the locality, "try[ing] to dictate things to the city and threaten that we could imagine moving operations elsewhere or something like that. We don't do such rubbish. Because we are also Schönauers and Black Foresters" (I2a: 175–178). This quote speaks to the strong local ties of the cooperative, despite its national outreach. To date, the leadership team includes founders and their family members.

Thirty years later, the political trenches of the early years are mostly forgotten or resolved. Said one interviewee, "I don't know who in Schönau is a member of the cooperative. No idea. I talk to everyone just the same." (I5: 190–191). The cooperative has come to be locally accepted and established and membership is no longer polarizing. One interviewee remembers a previous opponent who had campaigned against the project heavily, who later said "he had been completely wrong. There could not have been anything better for Schönau, than the EWS taking over the grid. Some show a certain humility, to say, I was wrong. It was stupid to be against it." (I5: 214–218). Prior opponents could say they had changed their mind. At the same time, another member underscores the importance of members' own humility in success, saying, "it was always important to me, that... there is a certain culture of winning... that you don't let an 'I told you so' stance or malice come up" (I3a: 207–210). Said another, "[there is] no stigma, not at all. I don't even know personally who is an EWS customer and who isn't. I know that more than 90% of Schönauers are our customers, though." (I5: 193–195).

Yet, some council members reportedly remember the conflicts of early years and will bring up their criticism of joint projects between EWS and the municipality. While the council and EWS cooperate often, "there are two, three representatives from the 90s left that haven't quite found their peace and when EWS wants to cooperate with the municipality, they immediately say 'always the EWS'. More or less. Well, then you just have to say, do you have a different partner?" (I5: 198–201). The small size of the village and the relative importance of the grid operator have created a new incumbent of sorts.

Regulatory reform on the national level in 1998, implementing EU legislation on market liberalization, helped the community project. As an established systems operator, the project could extend services to the national level once free choice of provider was established, and due to the discursive ties to the anti-nuclear movement the local company soon found customers nation-wide. Upon establishment as a cooperative in 2009, EWS had approximately 90.000 customers (EWS 2018a, p. 34). By 2019, the customer base had grown to 200.000 (EWS 2/27/2019), EWS ranking as one of five independent providers of renewably sourced electricity.[6] The cooperative recently introduced a tariff for members, underscoring the relevance of cooperative ties in commercial actions for the organization (EWS 11/16/2017). The project became established as an independent provider of renewably sourced energy, investing in own capacities for renewable procurement as well (see for example Winterer 2018). The majority of local customers did not switch suppliers following market liberalization, another indication of prevailing ties to the local community. EWS has established itself as a local partner to the municipality. Regional co-operations with energy utilities have been added to the portfolio of the cooperative (EWS 2018a). This leads to the last aspect of the project's innovation biography: Its hybrid character between an energy utility and a social movement.

---

[6]Independent refers to independence of market incumbents in ownership and decision-making structures. Renewably sourced refers to 100% renewable resources and the investment of revenues in additional capacity. Besides EWS, this list includes Naturstrom AG, Greenpeace Energy GmbH, Bürgerwerke eG, and Polarstrom GmbH (Stromauskunft 2019). Lichtblick SE was bought by eneco in 2018 which in turn will be privatized in 2019, and no longer ranks as an independent provider (Zeitschrift für kommunale Wirtschaft 2018).

5.4 Between Social Movement Organization and Social Movement Leadership

Members of EWS describe the organization as a hybrid of a social movement and a company, saying, "EWS comes out of a citizens' initiative. It is part of the anti-nuclear and ecology movement today." (I2: 26–27). This refers to the project's origins and biography, but also indicates the prevailing importance of the social movement today. An on-going commitment to its values frames commercial actions. One member proudly speaks of the project's "societal effect... of encouraging citizens to act" rather than feeling "helplessly at the mercy of big energy providers" (I1: 69–78). The empowerment of citizens locally through immediate project structures should then be transported to other communities. The grid takeover in Schönau was seen as "the basis for us to say, just look what is possible if you are just committed and courageous and stubborn and keep believing in and working towards your success" (I1: 82–84). The formation of an energy company as a provider of innovative and sustainable services is "nothing special. You simply have to do that" (I1: 88–89). What sets EWS apart, in this person's view, is "the mobilization of citizens and citizen support. We support other cooperatives. We have a support program. We really do very very much in many different ways on the level of the citizens. That is what makes us a special company" (I1: 89–93). While the organizational strategy of EWS is special, members emphasize that the underlying structure of community was not. Said a co-founder on the recurring question of what was special about Schönau to enable this development, "Nothing. If this inconceivable, unbelievable success... was possible in Schönau, it is possible anywhere" (Ursula Sladek, in Potthoff, n.d.: 104). This is an understandable phrase from an activist seeking to inspire. Yet, the project's innovation biography does point to distinct organizational qualities.

The cooperative emboldened its members and secured their continued engagement. "Somebody was always on the brink of resignation," one member recalls the early days of the project, "and then the group catches that and lifts them back up and that is very important." (I2: 190–191). Another recounted the anxiety of mobilizing for the second referendum, highlighting how in a small village this meant going from door to door in a social setting where everyone knew one another, "I stood there staring at the doorbell like a deer in the headlights" (I1: 250–253). Members motivated one another when one of them encountered a neighbor's opposition, which could be personal and hurtful: "if you had encountered this on your own, it would have probably been days before you were able to push another doorbell." (I1: 258–259). The diversity of the group also balanced individual strengths and weaknesses. One co-founder said on the challenge of campaigning for years while the utility kept adding hurdles, "when the going gets rough, that's when I flourish" (Ursula Sladek in Potthoff n.d.: 104), while her husband was better in maintaining commitment on the even track (ibid.). The ability to complement one another is seen as a key leadership quality. EWS was an organizational expression of the energy political beliefs of its members and the community requirements in fighting for them.

Citizen involvement is seen as a productive form of power, to "rather be able to be a supporter, than have to be an opponent" (Potthoff, n.d.: 102). This idea shaped the founders' ideal of participation. EWS was set up to counter what another interviewee described as a frame "signaling to the citizen that he is not allowed to play; he should just consume" (I2: 69–70). EWS customers are seen as collaborators in fostering change. "We see our customers not as customers, really, but as comrades-in-arms," one project member described the relationship (I3a: 21–23). EWS customers should "participate directly in what is happening. Pushing responsibility to the forefront. That is why we are an energy utility of the second generation." (I3a: 23–25). Responsibility concerns individual actions in consumption ("cancel your subscription to bad energy", I4: 209), and is immediately linked to direct ownership of energy assets by EWS. "I am responsible for what happens with my money and that's why I am also responsible, somewhat, for what happens to the electricity grid," one member explained this connection, "I think it is important... that people say, 'yes, this is the right thing to do, that I am putting my money into such a thing'" (I3a: 99–103). Responsibility directly connects political and financial involvement in this understanding; the responsibility to develop an alternative vision of technical infrastructures and invest both engagement (organizational resources) and money (material resources) in its respect. Self-reliance and responsibility also drove the mobilization of knowledge resources. Citizens giving their time developed new skills and capacities within the project to support its cause, "I knew nothing of corporate theory and actions, nothing of the energy industry" (I1: 328–330), and established networks, for example with financial partners (GLS 2019). The understanding of EWS as an 'energy utility of the 2nd generation' builds on concepts of direct participation and responsibility of individuals in energy system change (Figure 5.9).

The focus on responsibility and obligation, or duty, is linked to seeing customers as 'comrades-in-arms' and fellow citizens embedded in communities, rather than consumers: "Citizen to me always means that you are part of the community" (I3a: 83–84). The community refers both to the locality and to the community of intention built across the arenas of the project and the anti-nuclear movement. EWS engages in campaigns directly (EWS 6/23/2015, 11/27/2015), working on awareness-raising nationally and internationally, as well as educational work (EWS 2016, pp. 28–31).[7] Cooperative representatives also fulfill various functions within the social movement (Bündnis Bürger Energie BBE n.d.; (EWS 2015, 6/23/2015)), or support other organizations, for example as honorary board members (BürgerEnergie Berlin 2017: 10). Educational or institutional support of EWS is free of charge. Investments are not tied to an institutional role for EWS or its representatives in the respective initiatives. EWS' commitment to underlying values of anti-nuclear and participatory energy system design also speaks from its investments in renewable energy capacities. The cooperative partnered with Japanese organizations to install PV in the aftermath of the Fukushima nuclear accident (EWS 11/15/2014), for example, and introduced a 'double dividend' tariff for members, wherein energy efficiency is rewarded by investments in renewable generating capacities and revenues are shared between EWS and the customers that increased efficiency (EWS 8/28/2018). Corporate connections in this respect include EWS holding minority shares in three utilities that share its visions for energy system change (EWS 2017, p. 41).

As a social movement organization, the energy company provides energy services that express their political targets of a renewably sourced and efficient energy system as well as decentralized ownership and civic agency. In addition, its corporate activities provide financial resources to these social movement goals; this goes beyond immediate corporate action. Company resources are channeled towards ideological beliefs that originate in the underlying social movement, seeking to support others in realizing the accessibility and control over energy systems they themselves experienced (EWS 2016, 28–31, 41). The manifestation of political beliefs of the social movement through energy related services of the social movement organization is therefore a key function of the project. At the same time, the political dimension of energy services—like the payment of the 'Sonnencent' by EWS customers as part of the retail price—is often routinized. It continues without individual active endorsement, while the project ensures the institutionalization of its investment towards shared energy political goals. EWS customers trust the investment conditions without having to make an active decision on donating. Similarly, although membership numbers grow steadily (EWS 2018a, p. 30), not all members attend the General Assembly and make use of voting rights (I2b). The continued engagement of founders and their display of

---

[7]For reporting on financial investments supporting system transformation beyond core company activities compare: EWS 2016, pp. 28–31; 2015, p. 28; 2014, p. 25; 2013, p. 13; 2012, pp. 13–14; 2011, p. 10.

commitment create legitimacy and trust among members and customers. The social movement provides the background against which the EWS performs. Fostering change outside the local community is seen with caution. Critics within EWS refer to the limitations: "A beautiful strategy in theory" (I4: 193–196), but difficult to implement. "You cannot initiate this from the outside… to try to turn lots of tiny wheels and then to hope that they keep spinning by themselves and I can go elsewhere and turn more wheels," (I4: 202–207) one member explained the difficulty of sustaining engagement in new locations after investing money. Consumers have direct agency. "To cancel your subscription to bad energy," this person described the option to switch to renewably sourced energy services of EWS as a household, "is [a] good [option], because it is persuasively easy" (I4: 209–210).

Beyond an energy political agenda, founders were seeking empowerment to overcome their own lack of knowledge ("wanting to do something, but not knowing how") and their exclusion from decision-making in the energy system ("seeking comrades-in-arms to no longer stand by", both citations from the founding ad, EWS 2018b, see below at figure 4.9). EWS provided an umbrella, enabling its members to directly engage in energy services (reduced consumption and production, later distribution), to directly invest in energy infrastructure, and to mobilize politically for their targets on a municipal and later regional and national level. While, for example, members of the cooperative EWS combine civic engagement and economic investment, its customers (many more than members, yet more indirectly connected to the cooperative) show their awareness through a considerably reduced annual consumption.

Being a community energy project does not result in community support in all contexts. After establishing grid and retail operations, EWS began more proactively pursuing larger installations of renewable energy generation. The first jointly realized PV installation of EWS in Schönau was set on the roof of the village church and nicknamed the "Schöpfungsfenster" ('window of creation', see figure 5.7). The church's slightly elevated location near the center of the village placed renewable energy production in a highly visible position within the village; a symbol of the significance of the installation, and the consensus required for such a visible change to the locality (Herrberg 2009). EWS has also faced opposition against installations. Focusing on PV, combined heat and power, and the reactivation of small scale hydro in the region in its early years, EWS began engaging in wind power installations in 2012. Since 2015, it holds minority shares in several wind farms, and in 2016 its first wind farm began operations at the local Rohrenkopf, after four years of planning. Current plans for future wind developments, however, are facing increasingly harsh opposition of citizens in respective areas, based not on the organizational form but on a systemic rejection of any wind installations (project meeting, June 2017).

Figure 5.10 gives an overview of arenas and connections in the research situation of EWS between 2009 and the present. Two aspects stand out.

First, members were able to forge strong interpersonal relations to sustain engagement within the project arena, and extended these throughout the community once political endorsement had been won and grid operations had ultimately changed hands. The formation of a community of intention beyond the limits of the village through campaigning, and later through membership and an inclusive understanding of customer relations as a challenge to do one's part in energy system transformation, point to the project's extraordinary ability to create interpersonal commitment. This also shows in the local energy arena, where most conflicts with the municipality and prior opponents are resolved and a mutually beneficial relationship has been established.

Second, the community energy arena formed as a fifth arena; the result of growing civil society engagement in energy production and consumption and corresponding policy changes. EWS holds a resourceful position in this arena, fostering activities in other communities. EWS provides organizational, knowledge and often even financial resources to community projects. "We are… sort of like the 'Big One' that you lean upon. Where you can find help on various questions, company law, project management, questions on future energy systems in general, and–I won't deny it–money," one member described EWS' commitment to fostering community energy development in other contexts (I2: 164–169). Smaller organizations indeed lean on EWS as a source of knowledge, support and even financial resources (BürgerEnergie Berlin eG 2018, p. 10; olegeno 8/22/2013). The cooperative does not advertise their willingness to provide support one member described the process, "because of course, all of this creates a lot of extra work and is exhausting and we have enough operational work as it is, so you won't necessarily go out and look for extra work" (I2: 205–208). In addition to creating excess work for EWS team members, external involvement in a community project can also create adverse feelings among locals. "If small cooperatives manage by themselves, so much the better," this person concluded (I2: 213). This relatively resourceful position of EWS vis-à-vis the community energy arena is illustrated in figure 5.10 by setting the supportive symbol of the checkmark within the project arena. The growing community energy sector in Germany also provided advantages to EWS, as community-owned energy system services were established in other contexts and provided further proof of concept.

Figure 5.9 Raising
awareness Above: 1986
founding ad. Below: 2012
campaign ad (EWS 2018b)

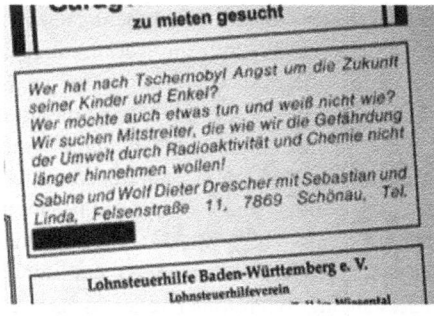

The formation of a community energy arena was driven not only by examples and supportive partners such as the EWS but also by structural conditions favoring community engagement in energy system services through the introduction of the German Renewable Energy Law, or EEG (see section 1.3). Technology specific feed-in tariffs for renewable energy installations reduced investor risks considerably and created strong incentives for small scale, often community led investments in renewable energy installations. On-going regulatory reform at the national level, however, has since withdrawn support from community projects and energy sector reform. From a regulatory perspective, this has created a mixed bag for renewable energy generation and community energy projects, as projects are not excluded from energy services by law, but receive little support vis-à-vis incumbent actors of the energy system. EWS has repeatedly campaigned at national and EU levels, challenging political decision makers, utilities and regulation. But while the project has created a community of intention of members beyond the local, and its position vis-à-vis local, and to some extent regional, political

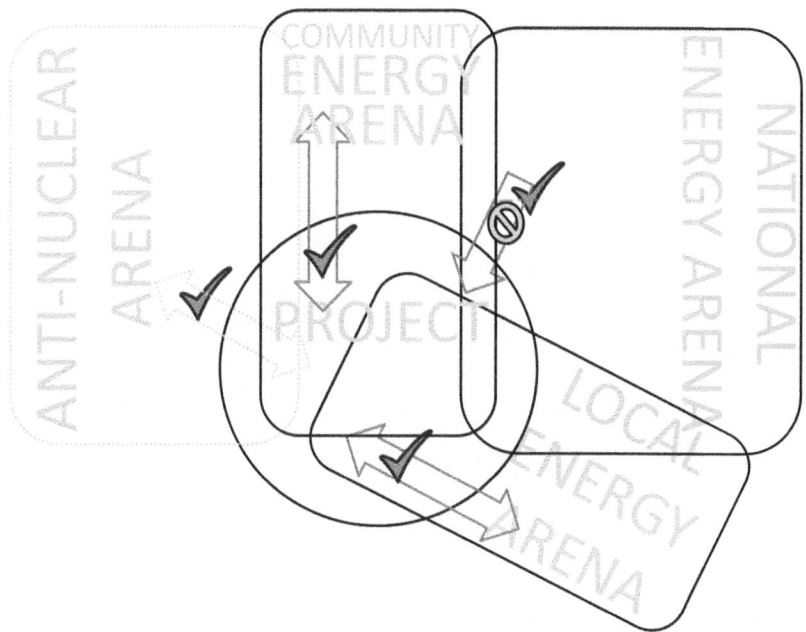

Figure 5.10 Arenas and interactions, EWS hybrid social movement / social movement organization. *Source* AC

decision makers has grown strong, influence on national decision-makers remained limited. EEG reform withdrawing support from community energy actors, for example, could not be prevented. Figure 5.10 also indicates that the anti-nuclear arena has faded. EWS will re-engage in anti-nuclear activism from time to time, for example mobilizing at European level (EWS 11/27/2015), but the anti-nuclear arena no longer provides active impulses to the community project.

Islands: Samsø and Mull

6

This second chapter of innovation biographies presents community energy projects on the two islands of Samsø, Denmark and Mull, Scotland. Beyond the basic similarity of geography, these cases share a pragmatic, rather than ideational, impulse for community energy engagement, and a strong role for project management. Shared features of the islands' innovation biographies also point to similar social functions of their geographies.

6.1   Samsø Fossil-fri ø: The Moderator

Nestled in the Kattegat, the island of Samsø was neither particularly environmentally conscious nor part of the vibrant Danish community of energy cooperatives when its mayor decided to enter the government competition for Denmark's "Renewable Energy Island" (REI). But it won. Within ten years, the island's energy system should convert to 100% renewables. Although not a bottom up initiative in the beginning, the island's Energy Academy has since succeeded in embedding energy system change in diverse citizen led structures on the island. It has moderated the ten year change as well as additional projects. The project's innovation biography is strongly characterized by procedural features of its institutionalization. The success of its leaders, in turn, has made them important figures of the national and international debate on energy system change.

© The Author(s), under exclusive license to Springer Fachmedien Wiesbaden GmbH, part of Springer Nature 2021
A. Colell, *Alternating Current – Social Innovation in Community Energy*, Energiepolitik und Klimaschutz. Energy Policy and Climate Protection, https://doi.org/10.1007/978-3-658-32307-3_6

6.1.1   Top-down Community Energy?

Denmark's most famous best practice example of community energy system change (SMILE, Smart Islands Energy System 2017) began with a mayor and a parliament representative planning to revive economic development in a community of the economically and institutionally outdistanced outskirts of rural Denmark (Papazu 2018, p. 203). The national government had opened a competition for municipalities to become the "Danish Renewable Energy Island" in the late 1990s. A local energy system should shift to 100% renewables to create an example of how the Danish commitment to emissions reductions of the Kyoto protocol could be brought to life, and showcase Danish technologies of renewable energy generation in the process. Communities could apply by presenting a master plan for a 100% renewable energy system within ten years. The program was not designed to promote existing community energy projects. Instead, its aim was to showcase system change in a community representing the Danish average. Covering approximately 12% of its energy demand locally, Samsø was "no greener" than the rest of Denmark (Papazu 2018: 204), its strong agricultural sector even performing below the national average of organic farming (Papazu 2018, p. 215). "It couldn't be an open process from the beginning," one project member described the top-down decision of entering the competition and devising a master plan for energy system change (Lundén 2003, p. 36). One reason was a lack of community interest in energy infrastructure potential for local development. "The mayor was convinced by an engineer from the mainland who was a consultant and who made the first draft for our master plan," one interviewee described the decision (I24: 15–16). When Samsø won in 1997, the competition provided funding for two full-time positions (an engineer and a communicator) creating the island's "Energy and Environment Office" (Papazu 2018, p. 203).

In 1997, a small number of actors shaped project development on Samsø. Island representatives made use of opportunities provided by national policies in a position of relative powerlessness. The island's mayor and its representative in the national parliament initiated the application process, supported by an engineer and advisor in drafting the master plan. These actors were powerful locally, among other things because of their ability to decide for the local community and their access to the government application process. But the community's eligibility for funding lay in its relatively weak economy, and its lack of local engagement or leadership in energy system services (Papazu 2018). The master plan connected to few existing local developments. One district heating system had been established in 1992–3 in Tranebjerg, the island's largest and main village, driven by local citizens that set up a cooperation with the regional energy utility (Lillevang and

Nielsen 2002). But while a community driven project in the more direct sense of citizens advocating for and participating in the process of establishing community oriented energy services, the Tranebjerg district heating system had not sparked imitation and its organizers were not involved in drafting the master plan (ibid.).

In the late 1990s, the previously strong Danish community energy sector was stagnating (Mey and Diesendorf 2018). Denmark's cooperative ownership of renewable energy generating capacities dates back to the 1970s (Maegaard 2013). Policy support for domestic coal and renewable energy resources and a ban of nuclear energy technologies, backed by a strong anti-nuclear movement, drove investments in renewable energy capacities. Wind energy was especially supported as an abundant domestic resource and an opportunity of national technology leadership and economic development (ibid.). Policies incentivizing investment in renewable energies in the late 1970s and early 1980s also included requirements for local citizen involvement (see section 1.3). The late 1980s and early 1990s saw the largest involvement of cooperatives in expanding renewables capacity in Denmark (Mey and Diesendorf 2018, p. 110). But while renewable energy capacities were still increasing in the late 1990s, the policy context had changed. Growing public resistance to wind energy installations in the early 1990s resulted in policy reform. Planning regulation impeded community ownership and commercial project developers increasingly entered the wind energy sector (Mey and Diesendorf 2018, pp. 110–111). Municipalities in charge of planning permissions, in turn, were weary of making decisions for local energy development amidst an increasingly unhappy public (ibid.). Fortunately, the REI competition provided a comprehensive frame for system change, with a fixed and ambitious target as well as a date for completion. This created important institutional lock-in on the municipal level: "otherwise, from case to case from project to project, you have to apply for permissions to do all these things. It is not embedded." (I24: 72–73).

If the masterplan should come to life given a lack of consultation and participation in planning and application, and the national context of declining community involvement in and support for renewable energy installations amidst continuously high and increasing levels of deployment, community engagement was required. According to one interviewee the project seemed like a "technical thing that **might** be a good idea. And nobody really knew if it was a good idea because they couldn't see the consequences of it… the perspective was really like very very [sic!] theoretical of what this could be." (I24: 26–33, original emphasis). The primary task of the communicator was therefore to engage the community and empower it to define subsequent project conditions. In the words of one member, to make "this [community] ownership for real and go out and talk to people and define with them, what do **they** see in it and what does **this** project do for them and the perspective or the ideas of the community." (I24: 34–36, original emphasis). The first challenge of the project was therefore to connect the island's new title, the master plan, and the community.

Project initiation on Samsø differed strongly from that in Schönau. Three arenas made up the research situation in 1997: a national energy and a community energy arena, and a project arena created by the REI competition. The energy and community energy arenas formed the backdrop for project development (figure 6.1). The institutions of Denmark's national energy arena included multiple references to renewable energies and community involvement through policy and regulation. In 1997, policy support for decentralized renewable energy structures had peaked, however, and increasingly large scale commercial developments were dominating the RE sector (Olesen et al. 2002, see section 1.3). The community energy arena had grown since the late 1970s, driven by citizen and community leadership and corresponding policy design. Beyond policy support, institutions of the community energy arena refer to the ability of community actors to uphold own interests vis-à-vis commercial or political opposition and their importance for technology development as early adopters consulting on functionality (see section 1.3, also Maegaard 2013, pp. 69–73; Jørgensen and Karnøe 1995). The energy and community energy arenas shared key argumentative positions, including an anti-nuclear and pro-renewable consensus, and general support for community involvement in energy infrastructures. The project arena on Samsø was not created by but for the community. The argumentative positions of project initiators referred predominantly to economic opportunities on the island. Argumentative positions shared by the island community connected to its agricultural tradition and established practices of neighborhood support. Consequently, the connection to the community energy arena was inactive in 1997 (figure 6.2).

Project organizers largely concentrated on negotiations within the project arena. The project actively connected to the national energy arena as national institutions provided funding for staff and set a deadline to the implementation. But otherwise the national level was neither invoked by the community nor offered further support. There were no active connections to the community energy arena (figure 6.4). Søren Hermansen, the project manager hired in 1998, was a local teacher, activist, and musician, son of a family of Samsø farmers. He had a sense of locality and an immediate connection to a core argument of the local identity: farming. In the words of one interviewee, "he has got credibility if anybody has. That's **really** street cred or what you would call it in a rural area. He as a person [was] very very important." (I27: 56–58) Hermansen himself had initially been doubtful about applying as he was not an engineer. But his credibility

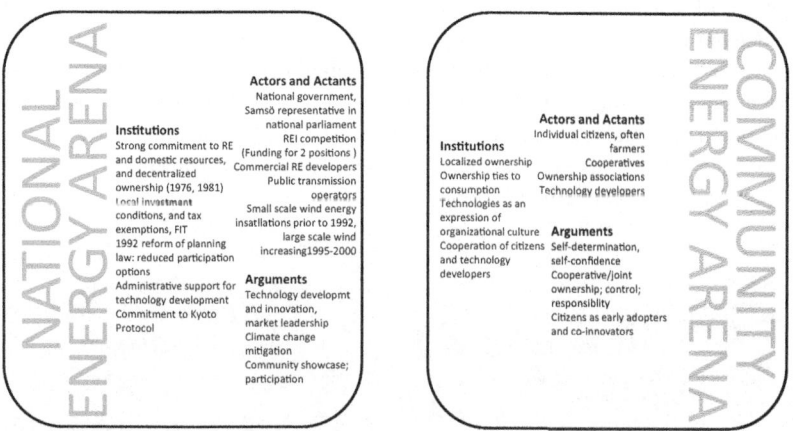

Figure 6.1   Energy and community energy arenas in the research situation on Samsø, 1997. (Source: AC)

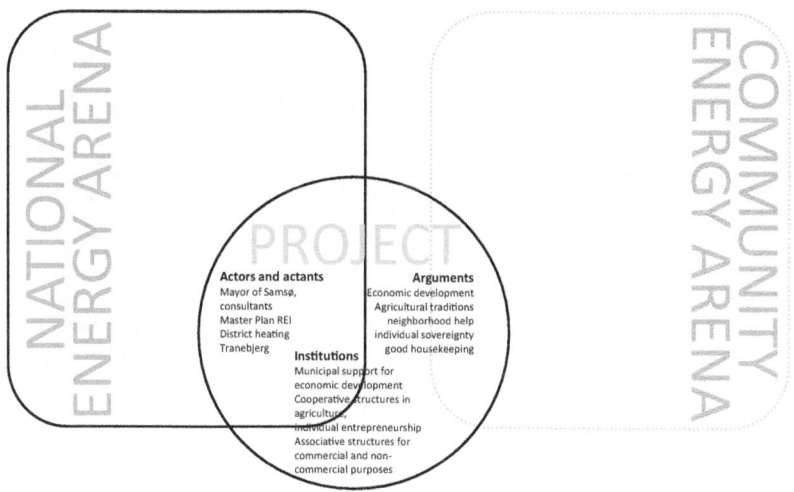

Figure 6.2   Project arena Samsø, 1997. (Source: AC)

countered a lack of technical expertise. A reliable and trustworthy person see-med more important, a person "that people know the parents of, the story. There

is a lot of info in that." (I25: 66–67, figure 6.3). The relevance of locality for
leadership is common to many community energy projects (Walker et al. 2007).
However, locality alone is insufficient, as exemplified in this quote pointing to
the individual abilities of community project leaders, "If you do not have this
ability to speak and create trust, then forget it." (I25: 71). The importance of such
leadership skills is also emphasized in the literature (Martiskainen 2017; van der
Schoor and Scholtens 2015).

Figure 6.3 Article in the local newspaper introducing Hermansen on March 10, 1998,
entitled "My task is to open doors" (Min opgave er at åbne dørene, Samsø Posten 1998)

The primary challenge after project initiation was embedding the goals for
energy system change into community narratives and established organizational processes. Technological features of the energy system were influential for
convincing the island's mayor, who had consulted predominantly engineers, but

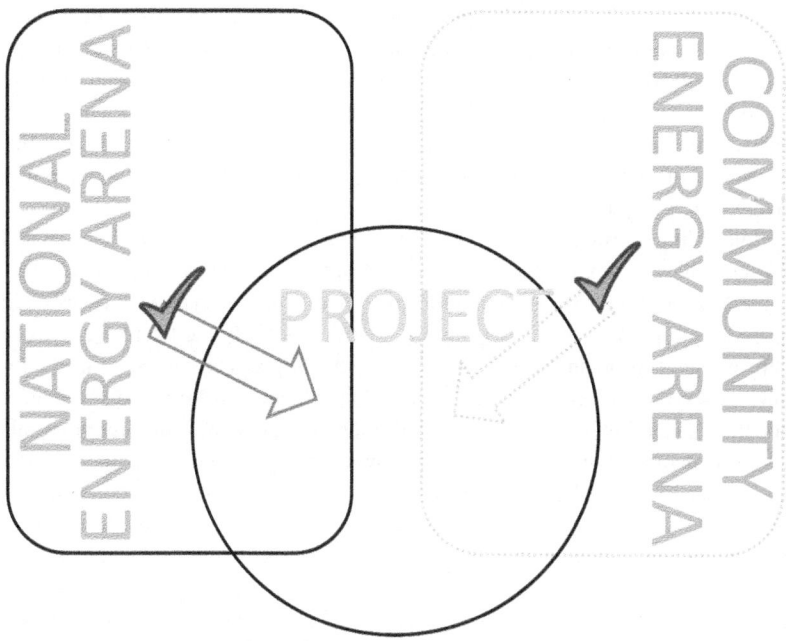

Figure 6.4   Arenas and interactions, Samsø 1997. (Source: AC)

they did not determine implementation. Project leader Søren Hermansen jokingly summarizes the importance of immediately grounding technology decisions in community values: "Engineers… know everything, but we shouldn't let them do everything. We should interact with them and talk about culture and beauty and fiction" (Hermansen 2013). Early stages of project work were determined by negotiations on different options of technologies, siting, ownership and financing. Ultimately, established and trusted technologies were chosen; wind, solar thermal and photovoltaics, and biomass (Lundén 2003: 96–97). The number of installations suggested in the master plan purposefully exceeded those necessary to achieve the target of 100% renewables locally (Papazu 2018). Proposals that found no support could be discarded without endangering target completion. The diversity of suggestions in the master plan acknowledged that realization of energy system change would depend on the community enacting changes, but also choosing which changes it was prepared to make. The master plan was publically available but not widely published to avoid an impression of centralized roll-out.

One project manager of the Energy Academy commented that this tactic created an impression of bottom up development, saying "it was made to seem like the project evolved from the bottom up" (Papazu 2018: 211). But beyond creating the impression of participation, this approach enabled organizational choices by local citizens and communities. Decentralization, again, referred to organizational aspects of implementation more than to technological features of a renewably sourced energy system.

Embedding transformative processes in established patterns of cooperation and cultural values was more important than determining how change should be achieved technologically. Cooperation and associative structures had been established on the island prior to the energy project. As one said, "In the telephone book… there is like 120 organizations for all kinds of purposes from soccer to gardening… everything is organized in a membership structure" (I24: 122–124). Project moderators could invite representatives of these associations to discuss proposals within their membership structures. They would then represent their members in larger discussions on the island. By inviting these organizations to cooperate, the Energy and Environment Office acknowledged existing organizational structures, "they were very aware of the importance of different groups being represented in the different projects" (I27: 92–93), and created additional streams of project communication. The underlying organizational paradigm, as another member points out, is one of cooperating within a group of people confined by geographical boundaries, "once you get to the water, that's it," one islander drily comments (I27: 73). One workshop participant points to the island "defining us", indicating that the geographical boundaries "throw us back on ourselves" (W1). The energy project could build on this shared sense of community and dependability. As one person points out, "we know who **we** are… It was not something we had to learn during **this** process. We live with it every day that we **have** to work together to make things happen… we know that we depend on each other" (I27: 74–78). In the local dialect, "skafning" is a standing term and roughly translates to "neighbor help" (Richardson 2013). This also speaks to the vulnerability of island life and the immediacy of experiencing natural resources. One project member recalls a heavy storm on the island and its influence on community life. "We were brought closer to each other and to nature," this member describes, continuing, "The wind has a special place in my life as I live so close to it. I have to be its friend" (Lundén 2003: 10). To a project member who moved to the island from Copenhagen his investment in an individual renewable heating solution expresses his commitment to the island, "I mean it. I am not just some clever guy from Copenhagen" (W1). While visitors would sometimes be surprised at individual investments that would not pay off economically, one project member emphasized how one had to "work together and stop putting money before everything else" (Lundén 2003: 100–101). Cooperation was an established experience on the island. Still, other members also point to the challenge of convincing islanders to surrender a sense of independence in favor of cooperative solutions. One interviewee said about the process of building cooperation, "This is a trust thing also. That you have to give up your individual leadership and then join the membership" (I24: 101–102). This person continued, "This local, individual leadership has been very interesting in a cooperative structure. Because this is very old; this is kind of the Viking leadership, how it has been for many many years [sic!]... embedded very much in the way we organize ourselves" (I24: 120–123). Responsibility and self-reliance therefore feature in many debates of project development. At times, "responsibility" refers to the commitment of the ten year time frame of the competition (Lundén 2003: 36). But more often it is phrased in reference to shared values. While energy system change could prove a strenuous topic of debate, one interviewee recalls how they could talk about 'good housekeeping' as a shared value of islanders: "'That's right, good housekeeping, that is what we do.' Alright, so we are on the same track now. Good housekeeping is also about making your own energy... So you have to kind of talk yourself into consensus about, what does it mean." (I24: 285–290) This sense of self-reliance as well as an almost irrational pride in alternative technology solutions also shows in individual realizations. Said one islander on his investment in solar heating, "it would take 410 years before the installations has paid for itself, but I haven't really thought about that, I just wanted solar heating" (Lundén 2003: 39).

Working on an account of the island's journey five years into REI process, one project members recalled the sense of needing to emphasize the social fabric of the technical installations that had begun appearing in the island's landscape, stating how this book "need[ed] to be about poetry,... need[ed] to be about art,... need[ed] to be personal" (I25: 29–30). Beyond discussing potential energy system installations with the various island communities and associations, the Energy Academy needed to ensure the islanders' lasting engagement with the overarching storyline (Papazu 2018). The Academy employed different communication formats, often convening large conferences for islanders. An early meeting of this kind commenced with an exhibition of local dreams for the island's energy transformation that the Energy Academy had asked participants for upon registration. "When people arrived... they were seeing all those dreams. And they were like, oh that's mine, what a nice dream you have... pretty basic, pretty banal. But, you know, people find themselves met, seen and mirrored," one remembers the effect of the conference on attendees (I25: 134–137). This person continues to emphasize the importance of this kind of work in community projects despite its sometimes limited recognition by members, fondly commenting "afterwards... some said, oh, it was so easy, we could actually have done it by ourselves... and I was like, okay, good art–you... feel... so invited into it, you are just moved into it." (I25: 137–140, original emphasis).

By connecting to the island's existing associations and communities as well as shared values and narratives, project moderators succeeded in filling the project arena with life (Figure 6.5). Villages, associations and individuals engaged with the idea of energy system transformation and began discussing suggested modes of implementation. This both reinforced existing structures of cooperation in reference to a new subject area and created new relationships as associations came together. This strategy also strengthened established community leaders, who had not been part of the application process or the newly created Energy and Environment Office, which avoided conflict on local leadership. Shared values such as 'skafning', 'good housekeeping', or 'leadership', were invoked and connected to energy system design. These measures actively embedded energy system change in established patterns and values of cooperation. The local project arena grew in relevance compared to the arenas of national energy politics and community energy. Previous connections between the municipality and the national government in the national energy arena during the application and decision-making of the REI competition were no longer key drivers determining implementation. Focusing on interactions with islanders within the community energy project arena also prevented a dominant narrative of powerlessness, based on the application of Samsø as a community in decline. Instead, islanders became empowered through their involvement in the process of decision making for and implementation of energy system change.

6.1.2   Success. What now?

Energy system change was successful. By 2007, the island's energy system relied on eleven onshore wind turbines covering 100% of the island's electricity demand, four district heating plants, as well as a number of individual installations for solar heating, heat pumps or wood burners outside service areas (Hermansen 2007). Transport emissions were reduced (targets set at 20%) and offset by ten offshore wind turbines (ibid.). Diverse ownership schemes were devised (Energy Academy 2011; Hermansen 2007). The master plan suggested technologies, but did not prescribe organizational terms. Ownership models varied depending on the size of the investment as well as organizational preferences. Offshore wind installations

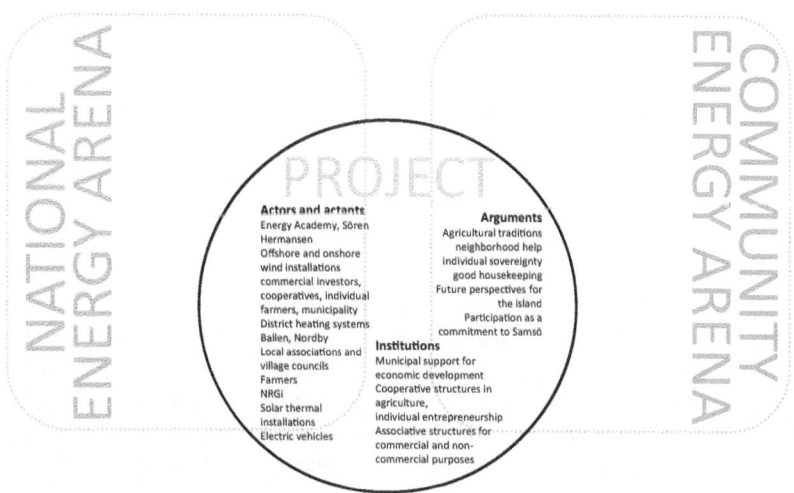

Figure 6.5   Project arena Samsø, 2007. (Source: AC)

required higher investments, as well as more institutionalized investment. Cooperatives were an established organizational model for joint investment on the island. They engaged with diverse technologies, including offshore and onshore wind, as well as building one of the district heating systems. Cooperative shares were popular. Many islanders wanted to invest and not all were ultimately able to do so as the number of shares was limited (Interview Samsø, June 2017). Onshore, nine turbines are owned by local farmers or small associations of farmers, and two turbines are cooperatively owned by various local shareholders. Samsø Municipality is the largest investor offshore, owning five turbines, with another three turbines owned by private investors, one cooperatively owned by local shareholders, and one owned by a professional investment fund. Both on and off shore, the turbines generate income via the Danish feed-in tariff. Of the district heating plants, two are commercially run by the regional energy utility NRGi (Denmark's fourth largest energy utility, cooperatively owned and based in Jutland, NRGi 2018), one is collectively owned by consumers, and one is owned by a locally based company. To realize the target of 100% renewables in 10 years, the island relied on nationally established technologies: Wind turbines (on and offshore), district heating, and individual solar heating systems. Installations included both units of power generation (wind turbines, biomass boilers, etc.), and distributive infrastructure (mainland grid connection of offshore windmills, pipes of district heating sche-mes, etc.). Local negotiations within the different island communities determined technology configuration and siting, resulting for example in three district hea-ting schemes based on local straw and one combining solar heat and local wood pellets.

Installations themselves became symbols transporting project messages. Simi-lar to the PV installation on the village church in Schönau, the onshore wind turbines on Samsø are among the most visible landmarks of change. Offshore installations are close to a major marine route around the island, project members comparing the turbines' iconic character to a lighthouse indicating land (Lundén 2003: 81). Visitors to the island immediately come into contact with its ambi-tious energy agenda as they travel on one of the country's first ferries powered by liquid natural gas (LNG). At the same time, as the turbines approach the end of their life cycle and blades are in need of renewal, investment conditions are being re-evaluated. Not all members share the understanding of individual ownership as a commitment to the island. Some individual owners, sensing adverse investment opportunities for refurbishment of turbines in the future, sold their windmills to outside investors from the mainland (W1). Others considered dismantling. This could endanger the net balance of energy system reform on the island. But dis-mantling of hardware installations is also seen as a threat to the project's narrative and success. "I never thought they might be gone someday," one member descri-bed her fear of this signifier of success and continuity being dismantled (W1). Islanders could make the same experience as community wind energy projects across Denmark in the late 1990s. In many cases, commercial investors took over ownership, offering attractive conditions of refurbishment to the cooperatives when turbines required renewal (Mey and Diesendorf 2018, p. 113).

The relevance of the project arena further increased upon achieving the REI targets (figure 6.6). The master plan had successfully been negotiated and imple-mented. This created a shared sense of pride and accomplishment among islanders (Hermansen 2007), which was independent of asset ownership. One workshop participant, an active participant in the negotiations for continuing sustainability transformation on the island today but without direct shares in an installation, said she still owned the system change of Samsø: "we own this project. This is us." (W1, original emphasis). Another islander referred to this as a sense of indirect ownership, which unlike direct ownership (for example through financial shares) would lead to "a greater and greater acceptance of the idea and people say[ing], publicly, this is a good idea." (I24: 65–68). In addition, completion of energy system change within the set time frame changed power relations in the national energy arena. Samsø had advanced to become Denmark's best known example of community-led energy system change. Energy Academy has since expanded upon this discursive position through extensive public relations and communication work, as well as its engagement in education and sharing of experiences (Papazu 2018).

Project representatives were very aware of this power. The sustained engagement of citizens for energy system change on Samsø, similar to Schönau, lent credibility to their claim. "I try to reconnect [to energy politics]," said Søren Hermansen, "because I have grown, my position is strong now, because I am the most famous energy person in Denmark." (Interview Samsø, May 2016). The credibility and (international) visibility of its success have created a mutually beneficial relationship with the national government. National representatives can present Samsø as a best practice case. On the other hand, project organizers also actively connected to the international level themselves, creating an international community energy arena reinforcing the project's position in the national energy arena (figure 6.6). On the national level, this reinforces the position of the Energy Academy when lobbying for continued support. One project member reports, "We have been kept on the national budget, this institution, even when it changed from a social-democrat government to a very liberal government and they cut **all** the cost on green innovation in Denmark, except us. We are the only one still on the national budget. And that is only because we serve them good." (I24: 432–436). Long lasting personal relationships both on community and on policy level are valued importantly: "most people change their jobs every 5 to 7 years… and then your network will crumble." (I24: 429–431). That the Energy Academy was able to maintain funding is ascribed to its lasting relationships to policy makers and its visible success. A story one workshop participant recounted of a visit by members of the royal family to the Academy indicates the self-confidence islanders draw from their achievements. The event was held in a locally established fashion, wherein a 'talking stick' would be passed among participants sitting in a circle and only the person holding the stick would be allowed to speak. "The king could only speak if he had the stick, he is no different from us," this person remembered, smiling (W1). Similarly, a local teacher recounted his students proudly telling a museum ward in Aarhus on a school excursion, that they were "from Samsø, you know, **that** Samsø" (interview Samsø, June 2017). This provides the local project with leverage in the national energy arena. It reflects positively on the community energy arena, wherein the project is now a positive example. But it also backs efforts of the Energy Academy on the island itself. Endorsement of the national level would also reinvigorate a sense of responsibility to lead, as Hermansen said, "I can then say to the people here, 'we have the support of the government, so we need to keep on working.'" (Interview Samsø, May 2016)

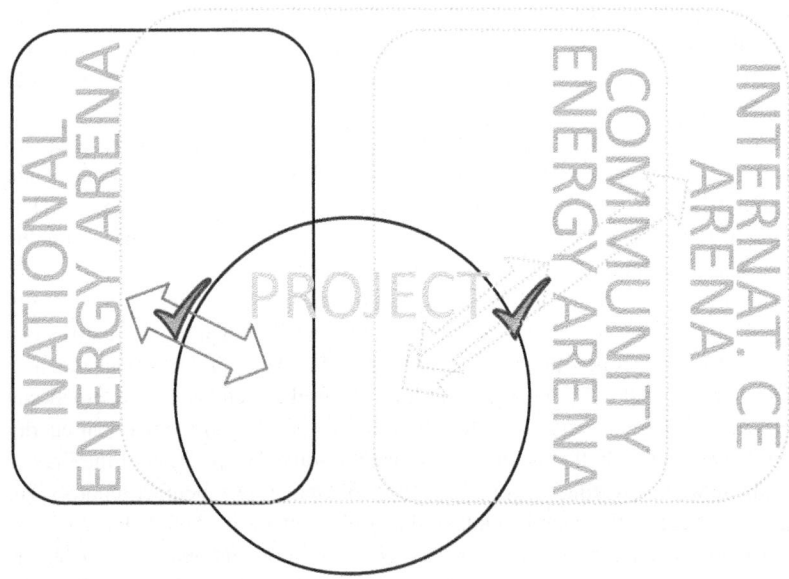

Figure 6.6   Arenas and connections, Samsø after 2007. (Source: AC)

Most installations on the island are owned by citizens and many islanders report a shared sense of ownership independent of actual ownership. Still, ownership of installations typically does not result in collectivization of financial benefits across community members. Ownership structures were shaped by the opportunity and ability to invest. Farmers feature prominently among investors, underlining their importance for the local economy. Local ownership could also be surrendered in favor of professionalized services, as in the case of the district heating plant run by the regional energy utility although it had originated in local mobilization. Some community members feel this loss more strongly than others, one commenting "it [the district heating plant] is a bit our own. But not in quite the same way" (I27: 30–31). This person continues to describe ambivalent feelings of assumed and actual ownership, saying, "I can't really say how I would feel about it [the offshore turbine installation, author's note], if I just owned shares. I think, I would have… another connection with it, than I have with our district heating plant" (I27: 27–28). The sense of community ownership was developed—and is cultivated—more in the establishment and management of the projects than in its legal manifestation. Said one interviewee on the process of establishing the REI, after Samsø was nominated, "ownership" needed to be "[made] for real", by people jointly developing shared visions for energy system change, "and slowly you could say we actually gained ownership of the project." (I24: 34–37). But in sustaining the island's activities, ownership is also exercised through the institutional make-up of many projects, where representatives of different communities and community institutions are chosen to sit on councils and boards of different projects. Community ownership on Samsø thus relies, to a large extent, on participation in negotiations, representation, and ideational ownership.

Both the broad discursive empowerment and the unequal economic empowerment speak to an underlying challenge of the community energy project in Samsø: The project's ability to deliver on the hopes for local development that had spurred the application. 'Udkantsdanmark' describes those local areas characterized by ageing communities, a loss of jobs and institutions (Papazu 2018: 204). The story of Samsø's energy system change is closely connected to economic self-help. In the local narrative, project initiation and the closing of the local slaughterhouse, the island's largest employer, are directly linked. Said one, "Those people were **very** frustrated about the possibilities of the future. And then this project could potentially produce some jobs. So, I think that was the main reason for a high percentage of the people to look into this as a possibility." (I24: 57–59, original emphasis). Another remembered "ha[ving] to meet the slaughterhouse people where they were—in crisis—they were thinking they were going to need to leave the island." (I25: 119–122). Members point to the importance of creating opportunities for those who had lost their jobs and were afraid of having to move to the mainland. But in fact, the closing of the slaughterhouse occurred well after project initiation, the narrative surpassing the actual chronology of events (Papazu 2018). The energy project did create economic opportunities on the island, for example through the opportunity to invest in energy installations and partake in financial revenues, or through the creation of local district heating networks and respective organizations. Economic opportunity, however, often benefitted those already in positions of relative strength.

Farmers are a prominent example. Farming is strongly engrained in the local identity, islanders often pointing to their family history in farming (such as in the case of Sören Hermansen) or referencing their personal connection as one workshop participant did when he introduced himself as "a teacher, but also a farmer" (W1). Energy system change opened new opportunities. Farmers invested in wind energy installations, or provided bio fuels for district heating. One interviewee remembers negotiating prices for straw to fuel a district heating plant, saying that while farmers sought a good price for their fuels the project itself had only just created this market which farmers needed to acknowledge. "There was no market before for straw, for fuel. Now there's a market. Thank you very much for doing that. So, we will offer **this** service for the citizens" (I24: 109–111, original emphasis). Yet, farming is declining. In 1998, the local economy was based on dairy and pig farming; today, local farmers today are struggling to find successors (Papazu 2018). Investment in renewable energies also offered an agricultural perspective, with families now 'farming' wind for electricity or straw for district heating. "This determined the future and perspective for the next generation on many farms," one islander and co-owner of an onshore turbine underscores (W1).Although local businesses were included in project realization as much as possible (Hermansen 2007), and locals estimated that the project had immediately created roughly 30 jobs on the island, economic activity created by the project mostly referred to a steady and growing stream of international visitors. This further strengthened the tourist sector on Samsø, but the island struggles to attract new businesses to set up on the island or maintain diverse opportunities of employment. Young people leave to pursue a higher education after the age of 16; few return. Samsø today has fewer inhabitants than in 1998. "No, no we can't," said Hermansen on the question of whether the Energy Academy could turn around the decline of rural community life on Samsø, continuing "We can stop the downward curve from getting even steeper. We cannot change direction. But we can certainly point out the weak spots of national administration for the rural area." (Hermansen cited in Richardson 2013).

While the relative power of the Energy Academy in the national energy arena has not translated to more political engagement for rural development, the Energy Academy was maintained as an institution. Much of its work centers on communication, education and awareness-raising on energy system change. The Energy Academy became a somewhat abstract institution to many islanders once the ten year plan was successfully completed. It continued to work on smaller, related issues as well as 'project maintenance' and communications. Islanders would have a general sense that the Academy is doing something for the island, and trust that it is a good thing, as one member summarizes a general attitude, "they tend to forget about things [...] only if we are threatened or something is happening... people wake up again and come back to the meeting." The Academy is happy to provide this background function. Said one "we have to keep the torch lit until next time there is something to meet about," and continues to underscore that the Academy through its institutionalization has the resources—time, personnel, knowledge—to provide this service, "that's fine, that's fair. That's a community at work. They have a lot of other things to worry about." (I24: 216–227) By maintaining the Energy Academy, a moderator for negotiations on system change on the island was institutionalized beyond the initial project. By 2007, the Energy Academy staff included an organizer of community involvement and participation. "I was part of the staff and I started to ask my bloody question again, I said, okay now, we **are** the Danish Energy Island... We cannot just go on. We need to redefine ourselves in this moment... And we need to make a community process as we did the first time" (I25: 180–185, original emphasis). To engage more parts of island life, the Energy Academy has since commenced moderation of additional transition processes focusing on complete independence of fossil fuels and the creation of a local circular economy (Kristensen 2015; Mathiesen et al. 2015).

Trust in the community energy organization often exceeded trust in the municipality. Although not personally involved in the REI project, one Samsø islander said, "I know Søren Hermansen, and he has done a lot of good work. Everybody praises him and feels they get good answers when they need it. The project is quite a mouthful for the municipality... I hope [the municipality] doesn't spend the balance before the windmills have generated it." (Lundén 2003: 39). The municipality initiated the REI project, but is not seen as the provider of the functions valued in the community energy project. This continues, as the municipality takes on larger coordinating responsibilities in current projects, such as the erection of a biofuel plant on the island—a much larger technical, financial and managerial challenge than previous installations. The municipality is proud of its more active role in energy projects, but citizens do not feel the same support of 'getting good answers when they need them'. Instead, islanders indicate their frustration over a lack of communication, "I do not understand why there is **not a word** from the people working with [the biofuel plant]. Nothing, nothing, nothing." (I27: 191–194, original emphasis). Islanders criticize this approach as the community moves towards more difficult aspects of system change: "these **very** large steps, we have to prepare even more" (I27: 195, original emphasis).

Similar to the small town of Schönau, Samsø's experience of the island's geographic limitations and the resulting tightly knit community can shape a perception of a shared locality as a prerequisite of cooperation (Hermansen 2013). But while technologies and installations can serve to reinforce the relevance of locality for example by partaking in a district heating system or living in direct view of a wind turbine, organizational functions and services can extend beyond the local community. Membership of the cooperative in Schönau was opened to enable participation of those sharing political ideals but not locality (EWS 2011). One Samsø project member speaks of "extended communities", saying "if you... **feel** connected you are invited to be a member. But it doesn't have to be the place only. It can also be the network in a way." (I24: 174–176) This person referred to the importance of alliances beyond locality, specifically in the context of acting as a social movement organization. The Energy Academy has attempted to facilitate such networks, for example through the conference series "From Best to Next Practice", wherein international experts are invited to the island biannually to discuss transition experiences (Flemming 2013). Without an overarching membership organization or connections through market relations, such as for example EWS could establish, it has been harder for the Energy Academy to build long-term connections beyond the local.

6.2   Green Energy Mull: The Reluctant Subversives

Energy aspirations on Mull resulted from the local Development Trust's search for new opportunities for a vulnerable island community. A small energy generating installation was planned, which would provide locally produced electricity as well as financial revenues that could be collectivized throughout the community. But new regulation limited grid access of renewable energy installations, casting doubt on the realization of the modest run-off hydro generating scheme planned on the island. Consequently, the Development Trust partnered with the Scottish association for community energy projects and market incumbents to prove the power of supply and demand balancing in island settings. The goal was to secure the grid connection of Garmony Hydro, as well as enabling communities elsewhere to follow this example.

6.2.1   Garmony Hydro: Just "clean renewable energy for the next 100 years"[1]

Not unlike Samsø, energy aspirations on the island of Mull began as a community oriented development rather than an energy political commitment. Quite unlike the Energy Academy on Samsø, however, institutionalization in the Hebrides provided a clear organizational setting and strong value orientation.

Enthusiastic about renewable energies, a group of volunteers began planning a local project within Mull Community Council in 2010, calling themselves 'Bold Renewables', "the clue's in the name," as one interviewee chuckled (I20: 343). Very soon, the idea was integrated into the activities of the local development trust, Mull and Iona Community Trust (MICT). A Development Trust is a

---

[1] I20: 274.

membership-based organization with the purpose of supporting the "social, economic, and environmental renewal" of a geographically defined community, which
although following no prescription for its legal form should contribute to local
value creation without generating private profits (Development Trust Association
Scotland, DTAS 2019). Many register as charities (ibid.). Financial sustainability is to be achieved through independent, non-profit operations, with returns
reinvested in the Trust or associated organizations of the respective community
(ibid.). The Development Trust also provides a quasi-public structure, by limiting
membership to locals—constituents of sorts—and tying reinvestment of returns
to the local community. Representing approximately 2.700 inhabitants on Mull
and just 125 inhabitants on Iona, the MICT was established in 1996 as a "local development agency [...] to formulate strategies and provide practical support
to local projects [...] improving the social amenities and physical and economic
infrastructure on the islands" (MICT 2019a). The organization was registered as a
Company Limited by Guarantee in 1997, and was granted charitable status (ibid.).
Membership is open to local residents sharing the organization's views and objectives and willing and able to pay 5£ in fees for a lifetime membership and voting
rights in the annual general meetings (AGM) as well as elections of directors
(MICT 2019b). In 2016, the trust had approximately 300 members accounting for
a little over 10% of local residents (Education Scotland 2016: 3). Part of the Inner
Hebrides, Mull is challenged by ageing and declining communities (although
Scottish isles overall are seeing an influx of investment and part-time residents
as holiday homes are acquired, (Alison Campsie 2016), comparatively high costs
of housing, fuel, or food (ibid.; Bunting 2015), low wages with islanders often
depending on multiple jobs in crofting, farming, fishing or tourism to support
their livelihoods, and a lack of accessibility (Education Scotland 2016: 2). The
situation on Samsø is similar in many ways.

   An inspection of the trust by Education Scotland and the DTAS emphasized
two key characteristics: Its commitment to accessibility and empowerment for
community members, and its efforts to hear diverse community voices in its committees and steering groups, including representatives from Mull and Iona islands,
as well as both community councils (Education Scotland 2016, pp. 2–3).Key
issue areas of MICT include public transportation and community accessibility
on the island for example by re-establishing and extending public transportation
and accessibility for remote communities on the island (Mull and Iona Community Trust 2019i, 2019f), the establishment of "community swap shops" (MICT
2018a) as well as second hand and food sharing projects (Mull and Iona Community Trust 2019c, 2019g), measures of environmental protection and nature
conservation in accordance with local tourism and respective job creation (Mull

and Iona Community Trust 2019d, 2019h), as well as childcare and health services (Mull and Iona Community Trust 2019e). Said one member of MICT "we try... to make the projects that we take forward relevant to the community. To make sure that... we are... taking projects forward that the community has asked us for." (I20: 376–379). The umbrella of MICT immediately framed the orientation towards community goals over energy political aspirations. Community interest in the Trusts' activities was key, rather than the availability of funding schemes or financial income: "if we just **did** it [enter a funding operation], well, people would say, 'Why are you doing that?'" (I20: 382–383, original emphasis). All aspects of a potential energy project—technology choices, siting, ownership, etc. –needed to be planned in consensus with the island's community.

With over 400 organizations across the country, Community Trusts have become an established organizational form based on their economic and social benefits to local communities (Education Scotland 2016: 4–5). The role of Development Trusts was further augmented by reforms to land ownership of the Scottish Government, notably in the Land Reform Act of 2003, and again in the Community Empowerment Act of 2015, and the Land Reform Bill of 2016 (Scottish Government 2015b, 2016a). The Land Reform Act of 2003 installed a "Community Right-to-Buy" allowing communities to seize land abandoned, neglected or used in ways harmful to communities independent of a willing seller (Education Scotland 2016). Claiming to bring "one million acres of Scotland in community land ownership by 2020", then First Minister of Scotland Alex Salmond in 2013 declared that the Scottish Government aspired to strengthen community ownership vis-à-vis private landowners (Scottish Government 6/7/2013). This framed land ownership emotionally, and linked land resources to the 'common good' (Land Reform Review Group 2014). The Community Empowerment Act of 2015 extended the Community Right-to-Buy to urban communities, and the Land Reform Act of 2016 included the Right-to-Buy in cases of planned sustainable developments (Scottish Government 2015b, 2016a). Legislation was reinforced by the Scottish Land Fund, awarding £ 9 million between 2013 and 2016 to communities buying back land and community assets, and commissioning another £ 10 million to this cause in 2016. Land reform and related legislation of the Scottish Government was criticized for weakening private ownership, and the ambiguity of poorly defined terms such as 'community ownership' or 'sustainable development' (Maxwell 2016; Combe 2016, 2014). Still, legislation and corresponding financial aid enabled many communities to substantially increase their asset base through acquiring public land. Much of this grant money was awarded to Community Trusts (Scottish Government 2016a). Shared ownership of RE resources, mostly community-owned onshore wind installations, was often realized in form of trusts (Friends of the Earth Scotland 2014).

Upon project initiation on Mull in 2010, three negotiating arenas therefore set the context (figure 6.7): a UK energy arena, including actors, institutions and argumentative positions of the national energy system, a Scottish renewable energy (RE) arena, referring to actors, institutions and arguments developed around energy system features devolved to Scottish government, and a Scottish community development (CD) arena, involving actors, institutions and arguments of the Scottish strategy for community support. The UK energy arena was marked by predominantly market-based strategies for increasing renewable energy capacities, which relied on established market actors and structures and had largely failed to meet the targets set by UK climate change mitigation strategies (see section 1.3). The Scottish government had chosen to interpret the devolution of renewable energy supoprt and energy efficiency (energy policy-making resided with Westminster, otherwise) in the context of its far reaching competencies for community development. This resulted in extensive institutionalized and financial support for community based actors, based on arguments of developing domestic resources and supporting especially vulnerable communities of Scotland's highlands and islands, as well as a certain competitiveness between Westminster and Holyrood. Its strong presence in local communities emphasized the relevance of the Scottish government for local welfare (Bunting 2014).

Technology choices on Mull were determined by reconciling interests in energy generation, landscape preservation, and community revenues. Wind energy opposition was expected to be very strong on the island, shutting down early considerations for local turbine installations. "There'll never be a windfarm on Mull," said one member ruling out wind installations (I20: 254). Strong opposition would have bound project resources, "we would be spending more of our time dealing with objections and, you know, defending campaigns against us, than we would be spending constructively taking the project forward in a positive way." (I20: 270–272). Opposition hinged on wildlife protection on the island, such as habitat conservation for golden eagles and hen harriers (see Bright et al. 2008 for an assessment of bird wildlife and wind energy developments in Scotland), but also resonated with larger regional trends. Scottish communities in the Southwest initially showed favorable attitudes towards wind energy developments, their level of support increasing with proximity to the installations (Warren et al. 2005) and community ownership (Warren & McFadyen 2010, based on a 2006

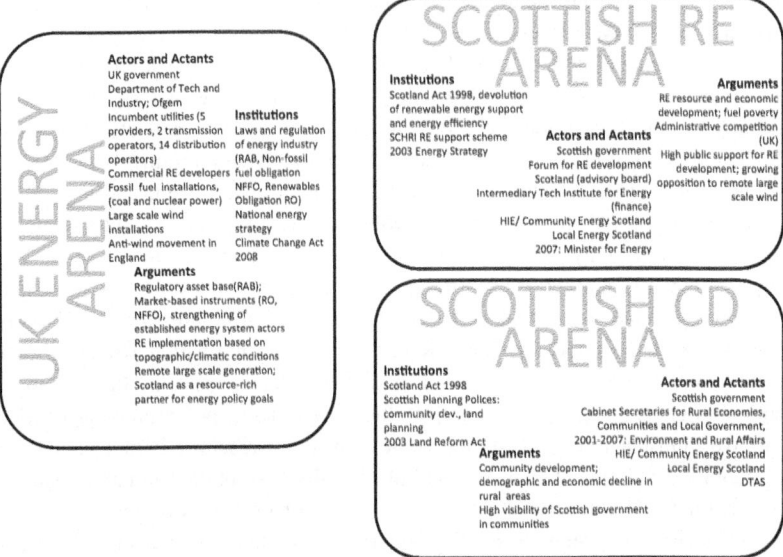

Figure 6.7 UK and Scottish energy and community development arenas, 2010 and after. (Source: AC)

study).[2] But when offshore wind development accelerated following the Scottish government's announcement to achieve 100% renewable electricity generation by 2020, however, opposition grew based on insufficient infrastructure integration and stakeholder participation (O'Keeffe and Haggett 2012). The 2008 defeat of a commercial wind farm development in the Outer Hebrides was still fresh on founders' minds on Mull, when starting project preparations in 2010. Technology rejection in community energy settings, like the anti-nuclear protests in Schönau, can also mean ex ante rejection of certain renewable resources despite abundant availability. Mull's selection of a river run-off hydro scheme speaks to its locality, similar to the connection of Samsø islanders to wind or agricultural resources. While minimally invasive with respect to wildlife and landscape alterations, the river run-off scheme acknowledges the island's natural qualities: landscape and abundant rain. The installation will be overgrown with time, vanishing into the surrounding hills and moors. "There's no visual impact at all. All you're doing

---

[2]Recent studies confirm relatively high levels of support for wind energy installations among communities in close proximity (Firestone and Kirk 2019).

is getting clean renewable energy for the next 100 years" (I20: 272–74). Mull's rain-laden climate exceeded expectations. "We've achieved in 10 months what we thought we would achieve in 12, just because it's been so wet. Every cloud has a silver lining," one interviewee affectionately joked (I20: 366–367). Like in other community projects, members make symbolic connections between the technology and hardware installations and the locality. At the same time, the 400 kW installation is relatively small responding to the wishes of the locality which largely opposed remote large-scale generation.

Once the technology had been chosen, implementation was determined in a two-step process, focusing first on feasibility and then on community decision-making. MICT successfully applied for a financial grant from the Scottish government's CARES (Community And Renewable Energy Scheme) scheme to conduct a feasibility study for hydro power locations on the island, with "Garmony Hydro" being chosen from the five locations suggested in the report (Local Energy Scotland 2015). As the location suggested was on public lands governed by the National Forest Landscape system, Garmony Hydro required a planning permission that included a proof of community support. MICT conducted a postal ballot among all 2.415 adults in the electoral roll which ultimately received a 70% (1.705 individuals) response rate with a 97% rate of approval (GEM 2014: 21). MICT had tried very hard to ensure participation in the vote by publishing the call widely and through face to face communication at community meetings. "We definitely spread the word very powerfully," one member describes mobilization for the survey, but also added that the issue of renewable energies mattered. "For other campaigns that we've had, where we tried just as hard, we had 30 or 40% [response], maybe? …It was something that people could sympathize with, relate to." (I20: 317–321) Mobilization of MICT staff for islanders to engage in the survey importantly increased levels of participation, but was supported by the resonance of the project plan with residents' values. Maintaining a sense of control and accessibility, and ultimately community ownership beyond share-holding, of the Trust is again connected to the relevance of MICT actions for community life. Beyond securing immediate support for projects planned by the Trust, local interest in and support of its plans ensure continued engagement of the community on other issues. One interviewee speaking about the danger of differing thematically from islanders' agendas said, "They [islanders] won't know in the future that if there's a problem, a thing that the community could address—they won't come to us." (I20: 407–409).

Organizational choices referred closely to the statutes of the Development Trust, although the energy installation is run by an independent organization, Green Energy Mull (GEM). Legally, GEM is a Community Benefit Society. The board of directors is chaired by Moray Finch, a co-initiator of Garmony Hydro and General Manager of MICT, yet GEM stands independent from MICT and directorship is tied to investment. This legal structure also allowed for funding from outside of the island. Funds were raised via community shares, reaching just under £500.000 in March 2015 and significantly exceeding the required £330,000 (Finch 2015), directly funding the installation at Garmony Hydro. "Two thirds of the investors are from the island and one third are from the mainland," explained one project member, noting that for the most part investors were locals but also included non-locals (I20: 31–32). In contrast, two-thirds of investment was coming from the mainland and one-third originated on the island. "We've got lots of small investors on the island and we've got some very large investors from the mainland" (I20: 33–34). While shares sold at £ 50, they can only be sold back to the Trust who would buy them back for £ 50 or less, depending on its financial capacity. Investors gained one vote, independent of their financial share, in the AGM of GEM (I20: 38–54). In 2016, shortly after the operations of Garmony Hydro went live, "The Waterfall Fund" charity was established to govern distribution of revenues according to the underlying values of MICT (The Waterfall Fund 2018). The revenues from energy generation which resulted from the UK feed-in tariff, introduced in 2010 (Ofgem 2010), are shared between investors of GEM that can receive direct revenues of up to 4% at the discretion of directors, and the Waterfall Fund charity. Investors are private individuals, who besides wanting to support the project and potentially receiving a small interest rate were attracted by the government tax incentives backing the investment. Waterfall Fund grants are given based on their community reference rather than topic, and include diverse community oriented initiatives (The Waterfall Fund 2019a).

The research situation on Mull during this first stage of project development, from the project idea to the commissioning of Garmony Hydro and fund raising through Green Energy Mull, shows many similarities to the early research situation on Samsø. A relatively small group of people initially conceptualized renewable energy operations on the island with strong references to community development, rather than sustainability concerns. The group initiating project considerations was embedded in organizations relatively powerful within the community. They began their deliberations within the Community Council and then moved further planning activities to the Development Trust. This initial group then began forging connections between the planned project and the larger community on the island, based on established local values such as environmental conservation and local value creation. A strong institutional embedding of the

project ensured organizational resources such as time and personnel, concentrated knowledge resources, and provided access to structural resources such as grant
funding.

At the same time, setting the project in established community oriented structures and connecting to argumentative positions such as nature conservation secured
trust and support of community members without previous interest in energy
policy issues. Project managers emphasized activities that would create ideational ownership for the project among community members. The Development
Trust functioned as a multiplier across the community, for example by awareness-
raising for the survey. The project arena had initially included founders and
their connections to the Community Council. The arena was extended as project
planning was assumed by MICT and shared argumentative positions on community development were incorporated. Access points for the community were
established, such as membership of MICT, participation in the survey, and later
membership of GEM. MICT was careful to base decisions such as the choice
of technology or location on community consensus, and create community oriented structures of collectivizing potential future benefits. This is summarized in
figure 6.8, setting the project arena in the context of the UK energy arena, and
the Scottish renewable energy and community development arenas. With respect
to the concept of social innovation, this first stage of project development predominantly reinforced existing power structures within the community, namely the
importance of MICT for conceptualizing and implementing local projects.

Upon initiation, the project arena on Mull connected to the UK energy arena,
as well as Scottish renewable energy (RE) and community development (CD)
arenas. Within these arenas, conditions for community engagement in energy
system services were relatively favorable in 2010. The UK energy arena had
softened its market-based approaches, impeding community-based energy projects, by introducing a feed-in tariff, and was acknowledging the importance of
policy support for alternative market actors to accelerate deployment of renewable
energy technologies. The Scottish RE and CD arenas were actively supportive of
community engagement in energy infrastructure services. Scotland (unlike other
parts of Great Britain, especially England) was developing a vibrant community
energy sector. This countered a concern of Scottish communities with UK strategies for RE deployment, which emphasized the Scottish potential for wind energy
generation to serve energy demand in the less resource rich south of the UK (see
section 1.3). This strategy, in combination with financial mechanisms of related policy instruments, resulted in large-scale commercial RE projects, but these
created strong opposition at community level. Smaller scale developments with
direct ties to local communities, by contrast, were seen more favorably. These

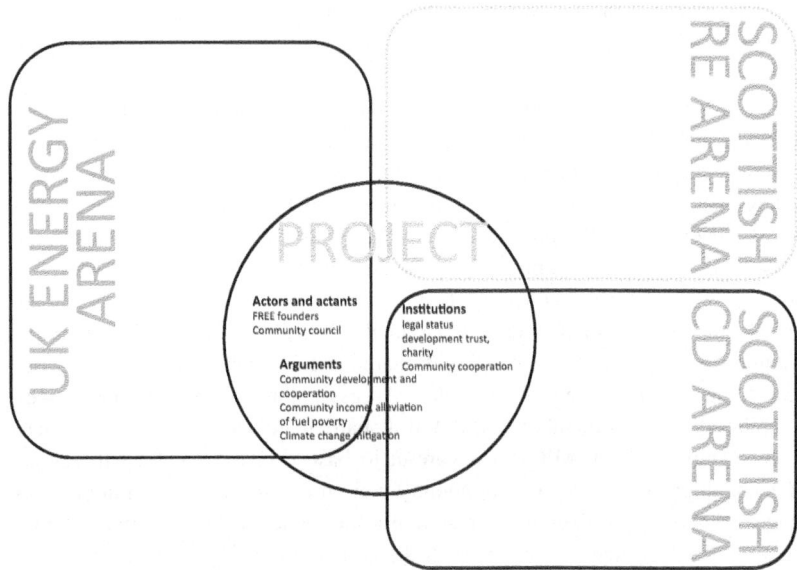

Figure 6.8 Project arena on Mull upon initiation. (Source: AC)

were embedded in a growing sector of community-owned trusts and cooperative organizations. These helped institutionalize community involvement in energy
installations within the Scottish community energy arena, as well as in the policies of the Scottish Government supporting community development. Beyond this,
they contributed to community ownership of land and infrastructures. The community energy arena also found institutionalized support in Community Energy
Scotland, an association that grew out of the Scottish national agency for community development in the Highland and Island communities. Project organizers
on Mull therefore strengthened argumentative ties to the Scottish community
energy and community development arenas, while distancing themselves from
the argumentative positions of the UK energy arena, despite benefitting from FIT
regulation. This is summarized in figure 6.9, wherein the UK energy arena is
shown as an inactive context arena as the project downplayed its relevance to
local choices of energy system alternatives.

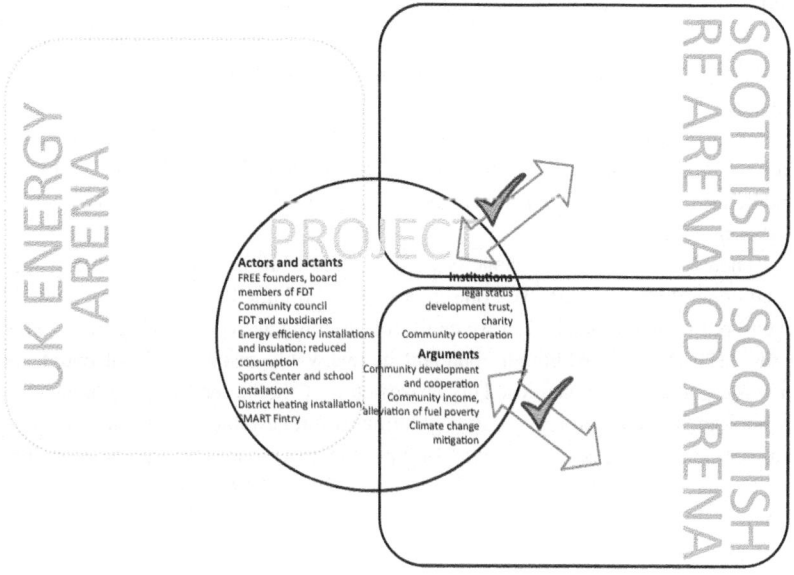

Figure 6.9   Arenas and connections, Mull 2010 (AC) Klicken Sie hier, um Text einzugeben

6.2.2   ACCESS[3] to the National Grid

The Mull community energy project started out as an energy generating scheme, seeking to combine the benefits of sustainable energy generation and a stable source of community income. This changed when grid constraints during construction called into question whether Garmony Hydro could be installed as planned (ACCESS 2015). At 400 kW, Garmony Hydro is a relatively small installation. This served a narrative of local production for local demand but was also a result of the island's grid connection. Transmission grid rules limit renewable generation on the geographic fringes of the national power grid, as local production can quickly exceed consumption and require grid balancing. In addition, grid constraints could also limit energy exports from Scotland to England. Despite national strategies to make use of Scottish energy resources for demand centers

---

[3] Assisting Communities to Connect to Electric Sustainable Sources

in the South, grid extension investments had not kept up with installations.[4] Operators of renewable energy installations received constraint payments to reduce production and support grid balancing (nationalgridESO 2014; Department for Business, Energy & Industrial Strategy 2012 (updated 2018)). This triggered protest as electricity prices were driven up to cover the cost of constraint payments and resulted in regulation to limit payments (Ofgem 2017). Dispute over grid balancing resulted in conflict over grid connections for planned projects, such as Garmony Hydro.

"There's no more [grid] capacity on the island Mull for any renewable energy to be generated above 50 kW," one interviewee explained, continuing "we would like to put in new projects. But we can't, for anything above 50 kW, connect to the grid." (I20: 67–73). Although abundant in renewable resources, Mull could not increase generating capacities. This drove setup of the larger project of balancing technologies for community based solutions of renewable supply and demand management, ACCESS (Assisting Communities to Connect to Electric Sustainable Sources, SSE Ltd 3/30/2015). Started in 2015, ACCESS seeks to develop a model demonstrating real time balancing of local generation and demand, as well as a corresponding system of local heat tariffs to commercialize services (Browne 2015). The goal is to enable communities to extend renewable capacities beyond current grid constraints. Project partners besides MICT include Community Energy Scotland as the consortium leader, SSE Ltd. (the UK's largest provider of renewable energy), VCharge (who designed the smart appliances set to balance Garmony Hydro generation and distributed demand), and Element Energy (working on commercial roll-out and system integration for other communities across Scotland). To balance supply and demand within the island's micro-grid, smart storage heating installations were made for approximately 60 participants (predominantly private households or small businesses) across the island and matched to generation at Garmony Hydro. ACCESS created a new arena of negotiation which tied actors of the Scottish renewable energy and community arenas and the project arena closer together in reference to the target of grid access, and created a new connection to actors of the UK energy arena (see figure 6.10).

Grid constraints created an impediment to Garmony Hydro that could not be overcome with established instruments of the Scottish RE and CD arenas (Figure 6.11). With construction of Garmony Hydro already underway, MICT

---

[4]Export of especially wind energy from Scotland to England nevertheless remains considerable, see Department for Business, Energy & Industrial Strategy 2012 (updated 2018).

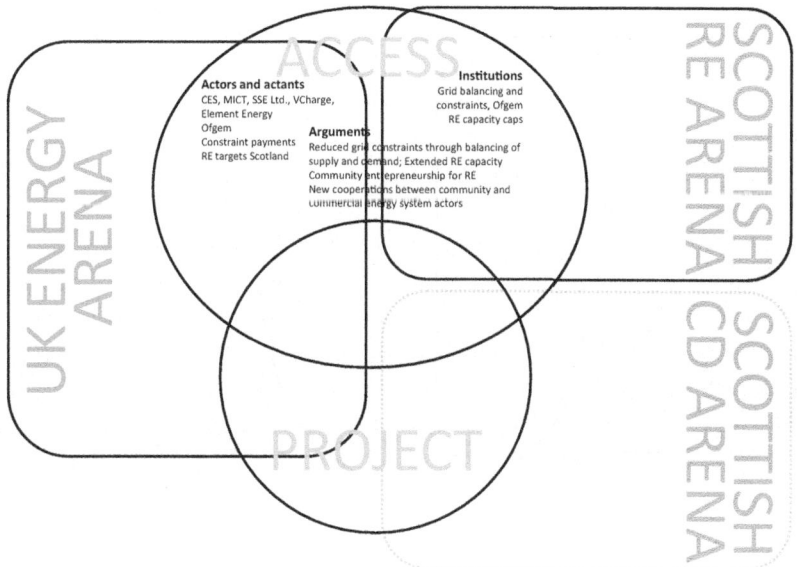

Figure 6.10   ACCESS forms new negotiating arena. (Source: AC)

responded to opposition by extending its project activities. Upon establishment, ACCESS was the first project including community and commercial partners to test local grid balancing of renewable capacities. It created a new negotiating arena, which found the support of the UK energy arena and integrated support for community energy actors available in the Scottish arenas, for example by establishing Community Energy Scotland as the consortium leader (figure 6.10). On Mull, ACCESS was seen as the key to realizing Garmony Hydro, an expression of community orientation ("With an eye to future energy projects on Mull, and to help support other community projects across Scotland", ACCESS 2015), and a way to strengthen MICT's own portfolio for future projects. GEM sees Garmony Hydro and ACCESS as stepping stones towards future engagements. "Will we use it for leverage for other projects?" one interviewee asked, immediately answering, "Of course we will. We'll use it as demonstration of track record. It might be for different and unrelated projects, we'll add it to our portfolio." (I20: 359–362). The establishment of community energy projects built expertise and credibility to enter new energy projects, but was also used to demonstrate resourcefulness for other project contexts (Figure 6.12).

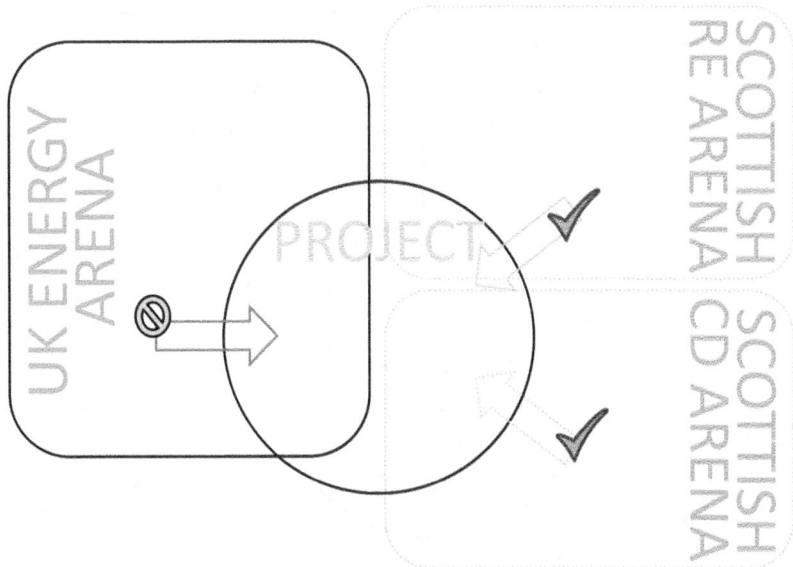

Figure 6.11   Grid constraint opposing Garmony (AC)

The establishment of ACCESS shifted the engagement of MICT in the energy sector from 'merely' establishing community owned generating capacities to becoming the first community based partner to demonstrate supply and demand balancing in a partnership including market incumbents, directed towards the national regulator. Said one interview partner on the project goal, "We believe it does work, we believe it will work, but… the whole point is to prove to the satisfaction of SSE and Ofgem that it does work." (I20: 290–91). Challenges to the project, from the perspective of the community energy project, are "mostly technical" (I20: 292). Cooperation between partners was described as productive, although at times cumbersome because of differences in corporate operations and requirements of documentation. One member described an example of communication with project partners, saying that a community energy projects might suggest "well, I'll nip up the road and ask so and so, whereas we actually need to write a letter to that person and it has to be formally documented and it has to go through the check procedure within these energy supply companies and it has to go through the partners and make sure everybody's happy and these have then got timelines and so on" (I21: 51–55).

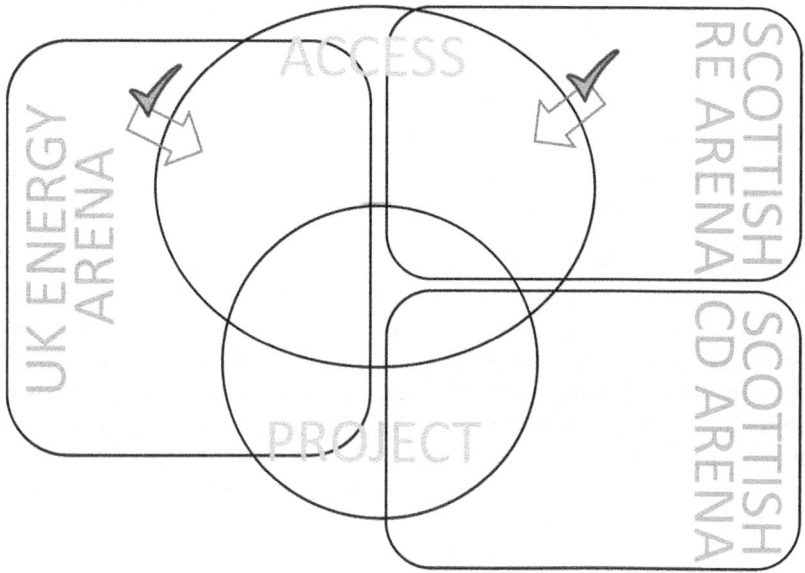

Figure 6.12   ACCESS connections to UK and Scottish arenas (AC)

More accustomed to, and more reliant on, working closely with their constituents, community energy projects reportedly struggled to accommodate more formalized structures of project partners. "It's been very very difficult," one project member on Mull acknowledged, describing the challenges of complying with making anonymous all consumer data while at the same time engaging in direct relations to those households involved in the project. "Our role is the community engagement, our role is to talk to residents, to explain the project to them, to help sell it, to answer any concerns and questions that they've got. As soon as we start passing their details to the other project partners, everything has to be anonymous. We have to allocate them a number, all the documentation is encrypted and password protected. We get questions coming back from the project partners and they talk in code. They talk about property number 33 that's got a problem and we think, well, who is property number 33, we have to go into the spreadsheet and remember what the password to get into the spreadsheet is and then look and see, oh yes, I know who they are, I'll phone them up." (I20: 152–158). While the community energy project does not question data privacy regulation, its role is to communicate with participants directly on all aspects of the project which can counteract privacy requirements. For a community energy project, this is also a resource intensive project setting which would require additional personnel if the project is successfully completed and subsequently extended to more participants on the island, "It's not been a fulltime job. But I think if we'd had more properties to deal with, the only way to cope would be to have more time." (I20: 168–170).

Setting the energy project within established structures of community cooperation, like on Samsø, supported community participation and consent. For example, project members reported not being concerned with data privacy compliance. They trusted the MICT to handle operations, and were focused on the advantage of having direct access to project organizers. Said one interviewee, "they just want to phone us up and say '[name], it's not working', or '[name], the guy hasn't turned up, he's supposed to come this morning, where is he?'" (I20: 174–176). The importance of MICT for the project as the direct link to participants was emphasized by other partners. This shows the general support provided by project partners. Because of the innovative character of the project, all partners were highly committed to its success. One described asking for a standard procedure, which was immediately sped as soon as it was made clear this was for the project: "Oh, it's for the ACCESS project, alright, I'll get that straight through." (I21: 59–60). But other partners involved were also specifically aware of the importance placed on meeting the expectations of MICT within the consortium. One reported that while project partners overall cooperated in a rather non-hierarchical manner, and organizational status was not translated into differences in power and influence within the consortium, special attention was paid to MICT. "There's definitely no sense of… that partner's opinion is more important than the other… If anything, everyone works very well to make sure that the community stays happy." (I21: 64–67).

Beyond the physical connection to national power lines, network integration created immediate economic and political connections. Economic connections are not only the financial subsidies for the feed-in tariffs, but also the contracts between energy suppliers and grid operators wherein the service and price of balancing is organized. Politically, grid integration affects the capacity of transmission. Distribution and transmission services are regional monopolies and as such closely regulated. Additional investments in transmission capacity that result from additional renewable generating capacities must typically be flanked by political decisions. This creates interdependencies between private service providers and policy makers, and between different levels of policy making. In Scotland, for example, energy policy was not devolved to the Scottish Government, and the Scottish Government seemed to have limited interest in this—despite its active role in encouraging renewable energies on community level. Said one interviewee,

"The Scottish Government… they want to take as much power as possible from London to Edinburgh. But on energy I don't think so. They say we'll carry on with the UK energy so that we can export into that market." (I20: 121–124). This project member observed that the Scottish Government's interest in energy sovereignty might be limited for fear of losing access to English markets for the considerable Scottish renewable capacities. Also, devolution of policy making power would increase financial responsibility. British investments in transmission capacity would be lost. The current limitation to increasing remote generating capacities that resulted in ACCESS therefore also resulted from conflict over political –and financial –responsibilities.

Within the project arena, ACCESS increased direct connections between the energy project and the local community. MICT functioned as a moderator, recruiting households and businesses that would change their heating system to be linked to Garmony Hydro and provide personal data to ACCESS. Participating in ACCESS forged lasting connections between islanders and the energy installation within people's homes, as Garmony Hydro was now powering the heater. ACCESS made significant changes to the community energy arena, rebounding on the Scottish as well as the project level. As the first cooperation between a community energy actor and diverse actors of the energy system, including incumbent utilities, to balance supply and demand, ACCESS underscored the potential for innovation within the Scottish community energy sector. The significant resources required to coordinate such a consortium and the involvement of Community Energy Scotland as a project leader emphasized the importance of institutional support to community energy projects provided by the Scottish Government. ACCESS created connections between the UK energy arena and the Scottish RE arena that had previously not existed, for example through a cooperation of SSE and community actors. At project level, it provided further proof of concept with respect to the capacity of community energy actors to forge lasting alliances in the community that create and implement new energy system designs, providing leverage for future energy projects and beyond. While not initiated as a challenge to the economic and political design of the energy system, Green Energy Mull nevertheless advanced to be just that.

An Unlikely Couple: Fintry and Oldenburg

7

Although quite different at first glance, Fintry Development Trust, the pioneer of community energy involvement in the Scottish Lowlands, and olegeno, the failed attempt of a citizen-led take-over of the power grid in a city of 165.000 in Northern Germany, the projects' innovation biographies show similarities. Both projects sought proximity between community members and energy system services in unusual organizational formats. In both cases, energy services of the community project were established without the project owning assets of energy generation or distribution. In Fintry ownership of material assets was not necessary to establish a diverse array of services. By contrast, in Oldenburg, the failure to attain ownership of energy infrastructure meant that subsequent energy services could not be embedded in community organizing.

7.1 Fintry Development Trust: The pioneer

In 2003, Scottish wind energy development was beginning its steep ascent but community involvement was still sparse. It was then that villagers of Fintry, Stirlingshire, suggested that a nearby wind farm developer might as well run a cable down the hill and power the village off the turbines they were building. Fintry Development Trust (FDT) was founded to realize their ideal of energy services that would benefit the whole community independent of the power to invest. Although ultimately not connecting the village's power supply to the turbines, a co-operation with the commercial developer channeled revenues into local energy system services. Since then, Fintry Development Trust has realized multiple projects of energy efficiency, reduced consumption, and improved service designs

A. Colell, *Alternating Current – Social Innovation in Community Energy*, Energiepolitik und Klimaschutz. Energy Policy and Climate Protection, https://doi.org/10.1007/978-3-658-32307-3_7

in the village. FDT has come to spearhead a growing number of organizations throughout the country developing energy resources in support of their communities. Fintry project organizers started community energy designs, providing the basis for projects such as GEM on Mull to create more complex community energy designs.

7.1.1 "Just a little bit awkward"[1]: Scotland's First Community Energy Wind Project

Scotland's first community energy wind project was a coincidence. Villagers of Fintry had begun investigating energy options, when a commercial project developer approached the village with its plans of erecting 14 wind turbines nearby. The villagers sought co-operation with the project developer. The arguments for a community energy project related to support for climate change mitigation and sustainable energy services, as well as fairly distributed community benefits. Villagers founded Fintry Renewable Energy Enterprise (FREE) in 2003. Expecting community opposition, the commercial developers were surprised when villagers instead demanded participation (Scott 2009). Locals wanted to "do things differently" (I19: 31). Timing was crucial: commercial developments had not been planned out when the community was approached, so the windfarm developer was able to accommodate community ideas. Developers "weren't that convinced by [community participation] being a good idea," one member recalled (I19: 16). The developer's initial opposition was not for technical or financial but rather for sectoral reasons. Twenty years of respective policy making had discouraged direct community involvement in the energy sector. One member recalled negotiations, summarizing the commercial developer's skepticism: "Why would a community want to have a stake in it, nobody had ever done it before, what a bizarre idea, what's that for?" (I19: 20–21).

Villagers initially sought a direct connection to the wind park as a local energy source. This was neither technically nor economically advisable. Instead, the developer suggested adding a 15th turbine to the initially planned 14 installations. "They would lend us the money… And they would retain ownership of the wind turbine but we would have a right to the income it generated," one interviewee remembered the negotiations, "but we would have to pay back the loan and interest on the loan and our share of the operating costs out of it. So, we said, okay, we'll go with that. That's a reasonable deal." (I19: 33–36). Turbine

---

[1]I19: 31.

ownership would be retained by the developer but revenues would be directed towards the community, reduced by refunding and interest for capital costs. "By that time we had kind of realized that we couldn't have a cable coming down to us to the village," this person continued to explain the community arranging itself with a more indirect connection to the energy installation, "what we **could** do was earn an income from it." (I19: 37–38, original emphasis). These revenues could then be directed towards local energy system actions, "we always had the view that that income should be used for other energy and sustainability and climate change projects in the village" (I19: 39–40). The energy project found broad community support (Scott 2009).

The windfarm developer suggested a cooperative investment model for the community based on Energy4All, a co-operative co-constructed by a Swedish company and UK investors which had successfully realized the first co-operatively owned wind installations in the UK in 2011 (Energy4All 2014). But Fintry villagers disliked the organizational pretense. "It would favor people who had money to invest," a member described their objection (I19: 60–61). The community opposed the idea of excluding villagers who were financially unable to invest, as well as unevenly distributed revenues: "that just seemed… a bit odd." (I19: 62–63). Instead, the Development Trust provided a citizen-owned and controlled organization wherein returns would be distributed throughout the whole community. "In terms of the relationship to the wind farm and also in terms of how we were structured later on, we wanted something that was just completely flat. Had no preference," one member explained (I19: 63–65).

Formation as a Development Trust in 2007 was, again, partially windfall. Unlike on Mull, there was no community organization in place. FREE members met with the Development Trust Association Scotland (DTAS) and found that the organizational structure fit villagers' ideals for "democratic" institutionalization: "anybody in the village who agrees with the views and objectives of the organization can become a member." (I19: 46–47). The trust currently has around 250 members; membership numbers began increasing slowly but steadily over time. Recruitment within the village is no longer important. As one member pointed out, FDT was established as a community actor: "people know we're here, if they want to join, that's fine" (I19: 226–227). A lifetime membership is open to locals willing and able to pay 1 £ and includes voting rights on organizational goals and directors. Members' oversight is exercised in Annual General Meetings (AGM), and online documentation of meetings secures transparency (FDT 2018b, 2018c). FDT has charitable status which "in terms of credibility was quite important" (I19: 51), both for reasons of general trust in the organizations' aims and for reasons of oversight (I19: 52–54). FREE was integrated as a company subsidiary, fully owned by FDT with all revenues channeled into the trust.[2] Negotiations between the developer and the community resulted in cooperation instead of opposition and the creation of a permanent income stream collectivized by the trust.

As with cooperatives in Germany, development trusts have become an established model of community energy organization in Scotland (Usmani 2017). Of the 666 MW operational renewable capacities installed in Scotland by June 2017, about 40% were built on Scottish farms and estates, 18% were held by local authorities, and 12% by community groups (Usmani 2017, pp. 2–3). The remainder of shares was held by housing associations, local businesses and public sector or charity actors. The structure of the trust ensures that returns are collectivized within the community without reference to individual financial capacity. While in cooperatives sharing returns beyond members rests on a voluntary basis (such as through the 'Sonnencent' grant scheme of EWS), it is a required feature of investments of a trust. Revenues in Fintry are predominantly invested in energy related projects, with some exceptions such as fencing for a local orchard (FDT 2018e). On Mull, as financial revenues are not collectivized by GEM but through the Waterfall Fund, there is no requirement of energy related investments, but funds must be channeled to the community.

The research situation upon project initiation was shaped by a relatively small group of volunteers in the community that shared a commitment to community-level climate change mitigation through energy system change in combination with collectivizing financial benefits. The group established communication with the village council, but was not mandated by municipal structures. Fintry is a small village struggling with many challenges typical of rural environments such as an aging population and a lack of infrastructure. But it had (and continues to have) a very active community. "Historically, this has been quite an active community, anyway," one member described the community setting within which founders began exploring opportunities for a community energy project. "This building we're in," the person continued, referring to the community sports center wherein FDT's offices are located and which at the time of the interview (a weekday in the morning) was full of elderly people bowling and mothers and small children eating lunch, "was entirely built by the community... And it's full of people. So this is kind of a very active community here anyway and we are just part of that active community." (I19: 207–209, original emphasis). Members

---

[2]Today, FDT includes tw other subsidiaries: Fintry Community Energy which owns and operates a district heating scheme, and Fintry Renewable Energy Distribution which runs the SMART Fintry project (FDT 2018a.

of FREE could therefore draw on locally established values of community involvement. The project arena created by founders was embedded in a closely knit community.

Just as on Mull, the project arena connected to the UK energy arena, as well as the Scottish renewable energy (RE) and community development (CD) arenas (figure 7.1). But the configuration of arenas was different in 2003. The project was set between the UK energy strategy, focusing on wind energy installations which preferably would be located in the resource-rich and scarcely populated North of the country, and the Scottish strategy for community development, which was directed at community welfare including references to sustainable energy developments.

The Scottish renewable energy arena was beginning to form when Fintry project organizers commenced their work, but was predominantly shaped by advisory bodies. Projects like the Fintry Development Trust forged a strong connection between the RE and CD arenas that project initiators on Mull could build upon in 2010. The Scottish RE arena was therefore supportive but inactive from the perspective of project development in Fintry, and interactions with the UK energy and Scottish CD arenas actively shaped project formation (figure 7.2). Public opposition to wind energy installations had increased over the course of the 1990s, especially in England, corresponding to energy and planning regulation that favored energy incumbents and neglected community involvement. Energy4All (2014) was seen as an opportunity to overcome public opposition to wind energy developments and increase the previously negligible shares of civil society investment in British renewable energy installations (Toke 2005, pp. 302–303). Yet, Energy4All was not driven by civil society actors but by a Swedish commercial developer (ibid.).

Fintry's choice of a different organizational form was noteworthy. Community ownership of energy installations had few precedents in the UK, and the cooperative model had just successfully realized a project. The success of the development trust model in Scotland was, in part, the result of Fintry's stubbornness. Their model created a successful example of an alternative structure of community involvement. But it was also the result of the Scottish regulatory and institutional landscape. Scottish land reform and related policy strategies sought to support communities in promoting their economic and social welfare through community development projects, including renewable energy projects. The DTAS actively provided knowledge resources and support to communities. Although less favorable to community energy actors than in 2010, configuration of the UK and Scottish arenas nevertheless proved beneficial to Fintry. Wind energy installations were realized by a commercial project developer capable of navigating the regulatory environment of the UK energy sector. The development trust, on the other hand, could benefit from supportive regulatory and institutional structures of the Scottish RE and CD arenas.

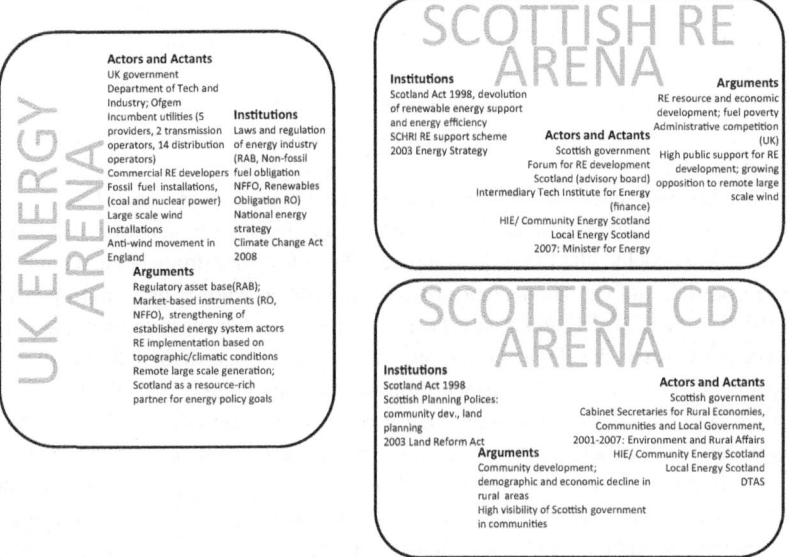

Figure 7.1 UK and Scottish energy and community development arenas, 2003–07. (Source: AC)

Subsequent technology choices of FDT pursued emissions reduction and climate protection through energy efficiency, reduced consumption, and improved distribution. Project activities were backed by revenues generated by the turbine. Most village households relied on oil or liquefied petroleum gas (LPG) for heating (Scott 2009). Therefore, FDT offered free loft and cavity wall insulation to villagers in 2007 and 2008 (FDT 2018e). That investments included individual homes as well as village institutions further supported the notion that benefits would be collectivized throughout the whole community. A professional energy consultant was hired in 2010, advising villagers on suitable installations and assisting with implementation (ibid.). Technology choices created close proximity between the community and energy installations, both literally, as many of them were

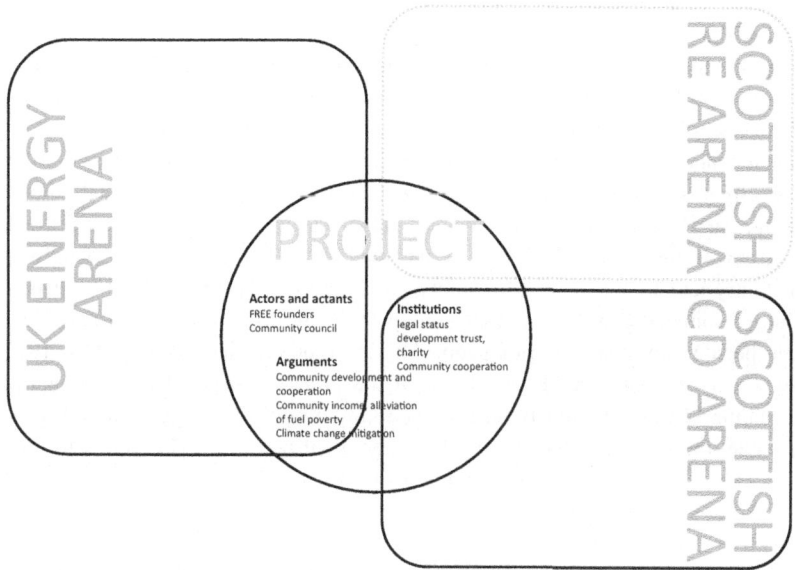

Figure 7.2   Fintry project arena 2003–2007 (Source: AC)

realized directly in people's homes, and figuratively. FDT emphasized the "emotional resonance and identification" of community members with its installations, which would require a close connection to installations (FDT 2018a). The 15th turbine, upon completed installation, was colored by children in a big festive event in the hills including musicians playing at its foot. "It was a big day out in Fintry," one interviewee remembered the emotional relevance of the completed installation and how the turbine provided "a sense of future for the village" (I18: 188–193). FDT members see the turbine, as well as installations throughout the village, as manifestations of agency. Climate change or energy system change could seem "big and scary and distant" (I19: 144) as one member put it. Installations in the community would ensure "that people don't feel isolated in terms of turning their concerns and fears into some form of action." (I19: 151). Projects have since branched out to address more fundamental questions of infrastructure design. The SMART Fintry initiative works on a tariff system for localized trading of renewable energy capacities via demand side management without establishing alternative grid structures (Smith 2018). Community energy engagement therefore includes technologically innovative solutions, but also organizationally innovative partnerships.

Ownership of the trust seems more important to project members than direct ownership of the energy assets. FDT does not own the wind turbine which forms the basis of its energy project activities, but project members underscore the emotional connection to the turbine. Said one interviewee, the festivities at the foot of the turbine were the community "taking ownership, we have managed to do something, this is ours" (I18: 190). Villagers' references to "sense of ownership", or a "sense of future" (I18: 191) indicate the symbolic dimension of asset ownership in Fintry and its effect on community building. "It's the idea of taking ownership of something as a community group. It's collective. It's not benefiting any one person, any one group of people. It's everyone in the village." (I18: 191–193). Connections were forged between the project arena and the community including spheres previously unconnected to questions of energy system design. To date, these connections often rely on informal networks. This refers to how new projects are chosen corresponding to village requirements or ideas, but also to institutional structures. For example, the village council is represented in the Board of Directors with one seat. However, this is an informal habit. When asked why this was not included in the Trust's statutes one member chuckled, "because we couldn't be bothered to formalize it. Just didn't seem worth the effort. We would have to write it into our rules of association and get it approved... much better to just keep it informal" (I19: 117–118).

Support for additional FDT projects within the community is most easily attained when financial benefits are transparent. "It's a kind of cynical view," said one, "but that's how it is... the easiest way to turn people off is to bang on about climate change" (I19: 188–191). While the attainability of energy services is a political goal for board and staff of FDT, this is less present for members. Unlike in Schönau or Oldenburg, members did not sign up for an organizational umbrella that would shield and fortify their efforts of system change. Still, FDT is certain that providing this organizational frame for members to engage with climate change mitigation secures this connection: "Even if they don't think about it every day it's kind of **there**. And if you provide... a way that people can take practical measures, they will" (I19: 154–155, original emphasis).

7.1.2 "If they have done their homework"[3]: Experts of Community Energy

The case of Fintry illustrates that empowerment through accessibility and responsibility is a key motive of community energy projects. Projects serve as challengers to the political and economic energy landscape, as well as moderators and facilitators of local processes of energy system change. Collectivization of benefits of the trust and its subsidiaries is more direct compared to an energy cooperative such as EWS where economic commitments to organizational values are prescribed voluntarily. It is also more direct than on Samsø, wherein a central moderator facilitates organizational ties and shared values across projects but does not prescribe organizational or economic forms of implementation. The regulatory requirements of the trust, including a charitable status, pose strict guidelines for project performance. This refers to a non-profit status and the dissemination of benefits throughout the community regardless of membership. Collectivizing benefits across the community is an important value mentioned across all interviews in Scottish projects. With the growing share of community involvement, representatives of community projects have also grown more confident. More recent partnerships between commercial developers and communities are critically reviewed: "There are examples were a developer has put 40 or 50 MW of windfarm, and the community benefit is on the benchmark... from the Forestry Commission of Scotland of 5.000 £ per installed MW per year as community benefit. That's **peanuts**." (I20: 197–200, original emphasis). The Scottish RE arena was extended by new actors, including community based actors, and policy instruments to support and increase community involvement. Yet, from the perspective of community energy organizers, regulatory requirements for commercial RE developments tied to land use and community ownership have garnered limited benefits for communities.

Projects in Fintry and on Mull have assumed distinct roles within the Scottish RE arena that exceed the local community. For FDT, this refers to their pioneering role as a community cooperating with commercial developers within the organizational model of the Development Trust. "If there is anybody looking to develop a commercial wind farm locally then they will always come and talk to us... as part of their community engagement process," one interviewee described the relevance of FDT in present commercial wind farm developments, continuing to ascertain their expertise: "And if they [project developers] have done their homework, then they know that we know what we are talking about." (I19: 149–153).

---

[3]I19: 104–105.

As with EWS and the Energy Academy, the long term engagement and success of FDT has turned into an important currency in counseling commercial developers as well as communities (FDT 2018d).

FDT's organizational model has come to be known as the 'Fintry Model' in national publications on community energy opportunities (Haggett et al. 2013, pp. 18–19). Consequently, public, commercial, and community partners of the ACCESS project were painstakingly aware of the pioneering character of this cooperation. CES created a full-time position for the first year of ACCESS, coordinating and following up on partners and ensuring timely delivery of working packages. All project partners showed particular attention to the community partner, "seen as almost like a client as well as a partner." (I21: 81) Project coordination was carefully tailored to MICT needs, as "obviously the delivery of the project is **for** the community... MICT is... the voice of the community and of the partners we are engaging. So we need to make sure that their concerns are looked on and are taken as very important issues" (I21: 85–89, original emphasis). For public actors of the RE and CD arenas, such as CES in the ACCESS case, performing according to community project standards matters not only for realizing energy political goals of the Scottish government. It also affects public perception of the Scottish government as a community oriented actor in the competition of administrative levels within the UK.

Across projects, management is characterized by the locality of team members, long-term relationships between team members, and unpaid engagement. EWS co-founders and long-term managers were from the village, and its management remains locally based today. The same is true in Fintry, where a majority of the board live in the village and its immediate surroundings. On Samsø, Energy Academy management and staff is also predominantly locally based. Similarly, on Mull, MICT and GEM staff is based on the island. In Oldenburg, there is comparatively more fluctuation in the team of voluntary staff but still a strong local base. In many cases, the locality of management—their recruitment from long-term community members—was a key asset upon establishment. EWS management, upon its founding and throughout the first 20 years of its operations, included the village doctor and his wife, Ursula and Michael Sladek, who were known and respected throughout the community. Upon winning the competition for Danish Renewable Energy Island, Søren Hermansen, the manager employed to breathe life into the organizers' master plan for the island, was a trusted local. The case selection therefore underscores the relevance of locality for leadership common to many community energy projects (Walker et al. 2007), and the importance of local leaders for establishing and sustaining community energy projects in the literature (Martiskainen 2017; van der Schoor & Scholtens 2015).

Case evidence also underlines the importance of long-term engagement at the local level for gaining more powerful positions in arenas of negotiation beyond the local. Support for new developments derives in no small part from what one founding member of FDT calls having a "track record" (I19: 116). FDT is established as a credible player in community energy projects vis-á-vis commercial developers and in grant applications. Over the years, this has resulted in more complex and also longer term energy engagements, such as a three year grant of the Climate Challenge Fund or the two year SMART Fintry project (FDT 2018e; Smith 2018). The active connection to the UK energy arena has faded after establishing the turbine. Connections to the Scottish RE and CD arenas have grown more important. Project activities relate more strongly to the growing regulatory and institutional support offered here and draw on the position of relative power established within these arenas as a 'pioneer' of community energy with a credible 'track record'. The relationship between the RE and CD arenas is not without conflict, for example regarding the limitations of community benefits in regulatory requirements for commercial project developers. Still, it is mutually beneficial. FDT gains support for extended project activities. The Scottish RE and CD arenas gain an experienced and credible actor showcasing the potential for community engagement (figure 7.3).

7.2   olegeno: The Catalyst

In 2011, Oldenburg citizens founded the cooperative olegeno to gain ownership and the right to operate the energy distribution system. Although the local incumbent prevailed in applications, the cooperative has since continued its work in projects of energy efficiency, reduced consumption, renewably sourced retail and electricity generation for urban environments. But while providing expertise and guidance in project development, energy initiatives were often realized without the cooperative's participation. It assumed a catalyst-like position, providing stimulus and initial resources to a reaction but ultimately not being part of the result.

7.2.1   A New Role for a Watchdog

In Schönau, on Samsø and in Fintry, community energy engagement started new projects, establishing new organizations in the process. On Mull, energy services were a new area of engagement for an existing organization of community self-help and support. In Oldenburg, the community energy project grew out of an

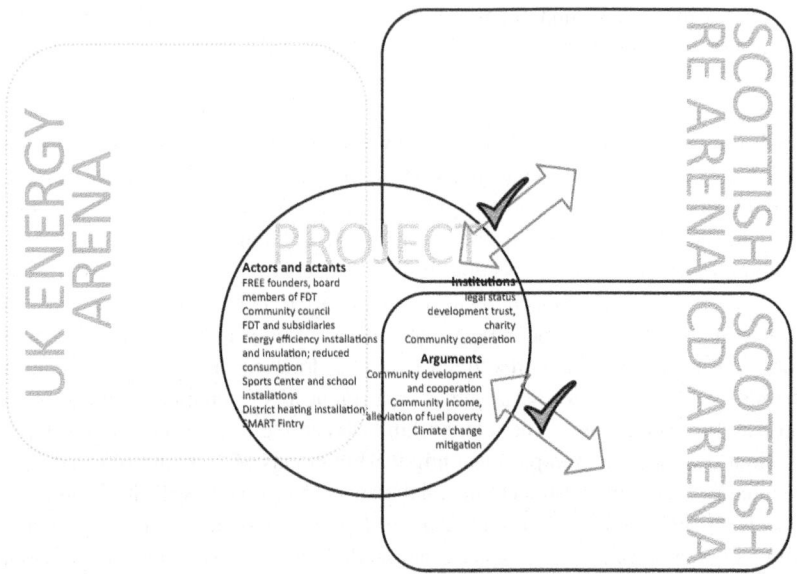

Figure 7.3 Current project arena Fintry and connections. (Source: AC)

existing citizens' energy association. For decades, these citizens had monitored
the city's energy political aspirations in the form of an "Energy Council" (Energierat), publicly commenting on local energy politics as well as actions of the
local incumbent, Elektrizitaetswerke Weser Ems (EWE). The Energierat had been
a critical voice for transformation and participation in the previous concessionary
process 20 years earlier and now sought to establish an overarching civic alliance
to create a new contractual basis for distribution system services. In the words of
one member, "the Energierat had monitored the concessionary process of the city
of Oldenburg. And out of this situation the necessity, the opportunity, arose to partake in the concessionary process of the city by establishing olegeno" (I8: 5–8).
The administrative process was seen as an opportunity upon which it was not only
possible, but necessary to act. It provided a clear target and frame of reference for
project actions. The cooperative was established as a vehicle of empowerment and
access to technology and local energy system design. Its political agenda focused
on citizen empowerment in energy system change, and the cooperative provided
the legal basis of applying or grid ownership and operations on this platform.

olegeno also framed its campaign as a rebellion against the established energy system, although less pronounced than in the case of EWS. Decentralization of energy systems was a key argument. For both projects, decentralization was not (predominantly) a result of the technical opportunity of smaller scale RE generation (Agora Energiewende 2017, p. 17). "Technologies play into this... but it is not enough by far," a member of EWS phrased the skepticism of realizing energy system change as a technological transformation, continuing "we need to learn and develop how we treat each other, how do we communicate... These questions will be decisive. And they are the Energiewende." (I3a: 54–68). Decentralization is not a technological paradigm of sustainable, small-scale, distributed generation but a socio-organizational claim. It refers to the projects' organizational ideal of energy systems owned and managed cooperatively by citizens, and connects to a narrative of control and empowerment of citizens in the energy system. This framing of energy technologies as fundamentally connected to community life is echoed in olegeno's decision to engage in distribution grids as the underlying structure of energy system services. Technologies and hardware, should not only deliver a renewably sourced and efficient energy system, but one that is "decentralized, small scale, close to citizens" (I7: 130–131). Technological and organizational design of energy system change is intricately connected. Energy system change is seen as a more "extensive... fundamental" transition including behavioral change (I8: 16–18). Decentralization presents a way of realizing accessibility, control and empowerment of citizens in the energy system.

Founding members of olegeno also mentioned anti-nuclear protest as a motivation for action. In 2010, shortly before olegeno's establishment, a liberal-conservative government had taken office at the federal level and revoked the nuclear phase-out installed by the previous red-green coalition Federal Republic of Germany 9/28/2010). In 2011, the nuclear accident of Fukushima reinvigorated protest across the country, mobilizing 250.000 participants in nation-wide demonstrations (Süddeutsche Zeitung 2011). "More than ever, we need to step up", an olegeno member recalled their reaction. "The situation after Fukushima provided an impetus to try different approaches in the concessionary process, but also to use this as a productive drive for energy system transformation" (I8: 63–68). Still, anti-nuclear mobilization did not dominate olegeno's campaign. Partly, this was because EWE was less directly tied to the nuclear energy industry than KWR in the case of Schönau had been. But more importantly, the German government reacted to the strong calls to return to nuclear phase-out that arose throughout the country following Fukushima. In a rather dramatic, and uncharacteristically abrupt change of heart, then German chancellor Angela Merkel had declared a moratorium on seven of the country's oldest nuclear reactors (Frankfurter Allgemeine Zeitung 2011). Shortly afterwards, the nuclear phase-out was reinstated, adding agreements on the immediate shut-down of eight power plants and a fixed date for the final shut-down of all nuclear capacities (Federal Republic of Germany 6/6/2011). Anti-nuclear protest therefore offered no platform to mobilize continued engagement for olegeno after 2011.

Central arguments of the cooperative, instead, were accessibility of and participation in decision-making processes of the city's energy infrastructure, responsibility and self-reliance. As quoted above, olegeno's founding members felt a 'necessity' to act. "It is about self-help, the classic cooperative goal," one interviewee summarized the cooperative's targets, and continued to explain how the cooperative differed from the German trend of cooperatively owned energy generation, "we want to make the energy transformation easier for people in Oldenburg as mutual economic help" (I10: 167; 173–175). The desire for easy and accessible structures was also built into the organizational structures of olegeno. The cooperative remained very close to the general requirements of the law in its statutes, summarizing its ownership and management structure in just one page (olegeno December 2014). Membership hurdles were kept as low as possible, reducing the price for one share to 60 EUR, which is comparatively low for an energy cooperative. Membership is open to anyone willing and able to pay for a share, and all members are invited to join management and operations. Members report a sense of joy and pride that comes out of the accessibility of cooperative structures, one saying "even a normal, small kind of [guy] in a cooperative gets an email from the head of Bürgerwerke [cooperative platform, author's note] ...to experience that. I think, that's a success" (I9: 123–125). This indicates a non-hierarchical approach to capacity building. Relying on honorary work also meant that anyone could partake in cooperative actions at all levels.

Establishment of the cooperative itself was seen as an expression of self-help. Prior to entering the application process, the Energierat had urged the city to commission an assessment of the local energy system to review the potential for municipal management. When the city council was unresponsive, the Energierat commissioned such an assessment itself (olegeno 2019a). When this report found potential benefits in alternative ownership structures for the electricity grid, the Energierat founded olegeno as a cooperative bidder in the application process in September 2011 (ibid.). In the fall of 2011, the newly elected majority held by the Social Democrats and Green Party in the city council commissioned a second, independent investigation into grid operations setting the concessionary application process on hold. Although the investigation favorably reviewed the political potential of municipal management and its economic implications, the city council rejected its recommendations. Social Democrats, Christian Democrats and Liberal Democrats all voted against the plan. In March 2013, applications were reopened. Indicative bids were requested of the two applicants in June 2013 (Kuchta 2013b). By this point, olegeno had mobilized approx. 230 members and 42.000 EUR in cooperative shares (Kuchta 2013a). Following negotiations with the city and the consulting law firm handling the application process, olegeno submitted a legally binding offer in November 2013 (olegeno 2013). The cooperative bid was rejected by the consulting law firm reviewing candidates' offers. In January 2014, the city council voted in favor of EWE's offer (Stadt Oldenburg 12/2/2013). Two political factions withheld their votes in protest of a nontransparent assessment process and the re-concessioning of EWE for twenty years without adaptations to the concessionary contract (olegeno 2019a).

The purpose of the cooperative's application was as much a corporate expression of their ideals for local energy system services, as it was an energy political statement. olegeno was established as "the challenger" as one olegeno interviewee said (I8: 42). "To be the alternative," another added, "to disturb, progressively, idealistically, to set the bar", especially in a sector often dominated by incumbents: "competition needs an alternative" (I9: 113–114). Following the loss of the concessionary application in 2014, the cooperative to date remains limited in terms of physical infrastructure change (change of grid management, number of energy clients, or PV installations, etc.). Still, members underscore that its success lies in showing that "there is a cooperative, they **have** a different concept, it **could** somehow be different than it is now" (I10: 179–180, original emphasis). By handing in a binding offer that withstood professional review, the cooperative proved it could compete "at eye level" (I8: 105)—that alternative energy system concepts were competitive. The framing of olegeno's activities was therefore a reaction to the configuration of energy arenas in Germany at the time of project formation. The project arena of olegeno was linked to the local energy arena, and national energy, community energy and anti-nuclear arenas (figure 7.4).

The national energy arena had changed profoundly compared to its configuration in1986, when EWS was formed. It now included powerful actors and institutions supporting renewable energy resources. The red-green government between 1998 and 2010 had mandated the "Energiewende", moving the Ministry of Economics and the Ministry of the Environment to collaborate in designing energy policy (see chapter 1.3). The introduction of the FIT and institutionalized support for community energy spurred development of alternative structures within the energy system.

The community energy arena had grown into a new active negotiating space, as the number of community energy projects and the share of RE generation increased. Community energy projects were recognized as relevant actors of deployment, and supported each other through the exchange of experience, knowledge, and at times even labor or financial resources (Becker et al. 2013; trend: research and Leuphana Universität Lüneburg 2013).

The anti-nuclear arena had not changed fundamentally with respect to actors and arguments, but its activities had. As an anti-nuclear consensus was negotiated with policy makers and the phase-out was agreed at the national level, mobilization within the arena subsided. After Fukushima, the strong and swift mobilization of anti-nuclear protesters to reinstate the nuclear phase-out showed that the arena still held important mobilizing power relying on established actor and support networks, as well as shared argumentative positions. When the government returned to phase-out policies, the arena resumed a dormant state. It provided no backdrop against which olegeno could have mobilized opposition to EWE.

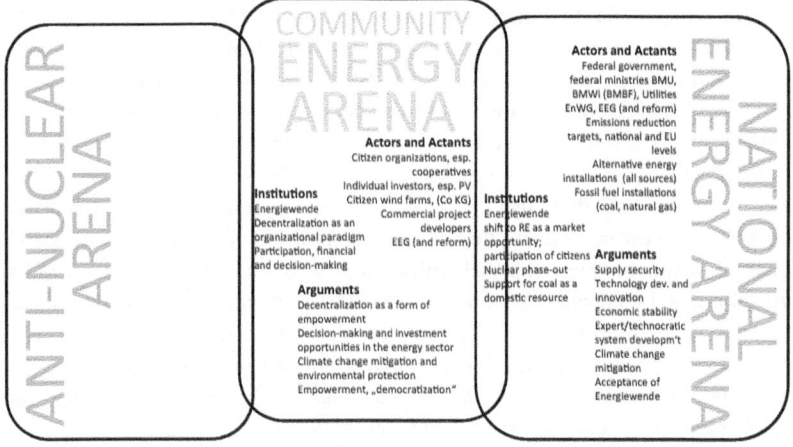

Figure 7.4 Configuration of anti-nuclear, community energy and national energy arenas after 2011. (Source: AC)

The relationships of the olegeno project arena and national arenas were inactive (figure 7.5). Negotiations on the community energy project took place in the local energy arena and the project arena, national arenas providing only a general background. The local energy arena in Oldenburg had included citizen activities prior to the energy project formation, as the Energierat had monitored and commented energy system performance. The Energierat was absorbed by the olegeno project arena. The local energy arena included the city council and EWE as the two dominant players of energy system design. The city council ran the application process for the grid concession. EWE had emerged out of a cooperation of multiple municipal authorities and was an important local employer. The relative power of the Energierat within the local energy arena had originated in its sustained and well-informed argumentative engagement with energy system services. EWE performed said services and the council was charged with providing the regulatory framework determining how services could be configured. The focus on consulting and knowledge-based arguments for energy system change was sustained as the Energierat established olegeno, and the project arena was formed.

Negotiations between the project and the local energy arena were again structured by the formal requirements of the grid application process.[4] And while according to the law the municipal authority must remain neutral throughout the award procedure, relations between the arenas were predominantly contentious (figure 7.6). As with EWS, this formalized structure included its own embedded power differences, not only between the municipal authority governing the concessionary process and those applying within it, but also between applicants. Arbitrary regulation still favored the incumbent over other applicants, a position frequently abused for example by current concessionaires withholding relevant information, or pressuring municipalities financially or politically (Becker and Templin 2013). Although the legal position on the price of infrastructure had been improved following EWS' lawsuit, regulatory ambiguities had not been resolved.

The power differences deriving from formal positions within the application process could be augmented or counteracted, depending on actors' positions within the public discourse. Unlike EWS which had successfully painted KWR in Schönau as an unresponsive and illegitimate concessionaire, olegeno was facing an energy company with favorable local reviews. Representatives of the city council had publicly declared their support for the current concessionaire before opening the application process, and locals felt informal, mostly favorable ties to the large employer (Kuchta 2010, 2011a). Ownership was a contested concept in the context of the concessionary process. EWE is municipally owned but citizens felt that public control of company actions was insufficient to implement political goals. The "energy utility came out of an association of municipalities... that is something really great," said one interviewee, "but we now have the situation that out of this association a company was established that works very independently and can no longer be properly controlled by municipal representatives and that is

---

[4] A third applicant can be neglected here, because of early drop out (Kuchta 2011b).

very unfortunate" (I8: 123–130). But the cooperative competing against the local incumbent was seen by many as an attack on the city itself. "You noticed in the whole city, in the whole region", this person continued, "that it was a taboo to compete against this company, because: That is us!" (I8: 131–133). Interviewees emphasized the audacity and stamina it took to enter and sustain the cooperative in the application process against the local incumbent. It took "a lot of courage to question things in an environment where one would otherwise not dare to do so," one member explained (I8: 138–139). And while regaling the story of multiple hold-ups in the application process another quipped, "don't hold your breath" (I9: 313). The local incumbent was closely tied to the municipal administration and the local community. Support could not be mobilized on a narrative of local ownership.

Within the local energy arena, the power to alter energy system designs of the Energierat had been purely argumentative. olegeno's power was also predominantly argumentative, as the concessionary decision rested on the reviews of individual bids, but was moreover inferior to the incumbents' argumentative position by regulatory design and concessionary practice. On the other hand, olegeno's embeddedness in the Energierat meant that members had access to knowledge and information (for example, members were aware of the concessionary contract expiring and could position themselves accordingly). Members had established communication channels with the city council through previous interactions. They had a credible track record in fighting for sustainable energy system services for example through local initiatives for renewable deployment (Solarinitative n.d.), or through the support of national campaigns (Klimaretter.info 2007). Also, a network of volunteers and supporters was aware of their work. Many supporters had been actively engaged for a very long time. A middle-aged member said about the starting point of their engagement with local energy politics, "ever since my college years ... it has been with me continuously" (interview olegeno, December 2015).

olegeno's organizational and argumentative positions resonated positively with the community energy arena. Still, connections were mostly inactive (figure 7.6). Community energy projects had become an established actor of the German energy system in 2011, most engaging in RE generation (Debor 2014). The organizational model had spread and gained acceptance. EWS had gained sectoral prominence and found imitations, for example in cooperatives in Berlin or Hamburg (Blanchet 2015; Colell and Pohlmann 2019). This lent support upon establishing management structures and strategies in Oldenburg. "We of course have partnerships," one member summarized, "with other cooperatives... especially those that also applied for grid management." (I10: 146–147). The

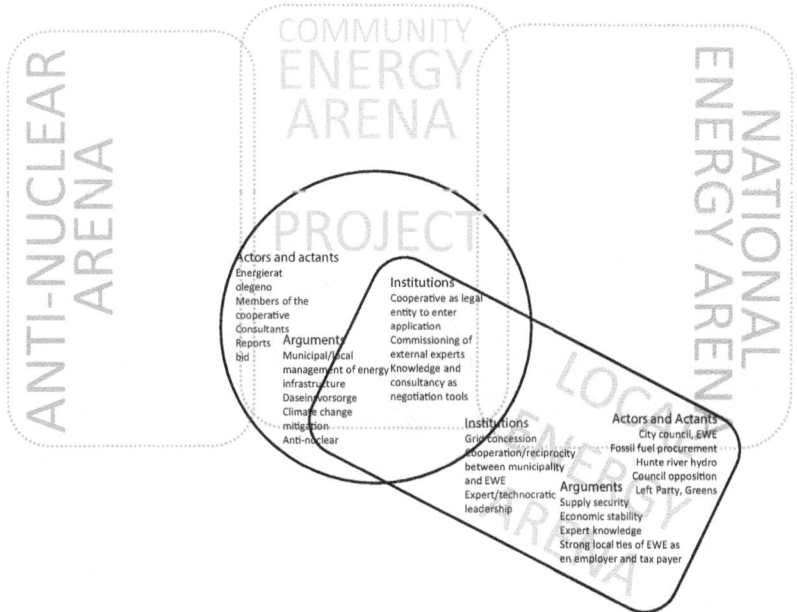

Figure 7.5   Project and local energy arena olegeno, 2011. (Source: AC)

cooperative could benefit from previous experiences of other citizen owned energy cooperatives. However, few citizen-led initiatives had successfully changed ownership structures of distributions systems (Stadtwerke Wolfhagen 2019). Most community energy projects focused on energy generating services which resulted in quite different organizational practices and financing structures.

Arguments of decentralization as an organizational paradigm, responsibility and self-reliance were connected to energy system services in narratives that referred to energy as a public utility and service. Members of the cooperative framed energy services as 'Daseinsvorsorge', as fundamental services of every-day life, in the words of one member "Energy I believe is, how do you call it, Daseinsvorsorge… completely fundamental part of life" (I9: 41–43). The cooperative was seen as a democratic, participatory structure organized in solidarity, "energy democracy… I would somehow think it self-evident that those belong together." (I9: 40–41). This suggests a quasi-public ownership and management structure beyond the municipality. On the one hand, establishment of the cooperative was seen as a result of public inaction. "For such a fundamental part of life, I would

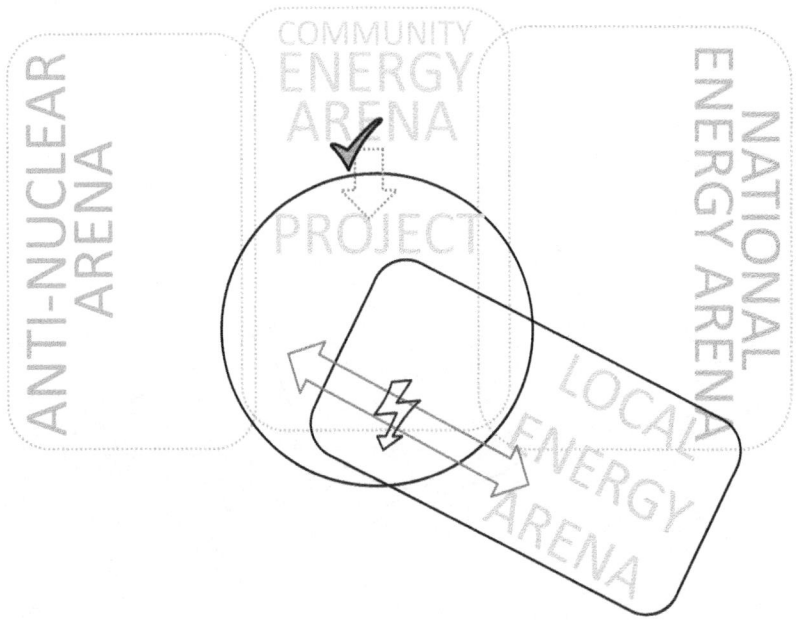

Figure 7.6   Arenas and connections olegeno, 2011. (Source: AC)

even say… this should be organized on community level, maybe even publicly," one member articulated his desire to move infrastructure ownership and management beyond profit orientation and towards macroeconomic considerations (I9: 45–48, also I8: 64–68; I10: 170–173). At the same time, municipal ownership and management was set apart from citizen ownership and management. One interviewee explained that joint ownership could improve many fundamental services of daily life, such as housing, and might even be organized in municipal ownership (I9: 218–229). "Yet," this person continued, "I have a strange reservation against calling municipal things citizen-owned." (I9: 229–230).

Civic engagement in community energy projects could also refer to the control of political institutions. The first referendum initiated by EWS interrupted an otherwise clandestine agreement between the utility and the municipality to prolong contracts for decades to come, and reopened a forum of public debate and control. olegeno in the course of grid applications also critically monitored the application process itself, and through the engagement of outside evaluation of the process strengthened its transparency and procedural credibility. The cooperative's commissioning of lawyers to evaluate alternative concessionary scenarios added complexity to the public debate of grid ownership in Oldenburg, and built pressure on the municipality that was ultimately moved to commission a second report. The publication of procedural details, if not protected by confidentiality requirements, during the application improved transparency for locals but also built pressure on the municipality and the competitor to be more transparent themselves (Kuchta 2013). This was extended by the cooperative publishing the offer and its evaluation following the final concessionary decision (olegeno 2014; during an on-going application, offers are confidential material).

After the municipality had decided on the concession, the regional cartel authority was moved to investigate the process. The cooperative suspected that it might have been facing undue opposition in the application process. Although the cartel authority found no misconduct, the cooperative felt recognized by the investigation: "that the regional cartel authority... made time for this, shows... that our concerns were taken seriously" (I8: 116–119). Overall, the public struggle for transparency and monitoring led by the cooperative and supported by opposition parties strengthened the procedural quality of the political decision (Kuchta 2014), and was defended even as higher costs for legal consultancy became public (Kuchta 2015). This indicates that procedural recognition of civic participation can be important regardless of whether civil society claims are ultimately supported or rejected (Grimes 2006; see also Tournhout et al. 2010). Although unsuccessful in their application, members of the cooperative felt validated in their actions.

The arguments and strategies of olegeno show its origins in a civic initiative monitoring performance and criticizing existing structures. The cooperative was the attempt to transfer political claims into entrepreneurial structures. Responsibility again features strongly in concepts of management: Of the municipality in exercising its rights of control and monitoring, of the utility in exercising its quasi-monopolistic market powers in respect to long-term considerations of energy system design, but also of citizens to hold both accountable—through the establishment of own political entrepreneurship, if necessary. A 'challenger' or 'alternative' position was not defined in protest of the incumbent but based on own experience and expert knowledge. The cooperative drew on established concepts published in the academic literature, examples from other communities as well as specifically commissioned expert reviews to build a strong case for its competitive position (olegeno 2019a). Its rhetorical framing was less emotional or based on locality, unlike for example the "Yes to Schönau" slogan of EWS, and rather based on arguments for the economic and environmental viability of its claims (ibid.).

7.2.2   A New Goal for a Company

Following the concessionary decision in 2014, the cooperative entered an extended period of reorganization. This process was on-going during the research project; various smaller steps had been taken to establish alternative services. The cooperative offered renewably sourced electricity and natural gas tariffs to local customers working with a nation-wide network of energy cooperatives (olegeno 2019e, 2019d), and engaged in local renewable energy generation, especially for renting tenants (olegeno 2019f, 2019c), increased energy efficiency, and reduced consumption (olegeno 2019b). The continued active engagement of both founding and new members in the cooperative can be called a success. Yet, the process of reorganization has been painful. While not wanting to become "just any old investment" in energy systems (I10: 173), the cooperative lacked an external frame of reference structuring its activities as the concessionary application had done. Key challenges of this process of re-orientation lay in determining which activities the cooperative would pursue, whom they would be able to mobilize within the larger community, and who amongst team members would engage in their implementation.

Without the concessionary application as an external frame of reference, the cooperative was now largely free in its operative orientation but remained tied to values of energy system change as a fundamental shift of services. Self-declared motivations of the cooperative remained close to argumentative positions of the concessionary application. These include "democracy" in the energy sector, realized through equal voice of members in the general assembly, "co-determination", as each member can vote on strategic decisions of the cooperative, "openness", as membership is unrestricted, "participation" in financial profits of the cooperative for all members, "solidarity" among diverse groups in civil society through an economic partnership of stronger and weaker members, and "limited liability" as financial liability is tied to the amount of shares (olegeno 2019g). Many founding members of olegeno had been part of the Energierat before establishing olegeno, which created a sense of responsibility after the application process ended. Members wanted to find new projects as well as honor the history of engagement. One interviewee said, "We absorbed many political [initiatives] here in Oldenburg… whatever there was in the energy sphere went into olegeno. That makes us responsible of following up on these things, at least that is the aim." (I9: 133–136). This indicates a procedural commitment in operations: To sustain an over 20 year tradition of critical civil society engagement in local energy infrastructure design and respective politics.

The attempt to translate social movement oriented values into an entrepreneurial organization could also be a burden. For example, although renewable electricity products abound in the retail market, olegeno wanted to offer a product governed in cooperative management and committed to directly increasing domestic renewable capacity. Yet members reported in meetings that product distinctions were hard to communicate to third parties, and customer numbers remained low. The cooperative's PV installations are designed to either cater to very small scales ("mini PV–solar power from your balcony", olegeno 2019c), or to address tenants who do not own their roof and therefore cannot decide on PV installations, as well as cooperatives in the housing sector. This, however, ties project realization to other, often honorary organizations with adverse effects on timely realization. In some cases, projects were planned with olegeno but implemented by someone else. Members' reactions to this varied, one pointing out how their daily bike path to work led past a solar installation that olegeno had planned but another organization had installed. "Every day I go past that," this person commented, not without resentment that the cooperative was not included in realization (W2). Another disagreed, saying how this was an example of "how visible the change is that olegeno has inspired" (W2).

Other projects lack feasibility, so far. For example, no business model has been found to directly combine electricity retail (or heating concepts) with reduced consumption. "[Our electricity product] has to be something that involves people in special ways, financially and conceptually, but that also incentives them to not only sell but also reduce", said one interviewee (I10: 188–191), but also described the struggle of calculating such a tariff within existing billing systems. This also speaks to the cooperative's aim to address citizens not yet involved in RE systems (I7b: 280–282), and improve accessibility of related services for low income households (I10: 205–206). These projects were in their infancy at the time of study, but market entry was slow or still pending. The cooperative sought to develop tariffs that would incentivize and reward reduced consumption, as well as looking to specify tariffs that would take special account of low income households' needs. "[The product] would need to involve people in special ways, economically and strategically… but also give special incentives to not only sell [energy] but reduce it['s consumption], too," one project member elaborated their understanding of an innovative, socially responsive retail product for renewably sourced electricity (I10: 185–191). "My idea was," this person continued to explain how they would have preferred to design incentives within energy tariffs, "anyone who saves 10% per year, gets a… bonus… we really want to bring the de-growth idea of 'less' into the mix" (I10: 191–195). So far, the cooperative was unable to design such a product in compliance with regulatory standards of retail and grid balance.

Members of the cooperative emphasize the agency of the individual, both as a consumer and as a citizen. Olegeno representatives highlighted the direct agency of consumers, similar to the power to 'cancel your subscription to bad energy' described by an EWS interviewee. "Consumers of course are those who can demand change... they can boycott products, they can demand products... And there are areas where **only** consumers can do something," one olegeno interviewee explained this perspective, summarizing their understanding of the power of consumers by saying they could "set [themselves] against growth." (I8: 177–188). The agency of consumers rejecting consumption, again, speaks to the potential of material participation as civic or public participation. While, for example, members of the cooperative EWS combine civic engagement and economic investment, its customers (many more than members, yet more indirectly connected to the cooperative) show the awareness of their active role as consumers by a considerably reduced annual consumption. olegeno attempts similar operations by offering an electricity product that is jointly realized by citizen owned cooperatives and committed to increasing renewable capacity (Bürgerwerke eG 12/19/2018, 9/13/2017, 3/11/2015). Civil engagement, by contrast, is understood as an act of empowerment not only for citizens themselves, but also to government and administration. Said an olegeno interviewee, "Those in powerful positions also need impulse and ideas from below... They need the public as empowerment; they can't decide something like the Energiewende by themselves." (I8: 177–183).

To reorient the cooperative, a participatory process was established open to all members of the cooperative. Said one, "we hosted a strategy workshop... with a membership poll beforehand"(I7a: 21–22). This workshop was open to all members, but mostly included those that had actively engaged in the application process. The first workshop involved mostly those "that were more or less still around from being involved in the application process... and then a few others, but many of them then pulled out of the circle of those actively engaged, and then of course management and board people were there. Roughly 10 or 12 people." (I7a: 56–62). A second workshop was hosted approximately one year later. Several topics were chosen that should then be explored further in working groups to define the potential for olegeno, which included energy targets as well as procedural challenges of the cooperative such as data management, administration and membership management (Project meeting, March 2016).

While the cooperative has managed to sustain its working structures and continued to grow in membership numbers (with approximately 300 members in December 2015, I7a: 74), only one operative project (electricity and gas retail) has been realized.[5] Members openly point to the limits to participation in their experience. This refers to projects' expectations concerning the engagement of customers or consumers, or the general public. One member of olegeno pointed out the limits of individual engagement, saying, "this idea of civic engagement… I don't expect that from everyone." (I10: 184–85). Managing expectations with regard to the involvement of others in decision-making processes echoes the sentiment of one Samsø interviewee, "that's fine that's fair, that's a community at work. They have a lot of other things to worry about." (I24: 216–227). But, the interviewee from Oldenburg continued, for those wanting to engage, and for those silently supportive, the cooperative needed to provide "coherence" (I10: 250–51). This is underscored by an assessment of regulatory structures preventing or impeding citizens' engagement. Said one interviewee on the experience with the regulatory environment of the concessionary application and subsequent project activities, "[regulation] cannot address the homo oeconomicus alone, people must find themselves seen in the content, find themselves socially, psychologically. And for that you need to strengthen the civic side… Give [citizens] space to implement ideas, economic space and planning space. So, that ethically motivated engagement becomes economically viable." (I10: 222–233).

The commitment and responsibility for sustaining energy engagement locally, and the diverse goals for the cooperative as an agent of energy system change compete with management and operations that rely almost exclusively on honorary work. The application was realized completely in unpaid labor, apart from external reports or investigations that were commissioned. After a resource intensive period of campaigning and fulfilling the various steps of the application process, the cooperative was now faced with 'defeat' and reorientation. In interviews, members recalled exhaustion and a sense of defeatism in the weeks and months following the city decision to resume a long-term cooperation with EWE without changes to participatory structures or sustainability performance. Several management members declared their wish to retire from their positions. As one remembered, "There was sort of a cut. And one or two members of management immediately said, well, we are most likely not going to go on." (I7a: 6–7). Others withdrew after the first strategy workshop. Team members were also frustrated. One recounted having made the experience that "participation, as a pretense, is officious. That was proven to me wonderfully in the working groups of olegeno. It will only ever be a few people doing that, and it runs itself dead if professionals don't come in." (I10: 288–290).

---

[5]While cooperative membership of olegeno is not confined to locals, it has remained much more local in its configuration than in the case of EWS.

This refers not to the degree of professional knowledge or expertise that can obtained in honorary structures but to individual exhaustion. For example, olegeno had attempted cooperating with students of the local university to offer practical experience to students but also gain consulting knowledge. As one member enthusiastically reported, the idea was to "bring together what belongs together, or at least what fits together." (I7: 70). The cooperative required additional input, while students welcomed the opportunity to work with a real-life case study. But while students worked for the course of a university semester and several daylong workshops were held, the cooperation was ambiguously. An olegeno member participating in a joint workshop between the university and the cooperative remarked slightly desperately when commenting on students' ideas for data management, "it is all very well, but I am sitting here wondering all this time who will actually be **doing** this in our organization" (Project meeting, March 2016, original emphasis). Students did not continue their engagement in the cooperative, and most ideas were not integrated in cooperative processes. This was also because students themselves had limited knowledge of the configuration of technologies, services, and guiding principles developed in the cooperative since its establishment. Some long-term members, instead, felt overrun by outside ideas with seemingly limited utility or respect for what was already established.

On the other hand, olegeno shared its concept for strategy workshops and membership polls with other cooperatives, which successfully applied the tools in their own settings successfully. "In Hamburg they picked up the idea of strategy workshops and then overtook us, they already have their first solar installation," one member recounted the experience of knowledge transfer between cooperatives (I7a: 24–25). This further underscores the idea of olegeno as a catalyst offering tools for reactions without being part of the result.

The limited success of cooperation with the university could be connected to members' claims that the cooperative maintained the character of a "civil society initiative, more than an economic association" (I10: 159). This frame relied on members' motivation and sense of ownership for not only the project but also its operations. One member said they had come to realize the relevance of physical presence in community activism. "Simply if you are there, attending a team meeting... for the... group that makes a huge difference. That I took the time to be present." (I9: 316–320) The importance of, even informal, institutionalizations for the group underscores the challenge of sustaining voluntary engagement over a prolonged period of time and the course of different, highly complex thematic group goals and services. Consequently, many interviewees in the cooperative explicitly stated the lack of paid labor as the key challenge in further developing the cooperative and thus hoped that electricity retail would develop respective financial resources. "There is a participatory aim," said one interviewee but summarized with a certain defeatism, that while the formal options of co-determination were quite distinct within the cooperative, many members were also glad to be able to delegate: "The [participatory] aim is there, but it is completely inflated." (I10: 73–83). olegeno is the only project in the case selection to sustain itself exclusively on unpaid operations. Yet, even projects that rely on paid staff identify "focusing on the relevant topics" as a key challenge. One interviewee joked dryly how "the quality of work does not necessarily improve if you are dancing on ten weddings where three would be plenty" (I5: 67–70).

During this process of reorientation, the cooperative's activities were centered predominantly on the project arena. This refers to the various working groups formed and activities attempting to establish local operations. Energy retail forms connections to the local energy arena, as the cooperative acquires customers locally. With respect to energy generating installations, connections to the local energy arena are consultative. More direct connections have formed to the community energy arena. The cooperative had consulted with other community energy projects that either had successfully obtained a grid concession, most importantly EWS in Schönau, or were applying simultaneously, especially cooperatives in Hamburg and Berlin. But no alliance formed in the community energy arena that visibly supported olegeno's campaign locally. This changed with olegeno entering into a co-operation with Bürgerwerke, a platform of German energy cooperatives operating electricity retail services (Bürgerwerke eG 3/11/2015). This cooperation explicitly formed a connection to other community energy projects, and created a continuous form of exchange for olegeno. The cooperative also began more actively engaging in national associations and membership organizations. But because REL reform had passed which limited financial support for small scale energy production, the cooperative could not create a steady stream of income from energy generation in the way many projects had done. "It feels as if we are late to the party," one member remarked on conversations with other community energy projects on branching out project activities (I9b).

Consequently, the assessment of power relations in the cooperative's interactions within the local energy and community energy arenas is mixed. The cooperative is predominantly visible as a consultant in local projects, despite its goal to be directly involved in operations. Its reputation as a knowledgeable watchdog of local energy politics was powerful in the local energy arena prior to applications. The failed attempt to gain direct access has weakened this position, despite the cooperative still performing as a monitor of local energy system actions.

7.3   Arenas, Interactions and Social Innovations

Three more general observations can be made on the development of arenas of negotiation and relationships. First, corresponding to methodological assumptions of situational analysis, membership within arenas of negotiation is dynamic and non-exclusive, and positions of an actor may differ across arenas. Differences show, for example, with respect to relative positions of power between arenas such as a relatively resourceful position of EWS in the community energy arena coinciding with a relatively powerless position in the national energy arena. Positions vis-à-vis actors of the same arena can differ. EWS could mobilize civil society within the national energy arena, creating a community of intention beyond the local, but could not alter political decision-making. Second, as project development progressed, active connections to more arenas could be sustained. The exception to this observation is olegeno. Lastly, arenas relevant to the research situation can ebb and flow between active and inactive connections. This may correspond to project development, the anti-nuclear arena actively providing support to EWS when the project required mobilization beyond the local. Deactivation of an arena, such as the anti-nuclear arena after 2009, could also correspond to dynamics of the arena independent of project development.

Three additional observations could be made regarding the creation of arenas. First, a project arena can be created by actors external to the arena without community contributions, but requires contributions of community actors, institutions and arguments to be established as a meaningful negotiating space. Second, if the project arena faces opposition of one arena that cannot be immediately countered by instruments provided in another arena, this can be overcome through creating a new arena which forges new connections between actors, institutions and the argumentative positions of previously involved arenas. And, lastly, entering international arenas of negotiation can reinforce a position on the national level.

The role of asset ownership deserves special attention. Direct ownership of energy assets was not necessary to create an assumption of ownership. But the assumption of ownership was important for projects to be able to embed energy services in processes of the community. When neither direct nor assumed ownership could be created, this coincided with a struggle to define and sustain project activities as well as honorary engagement of team members. This coincides with the inability of olegeno to sustain multiple relationships to different arenas as project development progressed.

Final observations turn to the characterization of projects' innovation biographies as social innovations. Again, three observations on shared features across projects are highlighted. First, all projects included social innovations referring to intentional changes in power relations and improved individual and collective capabilities. Community energy projects enabled members to partake in and provide energy system services in ways previously confined to energy utilities or political decision-makers. Second, this created interventions to established networks of actors, institutions and argumentative positions. Relationships within these established networks were altered by project actions, although these alterations were not necessarily permanent as indicated by the case of Oldenburg. Lastly, the way these interventions were framed corresponded to overarching narratives projects had developed in early development stages, creating an impression of consistency in project positions.

Part IV
Resource Mobilization in Community Energy Projects

The analysis of resource mobilization in community energy projects focuses on arenas of negotiation as spaces in which resources can be defined, mobilized or lost. Relationships between the various elements of the research situation are interpreted in reference to the kinds of resources required to alter power relations between them, as well as the origins of these resources. The innovation biographies of community energy showed that projects developed in distinct ways corresponding to their regulatory and economic surroundings as well as local preferences, narratives, and coincidence. Leadership features importantly, connecting project conditions such as the regulatory environment, and project characteristics such as narratives or organizational preferences.

The innovation biographies of communities attempting energy system change share three important features. First, arenas of negotiation differ between projects. This refers not only to which arenas matter for project actors to pursue their goals, but also to the actors and actants relevant to projects connected to seemingly the same arena. The relevance of context factors derives from the perspective of the niche actor. Second, although technology choices differ, projects are similar with respect to the social innovations—the kinds of changes achieved in power relations and improvements to individual and collective capacities—they entail. Comparison across projects can therefore provide indications regarding the conditions for social innovations across different innovation biographies. And third, the individuality of community energy projects in itself is a shared feature which could indicate how community energy projects navigate arenas of negotiation corresponding to the respective social innovations they entail.

This confirms results of more recent analyses of community energy projects pointing to the "situatedness" of energy services as social practices (Pohlmann 2018, 262, 273, comparing community energy projects in Scotland and Germany), and the importance of "intermediaries" moderating external and internal conditions of community energy projects (Sperling 2017, p. 886, analyzing the Samsø community energy project). The analysis of patterns of resource mobilization throughout biographies of social innovation adds to this understanding by providing focus and nuance in the understanding of power relations developed in the process.

Across projects, distinct patterns of resource mobilization indicate corresponding stages of project development. This section is structured accordingly. The first chapter (8) focuses on emergence and establishment. Emergence and establishment refers to resource mobilization after initiation (emergence) until the completion of the first project scheme (establishment). A project was considered established when the first scheme planned by the community had been completed, either by successful accomplishment or acknowledged failure. The second (9) studies maintenance. Maintenance refers to the resources mobilized in order to uphold and ex-tend project activities, sustaining the community energy project over time. The third (10) examines projects' reactions to challenges. Challenges refer to specific events or developments which altered project actions at a defined point in time beyond planned project activities. Such challenges could occur in both of the above named development stages and gave rise to specific patterns of resource mobilization. Within these stages of development, projects can be found to mobilize very similar types of material or non-material resources. Yet, what can be harnessed as a resource within these types of material and non-material resources differs between projects corresponding to their individual innovation biographies.

Figure IV.1 shows the development stages for each project over time, with emergence and establishment marked in green and maintenance in blue. Emergence and establishment of EWS occurred between 1986 and 1996, whereas this period lasted from 2011 to 2014 for olegeno, 1997 to 2007 for Samsø, 2003 to 2007 for Fintry Development Trust, and 2010 to 2015 for Garmony Hydro on Mull. Maintenance of project activities was analyzed for each project after establishment until the end of the inquiry period in December 2017. Challenges are marked as red lightning bolts across development stages.

Analysis considers material and non-material types of resources. Material resources include money, as well as technological installations or equipment, but also natural resources such as land or climatic and topographic conditions. Non-material resources include knowledge–information, skills, or tacit knowledge–as well as shared understandings of desirable (or undesirable) actions and narratives, and organizational resources such as time and labor, or networks. In addition, structural resources, or formalized mechanisms of resource allocation, were studied based on their availability within the projects' political and economic opportunity structures. Structural factors could provide different types

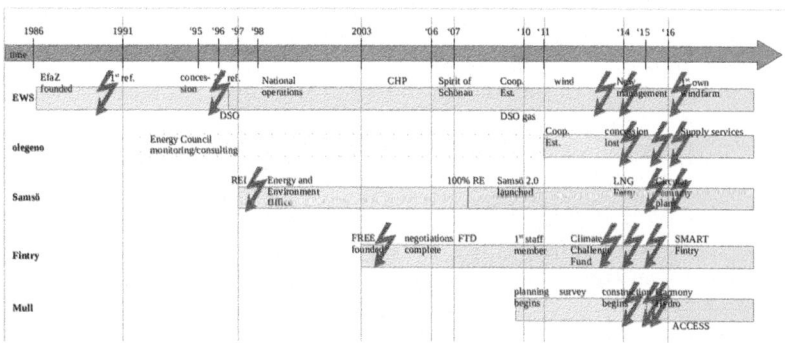

Figure IV.1 Development stages of projects over time

of resources, depending on how they were mobilized. For example, the regional availability of services for installations and maintenance of windmills on Samsø (features of the technical opportunity structure) provided organizational and knowledge resources to the project. But wind power was also established as a culturally supported practice in Denmark, symbolizing technical innovation and leadership. Structural resources, or "context variables" (Schreuer 2015, p. 64), cannot be "owned" by individuals or organizations (Avelino 2011, p. 70). Mapping of arenas and relationships shows how resources not 'owned' by projects nevertheless could be appropriated by them.

Emergence and Establishment: Harnessing Organizational and Symbolic Resources

8

In all cases, emergence created a project arena formed by founders that were either themselves part of the community energy project (in Schönau, Oldenburg, Fintry and on Mull), or interested in creating a community energy project (on Samsø). Across projects, this arena during emergence was characterized by negotiations on which energy system actions should be pursued and how these would be embedded in organizational structures locally, as well as on campaigning. Resource mobilization is dominated by non-material resources, namely the creation of shared narratives and corresponding symbols of campaigning for members and support, and organizational resources, namely time and labor, and networks. Money, as well as technological equipment and installations, or natural resources such as land were less relevant during this stage of project development.

8.1 'Talking yourselves into consensus'[1]: Schönau, Samsø, and Oldenburg

The nuclear accident of Chernobyl in 1986 formed the core of symbolic resources mobilized by Schönauers in the founding of EWS. "Chernobyl hit us like a bomb shell," (EWS 2018c) a co-founder remembered the effect on EfaZ founders. This overcame their lack of knowledge about energy or environmental questions. Community energy organizers in Schönau built on shared symbols of anti-nuclear protest ('grandkid suitability' and immediate danger) and civil society empowerment to construct a narrative of energy system change driven by citizens. While

---

[1] "So you have to kind of talk yourself into consensus about, what does it mean." (I24: 289–290).

199

A. Colell, *Alternating Current – Social Innovation in Community Energy*, Energiepolitik und Klimaschutz. Energy Policy and Climate Protection, https://doi.org/10.1007/978-3-658-32307-3_8

the nuclear accident of Chernobyl affected the national anti-nuclear and energy arenas, mobilizing civil society protest, KWR and the municipal authorities within the local energy arena saw no immediate reason to act upon the event.

This core of non-material, symbolic resources mobilized additional non-material and material resources. Non-material resources mobilized included knowledge (on energy alternatives, efficiency, reduced consumption, etc.), shared narratives of community and locality ('say yes to Schönau'), and civil society networks. It mobilized time and labor among locals, citizens organizing competitions for reduced consumptions, as well as raising awareness on energy system change, etc. Material resources mobilized on a narrative of alternatives to nuclear power included small scale technological equipment, for example additional electricity meters to educate locals on consumption of household devices.

Within the project arena, symbolic and organizational resources formed a self-enforcing feedback loop during emergence (figure 8.1). Shared beliefs and values mobilized voluntary labor as villagers organized to protest nuclear energy technologies. Villagers actively engaged in the community energy project supported one another by re-invoking shared values as well as sharing experiences. Self-reliance, civil society empowerment, and responsibility were established as key symbols of anti-nuclear protest. Immediate ownership and control of energy assets (the electricity distribution grid, as well as generating technologies) were defined as their expressions. When interactions with the local energy arena showed opposition of decision makers in the utility and municipality, the interpretation of self-reliance and empowerment grew more radical. Opposition to the actors of the local energy arena during emergence resulted from early negotiations with an unresponsive utility, as well as the regulatory framework of the national energy arena which did not allow free choice of supply without ownership of assets. Project campaigns began using a challenger-incumbent rhetoric, villagers calling themselves 'electricity rebels' opposing the local incumbent KWR and its strong alliance with the municipality. This was reinforced by regulation tying choice of supply to grid ownership.

Shared symbols of anti-nuclear protest resonated with the national anti-nuclear arena, which had been growing prior to Chernobyl (Schreurs 2014, pp. 11–12). The arena provided a sounding board for symbols invoked by EWS founders and created a sense of anti-nuclear companionship. In Wyhl, just 70 kilometers from Schönau, villagers had sustained protests for over a decade in the 1970s and early 1980s, ultimately successfully blocking construction of a planned nuclear power plant (ibid.). Protesters faced political denunciation and violent police operations (SWR 2016), but also rallied support beyond the local as French communities across the border joined their campaign (Mayer 2005).

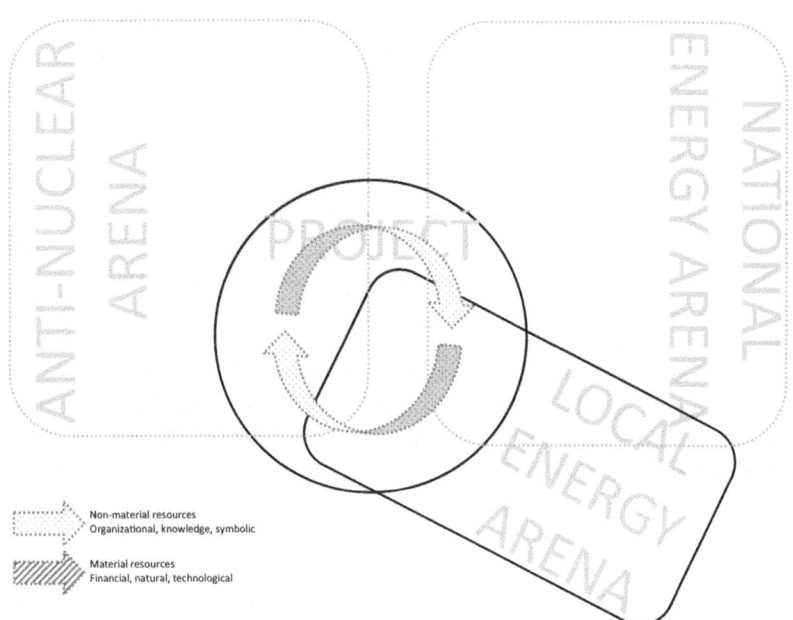

Non-material resources
Organizational, knowledge, symbolic

Material resources
Financial, natural, technological

Figure 8.1 EWS Feedback loop of symbolic and organizational resources during emergence. (Source: AC)

Strong political support for nuclear energy both on national and state (*Län-der*) level, disregarding public protest and realizing nuclear developments against public opposition, raised concerns over the accountability of German government with respect to the public opinion of nuclear power (Schreurs 2014, p. 12). Public debate on legitimate ownership of and decision-making in energy systems had begun all over the country (Schreurs 2014, pp. 12–13). The national anti-nuclear movement therefore not only reinforced technology rejection, but also symbols of self-reliance and civil society voice in infrastructure decisions. Self-reliance referred not to withdrawal from public infrastructures such as the electricity grid, but instead to their politicization (Colell and Pohlmann 2019). EWS co-founders coined the term "Schönauer Gefühl" (Schönau spirit, FUSS 2007) for this sense of empowerment.

The experience of association, on the level of the village and later as a symbol of the national anti-nuclear movement, importantly shaped EWS' ability to mobilize honorary engagement to run the campaign. Engagement was sustained through positive experiences of engaging the larger community. Especially in settings where personal connections between energy project(s) and other groups within a local context were close and professional and personal lives overlapped, mutual support of the group could be key. Responsibility, the third central symbol, referred to intergenerational effects of energy choices, "grandkid suitability" ("Enkeltauglichkeit", I3a: 61, 286). Responsibility concerned individual actions in consumption and was immediately linked to direct ownership of energy assets and participation in related decision-making processes by EWS. This definition of responsibility bridged political and financial involvement, calling on project members to create alternative infrastructure designs and invest non-material, knowledge and organizational resources, and material, financial resources.

Self-reliance and responsibility also drove the mobilization of knowledge resources. Citizens giving their time developed new skills and capacities within the project to support its cause ("I knew nothing of corporate theory and actions, nothing of the energy industry", I1: 328–330). Networks were established, for example with financial partners (GLS Bank 2019). Visible and credible commitment to these symbols was reinforced through the sustained activity of founders as well as local action such as village competitions to reduce household consumption.

When financial resources were required for the first time, the project had already been established for several years. Project members wanted to counter a financial offer of the utility to the municipality to prevent political endorsement. But the investment could be divided into relatively small amounts (250 times 100 EUR) and organizers in turn could build on established networks. Because grid operations required a political mandate (the concession), financial resources for energy operations were not mobilized until EWS had been politically endorsed, eight years after emergence. The project could draw on an even stronger base of symbolic and organizational resources: Shared symbols of protest, self-reliance, citizen empowerment and responsibility, as well as established working structures and networks. The image of citizens as a 'critical accident' ("Ich bin ein Störfall", EWS 1996) to the conventional energy system in 1996 raised 4 m Deutsche Mark (DEM) (approximately €2.5 m) in a matter of months (EWS 2018c). The energy utility attempted to exploit its strong position local energy arena as acting DSO. EWS had calculated a value of 3.9 m DEM. KWR was demanding 8.7 million DEM. KWR's demands threatened EWS but also provided a clearly defined target and opponent for mobilization.

Founders' commitment to their cause also lent symbolic resources. One member recounted how a personal response to a large donation resulted in the donor doubling their commitment, "This lady from France donates 20.000 DEM to support the 'Störfall' campaign. And [name] sends an affectionate thank you note. And she says, you sent such a wonderful letter, I am donating another 20.000 DEM… That's almost like a fairytale" (I3a: 233–236). For project members, this raised important financial support and provided an experience of solidarity. The member continued, "To allow yourself that experience, that something like this is even possible, that was a real learning experience" (I3a-236–237). Money was raised beyond national borders based on shared values and the project's strong ties to the anti-nuclear arena. Mobilization of financial resources relied on symbolic resources, anti-nuclear protest, self-reliance, citizen control and empowerment, as well as the organizational resources of founding members and a clearly defined target including a deadline.

Feedback loops of symbolic and organizational resources within the project arena were less important during establishment, when regime actors had endorsed the community energy project. The municipality had awarded the grid concession to EWS. This mobilized symbolic and organizational resources in EWS' favor in the form of public support of it claims, and a legal mandate for local infrastructure ownership and services. The project mobilized knowledge resources extensively to build capacities to assume grid operations, to gain credibility vis-à-vis political opponents in the energy arena, and to reinforce the project's standing in the regulatory arena. The court order confirming EWS' price calculations, for example, lent credibility to the knowledge and organizational resources mobilized by EWS. Endorsement of the municipal and regulatory authorities created new feedback loops reinforcing the credibility of engagement within the community energy project.

The nuclear devastation of Chernobyl provided an external impulse to community energy organizing, mobilizing non-material resources that shaped subsequent mobilization of material and non-material resources. On Samsø and in Oldenburg, project emergence followed an external impulse providing organizational rather than symbolic resources. Samsø's selection as the Renewable Energy Island (REI) provided a framework plan and funding for two employees to organize implementation. In Oldenburg, previously politically active citizens reacted to the municipality opening the concessionary application process, the regulatory condition to challenge an incumbent DSO. But while on Samsø the initial organizational impulse was quickly embedded in symbolic resources creating feedback loops of symbolic and organizational resources that mobilized a larger local community as described in the case of EWS, this did not succeed in Oldenburg.

The account of Samsø's innovation biography indicated that the project arena had been created for, rather than by, the project. Upon initiation, the arena was mostly empty with the exception of the two staff members of the newly created Energy and Environment Office. Project initiation on Samsø was driven by the municipality; and the municipality supported implementation but was not actively part of it. Again, a feedback loop of organizational and symbolic resources altered this setting. The staff of the Energy and Environment Office emphasized community ownership of the project processes (figure 8.2). Relationships to island villages and communities were formed, as technological and organizational options were discussed. Different technology choices took diverse organizational forms (Hermansen 2007). Of three onshore turbines, one was installed in a cooperative investment structure of islanders and two by investments of local farmers. Paludan Flak, the offshore wind park, included public, private and individual investors[2]. District heating systems were created as independent citizen led organizations or in cooperation with a local utility. In addition, individual solutions grew as islanders without the opportunity to partake in joint installations implemented their own installations. The municipality's master plan provided abundant but implicit knowledge resources for system change that were not advertised.

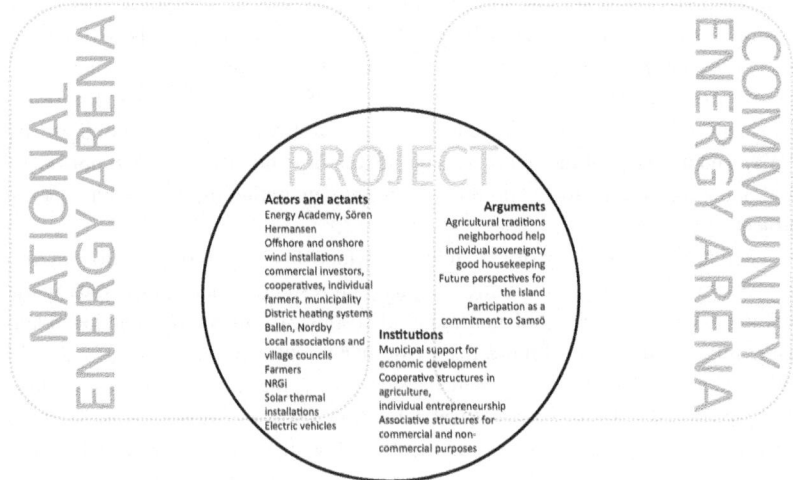

Figure 8.2 Project arena on Samsö, 2007. (Source: AC)

---

[2]Five turbines were realized by the municipality, three by a collective investment of citizen investors, two by individual investments of farmers on the island, and one by a locally established cooperative (Hermansen 2007: 21–22).

Connections between the project arena and other negotiating arenas were inactive upon initiation. The municipality's application for REI was not embedded in the community, or tied to material or non-material support of the community energy arena. Instead, material and non-material support was provided by the national energy arena, relating to Denmark's political commitments on energy and climate. Connections to the national energy arena were therefore strongest upon initiation; the island was clearly charged with the mandate and assignment to complete energy system transformation within ten years. The relationships forged between the project goal of energy system change and the local community by the Energy and Environment Office were extensive.

The Energy Academy, as the Energy and Environment Office was renamed, formed a hub of leadership for project development and coordination, mobilizing both organizational and symbolic resources. Academy staff moderated negotiating processes on the island wherein technology options, organizational models and ownership structures were discussed, and actively connected project activities to local narratives (Papazu 2018; Sperling 2017). Existing organizational resources (membership associations and established habits of assuming positions of responsibility within them) were mobilized to create new membership organizations around energy technology choices. Inviting these organizations to cooperate did not only show respect for existing organizational structures but also snowballed communications. One project member recalled being asked to sit on the advisory board of one of the local projects by a neighbor who knew of her community involvement in a different association through their own chairmanship of yet another association: "[Name] yelled at me when I was walking past down the street, '[name], would you consider coming to this meeting and be prepared, we might need somebody on the board, so could you think about it in advance?'" (I27: 87–90). The process of recruitment indicates the importance of shared knowledge in the island communities and informal modes of communication. Islanders could draw on experiences of cooperation and community: "That was there already" (I27: 79; see chapter 6). The continued mobilization of these symbolic and organizational resources reinforced the experience and expectation of community cooperation. Self-reliance, in the case of Samsø, was not an expression of frustration with the energy system as in Schönau, but of the geographies of communities and their socio-economic status.

Similar feedback loops were created based on the mobilization of established symbols of leadership for community development as well as cooperation on the island. One example is the symbolic value of the farmer's identity. The credibility of Søren Hermansen as a project leader was, in part, assigned to community knowledge of his family roots in local farming. The image of the farmer on Samsø is a romanticized connection to the land and an image of innovation. Denmark's earliest spring potato, nicknamed 'Samsø gold', was developed by local farmers, who (allegedly) were also the first to artificially inseminate cows and exterminate bovine tuberculosis (Papazu 2018: 7). Farmers were seen as steadfast and mindful partners, "They are rock steady in their way of things… they have been taught to think in long terms" (I24: 293–295). Many farmers associated investments in energy system change with economic opportunities more than values of energy system change, as exemplified in farmers' attempts to match oil prices when selling straw fuel for district heating which was refused by project organizers (see chapter 6). The symbolic and economic importance of farmers was acknowledged in the project design, but the economic agenda was counterbalanced by the agenda of energy system change in favor of the larger community. Farmers formed a center of mobilization for natural resources (straw, land for wind energy development, etc.), financial resources (their own investments in installations), and symbolic resources.

Other symbols of leadership were mobilized without connection to a specific group on the island. Most notably this refers to the narrative of 'good housekeeping' and ideas of individual leadership. Energy system change was actively moved into the concept of 'good housekeeping', a shared value of household and community management, in a process that one member described as islanders "talking [themselves] into consensus" (I24: 288) about what the energy transition should mean locally. The narrative of 'Viking leadership' was invoked to instill a sense of agency and responsibility in islanders, who were being invited into energy projects as community leaders in their own right (see chapter 6). Historically, Samsø was indeed an important location for Viking peoples gathering their fleets in the protected waters east of the island.[3] Both examples point to a practice of project organizers to build on established symbols and traditions. Said one community member, "we just used what was already there and kind of tried to respect the structure of democracy in the local context" (I24: 124–126).

The reinterpretation of associative structures, leadership and established symbols within the context of energy system change created ownership as a symbolic resource. Project ownership of islanders was assumed through actual ownership of assets or shares as well as ownership of processes, as exemplified by the above

---

[3] The Kanhave canal, dug 500 meters across the island's most narrow part, dates back to the years 726–729 AD, and proves the relevance of Samsø for controlling trade and warfare in the Kattegat waters (Danske Fortidsminder, n.d.). An art installation of a life-sized, stylized Viking longboat crossing the island in view of the main road to the island's north keeps this memory alive to date (Visit Samsö 2019).

mentioned workshop participant pointing out that islanders were the project independent of holding financial shares (see chapter 6). This collective assumption of ideational ownership reinforced the organizational mandate of the Energy Academy. Islanders feeling the Energy Academy provided 'good answers when one needed them' refers not only to knowledge but also to the structure project organizers provided for implementation. Institutionalization and funding of a moderator, the Energy and Environment Office/Energy Academy, through the REI competition created a position otherwise often neglected in community energy project setups. Organizational and financial resources were mobilized to hire a communicator and an engineer (Papazu 2018, pp. 7–9; Sperling 2017). Time and labor are often donated in community energy projects. By consequence, these often scarce resources are typically assigned to immediate implementation rather than social embedding of project targets. Financial support from structural resources mattered not for its overall sum, which was rather small considering the extent of system change intended, but for its investment leveraging organizational, knowledge and symbolic resources central to project establishment.

Regulation and installations of the energy sector provided structural resources that were appropriated as symbolic or organizational resources. Technology choices on Samsø resonated with nationally established practices of renewable energies. Over 60% of Danish private households are connected to district heating services for heating and hot water (Danish Energy Agency, p. 4), and islanders could draw on the local experience of the Tranebjerg district heating system. Wind installations were embedded in the strong support of wind power as a local resource and established technology in Denmark, similar to the embedding of Schönau's anti-nuclear sentiment in a national movement. Cooperative ownership proliferates in the Danish wind energy arena (utilities only accounting for about 10% of installations, see section 1.3), both historically (providing symbolic resources as well as knowledge) and through the regulatory requirement of local ownership as a reinforcement of collective citizen ownership (an organizational resource). The regional incumbent utility NRGi, directly involved in the district heating system in Nordby, is itself cooperatively organized. "They were expected to be helpful," one islander remembered the connotation of cooperative ownership in negotiations with NRGi about their involvement, but also pointed out that "they might as well have said no if it wasn't commercially alright" (I27: 40–41).

Project establishment was characterized by strong diversification of resources, referring to both type and origins of resource. Successful realization of energy system change rested on the mobilization of voluntary time and labor of island communities and organizations, as well as individual citizens in determining the organizational and technological parameters of individual installations. Knowledge resources were mobilized, as well as natural resources (such as land for turbine installations, straw for district heating, etc.). Shared values and beliefs of neighborhood help and good housekeeping were invoked to support investment in shared or individual installations.

Symbolic resources shaped successful establishment of projects in Schönau and on Samsø and referred to concepts of self-reliance, responsibility and ownership in both cases. Yet, mobilization of these symbols was distinctly different. Schönau citizens reacted to an external event which imbued energy practices with symbolic meaning. The desire for change was reinforced by a larger national movement. Samsø citizens reacted to an organizational impulse that required that they change; these were actively embedded in local symbols of cooperation and leadership through project organization. The cooperative in Oldenburg forms a counterpart. An external organizational impulse could not be embedded in shared symbolic resources to (sufficiently) mobilize local action. And although an external event activated project members, its symbolic power did not mobilize the community in their favor.

Two external drivers predated emergence of olegeno: the city council of Oldenburg inviting applications for the DSO concessionary process, and the nuclear incident of Fukushima in Japan. Members of the Energy Council had organized around local energy politics for years, which provided important knowledge resources. They knew about the concessionary process, knew many municipal actors involved and were familiar with the commercial practices of the competing applicant, the local energy incumbent EWE. This was the prerequisite of identifying the concessionary application process as a potential window of opportunity upon which the community could act to pursue its own political aims related to energy. Very similar to Schönau, Energierat members referred to symbolic resources of self-reliance, empowerment, and responsibility (see chapter 6). The nuclear accident of Fukushima in March 2011, predating olegeno's founding by just a few months, reinvigorated German anti-nuclear protests. Fukushima instilled members of the Energierat with a sense of urgency, feeling the need to try new approaches to energy system change (see above 6). Self-reliance and responsibility were again phrased as symbols politicizing energy infrastructure ownership. Political frustration referred not to an alliance of all political levels supporting nuclear technologies—indeed the national level had shown its responsiveness to high public pressure—but to the local level. The city council had chosen to ignore the Energierat's recommendation for a socio-economic investigation of alternative grid operation models.

The project arena of olegeno related to local and national arenas of energy and anti-nuclear protest. But while these are similar to negotiating arenas connecting to the project arena of EWS, the configuration of these arenas changed. The national energy arena was no longer dominated by vertically integrated energy incumbents, although energy market liberalization and subsequent consolidation of utilities resulted in four large energy utilities dominating the national market. Nevertheless, the market diversified. Customers could choose from an array of over 1.000 energy providers (1.100 in 2011, BDEW 2012, p. 32), several of which offered services sourcing electricity renewably.[4] Direct ownership of distribution systems, a narrative which had fueled the EWS campaign, was no longer a prerequisite for individual influence in energy system change. Instead, this created an explanatory burden for olegeno in determining why ownership of the power grid was necessary. Similarly, the comparatively successful development of renewably sourced and community owned energy systems in the national energy and community energy arenas were cumbersome for olegeno. Policy support of the EEG related predominantly to energy generation, not distribution, as did most projects of the community energy arena. Both arenas provided limited resources to the project, while supporting an overarching narrative of successful community involvement in energy system change which impeded mobilization among citizens of Oldenburg. The anti-nuclear arena had lost its mobilization power after forcing political commitment to nuclear phase-out. And EWE was a mostly favorably reviewed local actor. The cooperative challenging EWE could not rely on sympathies of a 'David and Goliath' setting, as EWS had benefitted from. This underscores the importance of understanding arenas of negotiation from the perspective of the niche actor, as their configuration determines the respective availability of resources (or lack thereof).

olegeno concentrated on mobilizing knowledge resources in the energy arena, building on the legitimacy of the Energy Council acquired in decades of critical monitoring of energy politics. The symbolic power of citizen knowledge was purposefully communicated, in combination with symbols of accessibility and more direct control. In the face of an unresponsive city council denying a formal investigation into concessionary potential in 2011, the Energy Council did not campaign to pressure the utility or the municipality to act, but commissioned external experts to conduct such an investigation. Once expert opinions supported alternative models for grid operations, the Council founded the cooperative as a representation of this alternative in the bidding process. While the symbol of the community energy project as a challenger to energy system structures was also

---

[4]The number of electricity providers has since grown to 1.300, BDEW 2019.

important to olegeno, it was mobilized within the economic and political conditions of the application process. Self-reliance as a symbolic resource was not framed as rebellion ('Ich bin ein Störfall') or local leadership qualities ('good housekeeping, that is what we do'), but as an expression of knowledge of alternative management structures which would enable civil society access and control. To submit a binding offer supporting its claims of alternative grid management structures and fulfill requirements of the bidding process, the continued mobilization and extension of knowledge was key to cooperative activities (Kuchta 2011a, 2011b).

The mobilization of knowledge and symbolic resources by the cooperative remained closely connected, even after losing to EWE in the application process. The endorsement of EWE by the municipality provided an organizational commitment but also constituted a strong symbol of political support for the company's actions. Project members therefore highlighted how having submitted an application 'at eye level' in the bidding process was an important constitutive success for the cooperative. In this respect, olegeno exemplifies how credibly showing an alternative energy system configuration matters as a symbolic resource to participants of community energy projects, even if this alternative is ultimately dismissed. This frame of "it **could** somehow be different" (I10: 179, see section 6.2) shows in similar ways across all projects. Those sporting a longer track record such as Schönau, Samsø, and Fintry mobilize this narrative as proof of concept, lending credibility and legitimacy to their projects. Younger projects stress how establishing a track record, for example in the case of ACCESS on Mull, would subsequently enable the pursuit of other project activities.

Financial resources were mobilized quite late in the process of olegeno's emergence. The time and work of cooperative members was donated, and campaigning relied on voluntary engagement. Membership shares were raised as part of the final bid in the application process to indicate the financial capability of the cooperative in the bidding process (Kuchta 2013). However, the communication of the project relied predominantly on its conceptual goals for grid management, arguing that financial endorsements would follow a political commitment (ibid).

8.2 Don't you go 'banging on about climate change'[5]: Fintry and Mull

For Fintry Development Trust (FDT) and Green Energy Mull (GEM) in Scotland, organizational resources played an even more substantial part in project emergence. In Fintry, establishment of the energy project was predominantly a function of strong mobilization of organizational resources. And while project founders and organizers saw organizational resources embedded in these shared symbolic resources, these were not necessarily shared by the larger community supportive of the energy project. On Mull, project realization was also shaped by organizational resources, namely the capacities of MICT staff. MICT, however, paid reference to shared local symbols during project realization (technology choices and implementation) to ensure continued community support for the energy project and for the trust.

In Fintry, founders' pitch of a local energy supply to the community council in 2003 was based on personal enthusiasm for a narrative of low carbon community development, embedded in an image of community pioneering (as captured in the founders' organization's acronym FREE, see section 1.2). Discussions coincided with a commercial project developer pitching a wind farm. This external action provided a technology selection, that is, wind power installations, and an organizational impulse to the community, which resumed its planning activities with a view of connecting the village to the developer's plans. Villagers suggested to "combine forces" (I19: 19) to make use of both the knowledge and organizational resources (labor, networks, etc.) of an experienced commercial windfarm developer, and community resources referring to discourses of energy system change and community pioneering (symbolic resources), institutionalized involvement of community decision making concerning siting of installations (organizational resources) or community based financial investments (financial resources). Falck Renewables' reluctant suggestion to create joint ownership with citizens organizing as a cooperative was rejected by FREE founders based on considerations of community ownership. The suggestion of the commercial developer induced a discussion of which symbols founders wanted to connect to community ownership, namely a 'flat', 'non-hierarchical' organization without individual financial benefits (see chapter 6). The choice of organizational form was again decided by an external impulse, as DTAS consulted with project organizers. One interviewee remembered, "All these things [were] a bit haphazard... we met someone from

---

[5] I19: 191.

the DTAS which had just been newly formed and they basically sold it to us as a structure" (I19: 53–56).

This account of the seemingly unstrategic mobilization of structural resources as organizational and knowledge resources to the project by FDT founders points to two important features of community energy projects. First, projects make use of alternative sources of organizational and knowledge resources, beyond their own frequently unpaid and laypersons' capacities (as in the case of EWS campaigning, or Samsø mobilizing through existing associative structures). And, second, projects develop argumentative positions (knowledge and symbolic resources) in the course of reviewing an external impulse.

It was only once commercial realization had been secured with the windfarm developer and organizational realization had been established with the help of DTAS advice, that FREE founders actively invoked symbolic resources at the community level to mobilize within the village and create a membership base for the Development Trust. Symbolic resources connected the project to larger debates of both energy regime change and civil society control. The virtual connection of villagers to renewable energy installations should engage and empower them in questions of climate change mitigation (I18: 53–54; I19: 35), while ensuring 'democratic control' by villagers in the structure of the Development Trust. Democratic control and charitable status of the development trust were connected to symbols of responsibility by founders and team members of the Trust. These mattered for their engagement, but did not necessarily extend to 'ordinary' members. Said one project member on their understanding of the responsibility of FDT, "We're privileged... that we can afford to think about climate change and we can afford to think about the transition of energy system... I think we have responsibility to try and make that happen in general." (I18: 67–71) At the same time, while villagers in general are highly supportive of the Trust and its activities, climate change mitigation is not a shared mobilizing symbol, as indicated by one member cautioning against "bang[ing] on about climate change" (I19: 230–231). The project arena formed gradually, FREE founders building knowledge and organizational resources. Organizational, knowledge resources and symbolic resources, namely the support of Community Council as elected officials for the continued investigations of FREE, circled within the community. The project arena was established once an external impulse indicated that energy developments would occur locally, and that the village's position would now depend on its diligence in negotiations. This intensified interactions between FREE and the Community Council, and connected the project arena to both the Scottish and the UK energy arenas as the relevant regulatory and economic environments.

Accumulation of knowledge resources within the project arena was key. This refers for example to the connection between the wind farm development and the village, which initially had been planned as a physical connection of electricity supply but was soon replaced by a financial connection upon realizing both the limitations of grid access and the benefits of selling to the national grid under UK feed-in tariff regulation. However, the concentration of knowledge resources within FREE and the subsequent expansion of negotiating powers vis-à-vis the project developer importantly determined the project's ability to press for terms of cooperation that matched their interests in community organization. This connected the project arena to the negotiating arena of Scottish community development strategies. FREE rejected the path for community cooperation established within the UK energy arena and chose the path suggested by the Development Trust Association Scotland instead. Through the establishment of a feedback loop of knowledge, organizational and symbolic resources between community energy organizers and the community council, and additional knowledge, organizational and symbolic resources provided to the local project within the Scottish community development arena, a new organizational model could be established in the Scottish energy arena: the cooperation of a Development Trust and a commercial wind farm operator.

In Fintry, the project was entirely released from mobilizing financial resources as the investment would be refinanced through financial securities provided by the UK energy arena, namely the feed-in tariff. Financial resources, however, changed the picture upon the project's establishment. The establishment of a steady source of income quite literally provided the ability to 'afford' to think about a pioneering role in implementing energy system change locally. The UK feed-in tariff was not immediately instigative to community energy considerations in Fintry, as is apparent from the project's innovation biography, but as a structural resource the FIT provided a narrative of financial stability to mobilize villagers' support for the Trust. This increased membership numbers and ultimately credibility (measured in representation by numbers) of the Trust. Although the membership shares raised upon establishment of the Trust did not fund installations, the security of future income secured establishment. Also, the FIT provided the financial security for the project developer to set up an additional turbine which would be paid off by the community during operations.

Garmony Hydro on Mull is the only example where the availability of material resources, namely natural resources for energy generation and financial resources for their exploration, featured more prominently during the stage of project emergence. These material resources were embedded in a strong organizational context. Although ultimately realized as an independent project, early project considerations were soon run under the local development trust (MICT). This provided organizational resources to the project, as coordinating work could be done by staff members who were also knowledgeable regarding grant application opportunities and other structural resources made available at Scottish and UK levels. The umbrella of the development trust, as well as the leadership of its director in the energy project also provided symbolic resources to the project.

This is exemplified in the process of technology selection. Natural resources for energy generation are abundant on Mull, offering different opportunities for community energy schemes. Wind, tidal and hydro energy installations were the predominantly discussed options. The community chose a hydropower option which in the words of project members had almost 'no visual impact' on the countryside and quite literally imbued the local clouds with a 'silver lining' (I20: 272, 367). In other words, the technology choice was framed by strong symbolic references to the locality. MICT then successfully applied for a financial grant from the Scottish government's CARES (Community And Renewable Energy Scheme) scheme to conduct a feasibility study for hydro power locations on the island, with Garmony being chosen from the five locations suggested in the report (Local Energy Scotland 2015). The setting within the development trust provided the important advantage of eligibility, as not all legal entities can apply for grant funding. Commercial organizations, such as cooperatives, are often excluded from such instruments or the application process exceeds knowledge and personnel resources of voluntary settings.[6] MICT again provided important organizational resources to the project by setting up the survey and advertising for participation across the island. The 'feedback loop' created during project and emergence in the case of Mull was therefore largely one of organizational resources provided to the project by MICT supported by the credibility of the organization (symbolic resources), mobilizing symbolic resources (technology choices, support, etc.) in the larger community which in turn provided the necessary grounding of the project scheme in community interests and support.

Upon initiation, the project arena for Green Energy Mull immediately included more references to established local organizations, and direct interlinkages to respective arenas of negotiation. The early link to MICT lent organizational resources in the form of personnel, as well as knowledge resources, credibility, and legitimacy to the energy project. MICT was an established community development actor, which gained community trust in the project. At the same time, the

---

[6]Eligibility for grant funding is often tied to non-commercial structures. In the UK, this changed with the establishment of associative funding such as the Community Shares Booster Programme (https://www.communitysharesbooster.org.uk/).

Development Trust could not risk community support on an energy project if this were to jeopardize future activities in other areas. This ensured that MICT carefully tailored the project to community preferences, as exemplified in the choice of technology or siting.

The close connection of the project to MICT set project preparations within the Scottish community development arena, characterized by regulation of the Scottish executive to increase community access to land resources and support community attempts of local economic development (Scottish Parliament 3/25/2003). In the Hebrides, specifically, this resulted in a rising number of communities buying lands from private landowners, often in combination with subsequently established renewable energy installations which significantly increased community income (Fletcher 2016; Bunting 2015). While Mull islanders were not looking to acquire private lands, project organizers were aware of developments on other islands of the archipelago and the economic potential of renewable energy developments for the economically vulnerable community of Mull. They were also aware of the limitations to community income in partnerships with commercial developers, as this quote indicates: "There are examples were a developer has put 40 or 50 MW of windfarm, and the community benefit is on the benchmark [...] from the Forestry Commission of Scotland of 5.000 £ per installed MW per year as community benefit. That's **peanuts**." (I20: 197–200, original emphasis). Key argumentative positions of the Scottish community development arena connected to community empowerment, access to land and other natural resources for communities, and community-oriented collectivization of benefits to be gained from such resources.[7]

Consequently, although the project connected to the UK energy arena through plans to benefit from energy installations financially (which meant that installations would export energy to the national grid under FIT regulation), this connection was down-played argumentatively. Instead, project ties to the Scottish energy arena were emphasized. This was an argumentative decision but also a practical advantage, as MICT could follow the examples of a growing number of projects establishing energy installations in Development Trusts or similar community-oriented organizational structures, and benefit from institutional support established by the Scottish executive, namely financial support offered within the CARES program, and organizational support of the association Community Energy Scotland during later stages of project development.

---

[7]Regulatory reform for community empowerment continued I the community development arena after project initiation with the Community Empowerment (Scotland) Act of 2015 and the Land Reform (Scotland) Act of 2016, which both further extended community rights to own and manage natural resources and community infrastructures (see section 1.3.3).

During project emergence, MICT provided important material and non-material resources. Its staff handled planning and acquired relevant knowledge on technological alternatives, siting, and organizational structure. Its organizational form offered access to institutionalized, financial support such as grant-funding and lent credibility and legitimacy to the project through its track record of community engagement. The Scottish community development and energy arenas provided non-material symbolic resources in the form of narratives of community empowerment, non-material organizational resources through regulatory statutes securing access to lands for community development and institutions offering knowledge and organizational support, and material resources through financial grants under CARES. The UK energy arena provided the regulatory basis for future financial resources to be mobilized for the project when the installation was conceptualized under the FIT.

Project preparations were complicated by regulatory constraints of the national energy sector, which limited energy generation at the fringes (see chapter 6). This resulted in the creation of ACCESS as a second project arena, overlapping closely with the first but connecting more strongly to the Scottish and UK energy arenas because of the project membership of key actors from both arenas. Community Energy Scotland, the leading public institution of the Scottish energy arena, took over project management, providing key organizational resources by hiring a project manager, the organizational and knowledge resources assembled in the association, and symbolic resources. When SSE Ltd, one of the dominant British energy utilities entered the project, the project became connected to the UK energy arena not only through the technological and knowledge resources of the company, but also through the credibility project results gained on the national level vis-à-vis regulatory bodies such as Ofgem.

The initiation of ACCESS provided additional knowledge and symbolic resources to the community energy project. It lent additional credibility to the Garmony Hydro project, as islanders reaffirmed their commitment to an energy installation in the face of structural opposition. This also supported the experience of self-help and self-reliance on the island central to the establishment of community trust organizations, especially in vulnerable regions as has been shown in chapter 6. Knowledge and symbolic resources were also mobilized vis-à-vis the energy sector. The implementation of Garmony Hydro had been challenged by actors from the national energy sector. MICT succeeded in realizing not only the original project, Garmony Hydro, but also a second, more complex project of energy system integration. This added credibility to the organization as a community energy actor in the renewable energy arena. Similarly, the establishment of the Waterfall Fund, as the foundation ultimately administrating the revenues

Garmony Hydro generates to community development projects, lent additional symbolic resources to the project. It was indeed 'there for the right reason' (I19: 66). Similar to Samsø project development, trust as a symbolic resource provided by the island community was mobilized through individual leadership, personal attributes of the people involved, and organizational processes. The director of GEM, a voluntary engagement, was known to islanders as the director of MICT, a paid position. While GEM operates independently of MICT—connections such as the directorship are not a result of organizational ties—its director was known on the island through GEM's work for MICT. This helped gain islanders' trust. Project organizers were careful to ensure close connections of project concepts and islanders so as not to jeopardize the institutionalization of cooperation the project was built upon. Said one interviewee about the long-term danger of losing islanders' support, "they won't know in the future that if there's a problem, a thing that the community could address—they won't come to us" (I20: 406–409).

Compared to the map of arenas in Fintry, the map of arenas in Mull points to more complexity within the project arena, and different configurations of the UK and Scottish arenas the projects connect to. Upon project emergence Fintry, the dominant energy sectoral blueprint for cooperation with citizens (if enacted at all) was financial representation through cooperative structures. The DTAS as an actor of the Scottish community development arena mattered for providing knowledge and organizational resources to FDT. Upon project initiation on Mull, by contrast, the island could build on the experience of community support through the Development Trust. Development Trusts had become an established tool of community energy engagement in Scotland. While the DTAS provided organizational background, it did not actively instigate institutionalization. Also, new tools had emerged to support projects financially (such as CARES in the case of Mull), and Community Energy Scotland offered knowledge resources, as well as organizational and symbolic support when necessary.

8.3   Summary of Resource Mobilization during Emergence and Establishment

Tables 8.1 and 8.2 provide overviews of resource mobilization by type for all projects during emergence and establishment. Material and non-material resources as introduced above were separated into six categories. Material resources include financial, technological and natural resources. Non-material resources include organizational, knowledge, and symbolic resources. A " + " indicates that resources of this type were mobilized, while a "-" indicates that this resource type was not mobilized during this stage. The indication of mobilization in brackets points to an indirect mobilization.

All projects mobilized organizational, symbolic and knowledge resources during emergence. Material resources were less important. The general availability of natural resources was important for project conceptualization on Samsø and in Fintry. But it was not the availability of natural resources that initiated project activities in either case. Rather, organizational resources were important. By comparison, the availability of different natural resource types and the symbols attached to them shaped project decisions on Mull. Financial resources were not mobilized during emergence, with the exception of Mull. Security of future investments also influenced campaigning in the cases of Mull and Fintry. Technological resources were not instigative for project emergence but could support campaigning. In Schönau, the experience of efficiency competitions, realized in part through the installation of old metering technology to create awareness about consumption provided a sense of self-efficacy to members. Similarly, technology availability was not relevant to mobilizing community involvement on Samsø, but it was important for convincing the mayor to apply for the REI competition as a measure of community development.

All types of resources were mobilized by all projects upon establishment (the only exception being EWS and olegeno that did not mobilize technological or natural resources for establishment/rejection of the projects as local grid operators).

Table 8.1 Mobilization of resources by type during emergence

Project	Mobilization by resource type					
	Financial	techno-logical	natural	organizational	knowledge	symbolic
EWS	−	(-)	−	+	+	+
Olegeno	−	−	−	+	+	+
Samsø	(-)	(-)	(+)-	+	(+)	+
FDT	−	−	(+)	+	+	+
GEM	+	−	+	+	+	+

Analysis of the configurations for each resource type within the projects shows that symbolic resources mobilized during emergence and establishment refer most strongly to symbols and values connected to technology choices, community functions of energy infrastructure, and ownership and returns. Organizational resources also strongly reference the dimensions of ownership and returns, as well as management and operations.

Table 8.2 Mobilization of resources by type during establishment

Project	Mobilization by resource type					
	Financial	techno-logical	natural	organizational	knowledge	symbolic
EWS	+	−	−	+	+	+
Olegeno	+	−	−	+	+	+
Samsø	+	+	+	I	+	+
FDT	(+)	(+)	+	+	+	+
GEM	+	+	+	+	+	+

Three lessons can be drawn across cases. For one, symbolic resources and organizational resources, that is, the stories that were told and those telling them and building the organizational structures to bring them to life, were key to project emergence. These symbolic and organizational resources could either originate in the project itself, or be mobilized on the basis of structural resources that were appropriated and reconfigured to become symbolic or organizational resources. Variation refers to the order of appearance of symbolic or organizational resources as well as their local specification, but not to the types as such. However, what could be mobilized as a symbolic resource differed across cases, and also shaped subsequent institutionalization of the projects. For example, anti-nuclear protest and challenger-incumbent frames were mobilized by EWS in Schönau and olegeno in Oldenburg very differently.

Secondly, while other resource types may have also been mobilized, they alone were insufficient to instigate actions. Financial resources, for example the capability of project members to invest private money or the availability of subsidies to be mobilized by projects, although important in later stages for realizing installations and providing respective energy services as will be discussed in the following section, did not provide similar initiating impulses. Similarly, knowledge resources such as awareness of policy support mechanisms for community engagement in energy systems or expertise in renewable energy technology, as well as natural resources such as climatic conditions favorable to wind energy generation or space to establish installations were insufficient to create enough momentum for project establishment without their embedding in symbolic and organizational resources.

Lastly, the mobilization of all types of resources for establishment underscores the broad range of activities necessary to anchor an energy project within a community, and the diversity of discourses projects must engage in within the community.

Maintenance: Growing Relevance of Financial Resources

9

During project emergence and establishment, mobilization of financial resources typically occurred in connection to either a specific activity or event. Examples are the mobilization of grant funding for a feasibility study of hydro-power siting on Mull, the mobilization of donations to counter the energy incumbent's offer to the municipality in Schönau, or the mobilization of membership shares at the time of establishment. Financial resources were instrumental rather than instigative. When analyzing project maintenance, by contrast, cases can be roughly categorized based on their ability or failure to mobilize a consistent flow of financial resources. Sustained influx of financial resources could be established based on energy services (generation, distribution, or retail), or through public funding. Mobilization of financial resources for project maintenance was still instrumental, but took a more prominent role in upholding and extending project activities.

9.1   Feedback Loops of Material and Non-material Resources

Both Scottish projects and EWS in Germany successfully mobilized a steady income based on energy services, although mobilization of material resources through these services was partially supported by structural resources. In the Scottish cases, energy generation creates a steady income stream regulated under the feed-in tariff. Revenues generated by the turbine in Fintry are channeled to the Development Trust, which decides on reinvestment within the community. Membership shares are raised upon entry into the development trust, but do not contribute substantially to FDT's assets. Revenues are not distributed among

© The Author(s), under exclusive license to Springer Fachmedien Wiesbaden   221
GmbH, part of Springer Nature 2021
A. Colell, *Alternating Current – Social Innovation in Community Energy*,
Energiepolitik und Klimaschutz. Energy Policy and Climate Protection,
https://doi.org/10.1007/978-3-658-32307-3_9

members but invested in services of the Trust within the community. Community redistribution of revenues, as well as membership, is limited to the village of Fintry. By contrast, shares raised for the establishment of GEM contributed directly to the installation of Garmony Hydro on Mull. Revenues of energy generation, secured by a feed-in tariff, are shared between investors of GEM that can receive a direct revenue of up to 4% at the discretion of directors, and the Waterfall Fund charity. In its first year in 2016, the charity received £ 25.000 from GEM (Waterfall Fund 2018). Apart from potential revenues from generation, investments in GEM were supported by a tax incentive. Investments of the Waterfall Fund are tied to the communities of Mull and Iona and immediately adjacent isles (Ulva, Gometra, Erraid, Calve and Inch Kenneth, (The Waterfall Fund 2019)). Membership in GEM and corresponding redistribution of revenues according to investment is open to non-locals.

For EWS in Germany, the first steady source of income was established by winning the local grid concession. At the time, this included income from distribution systems operations, as well as retail and generation. Assets of the cooperative today include membership shares, as well as direct revenues from its five subsidiaries dealing with distribution systems operations, retail of electricity and natural gas, direct marketing of renewable energies, income from installations and operations of energy production, and a windfarm (EWS 2018b, p. 51). While the cooperative has realized energy installations under the feed-in tariff, these account for a smaller share of overall financial assets. Revenues of the cooperative are divided between members, who earn a maximum dividend of 3.5% annually, and reinvestments in energy projects, either directly or through grant programs. Financial income streams stabilized operations. The cooperative could hire permanent staff to run energy system services, provide knowledge resources and ensure that energy operations were coordinated according to shared symbols established during emergence. This created feedback loops of non-material and material resources between the project arena and the local as well as national energy arenas (figure 9.1).

On Samsø, income generated by energy installations under the feed-in tariff is distributed among investors depending on the organizational structure of the installation. The Energy Academy as the organizational hub of the project is funded by the national government. Mobilization of financial resources for the Energy Academy relies on political support. Respective networks, established and sustained by leadership of the Energy Academy over 20 years, have created a system of reciprocity. Funding for the Energy Academy is sustained, because its activities are a successful example of government-supported projects. "I am the most famous energy person in Denmark," said Søren Hermansen in an interview, and

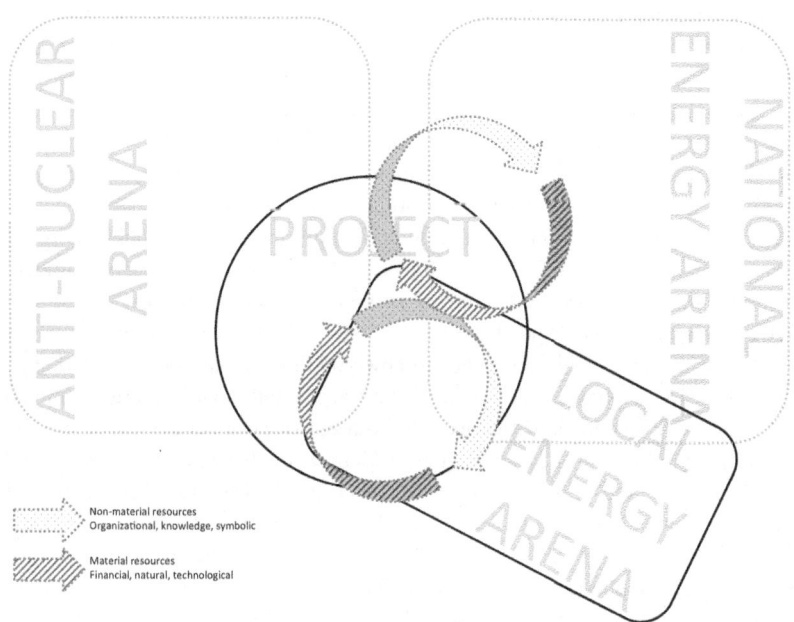

Figure 9.1   EWS feedback loops of non-material and material resources during maintenance. (Source: AC)

continued that while it was "not always interesting to be that person" he had now developed a network of politicians which would "try to keep me in their network, also, because I am good for them and they are good for me." (Interview Samsø, May 2016). The Samsø Energy Academy is frequently referenced as a 'best practice' example for renewable energy and community development by EU representatives (Buffetaut et al. 2017; Ankjaergaard 2017). On the final day of the United Nations Climate Change Conference COP21 in 2015, US national television reported on Samsø. "We had ten minutes on CBS Morning News and ten minutes in the CBS Evening News. That is **big**!" one interviewee remembered (Interview Samsø, May 2016, original emphasis), adding with a laugh, "that is where we are paying off the debt" (ibid.). This dependence on government support has gone unchallenged so far, the Energy Academy even surviving budget cuts affecting other organizations in the sector. Still, it has given rise to concern in the organization and led to new activities to diversify public funding, for example through project grants (Christensen and Friis 2017), or independent income through consultancy work (Energy Academy 2016). Like in Schönau, feedback loops of material and non-material resources were characteristic of project maintenance. Continued funding for the Energy Academy enabled its staff to provide background functions ensuring project maintenance, to 'keep the torch burning' (see chapter 6), and develop new ideas for project activities.

Across cases, the mobilization of financial resources matters most for the ability to invest in the realization of project goals locally, increasing the projects' credibility and support within the community, and for the mobilization of organizational resources. To Fintry project leaders, more than the actual amount of money the knowledge of the amount of money recurring annually fueled project development: "The knowledge that it is every year and that it's going to keep going… makes a big difference to how you think about what you're doing" (I19: 317–318). Money enabled project investments, and mobilized organizational and knowledge resources. If one were to "run out of personal energy, at the end of the day you're almost faced with a problem, actually: what would we spend the money on? And you have to just get back to it" (I19: 318–321). Financial investments in wide-ranged projects throughout the community strengthened local support for the trust (FDT 2018). Installations provided a material expression of the legitimacy implied by the organizational status as a charitable organization; the Trust was indeed "there for the right reasons" (I19: 66). Each new activity also built FDT's track record, increasing its legitimacy and credibility in negotiating arenas beyond the local and underscoring its image as a Scottish community energy pioneer. "Because we have been pioneering in that field and continue to be pioneering… it's like a snowball effect. Once you do one project, the rest is a lot easier. So, it's building up that momentum and that momentum has just kept going." (I18: 114–120) The project's track record in combination with the security of a second income stream underscored FDT's application for funding through the climate challenge fund which financed two permanent members of staff in 2010 (FDT 2018). Administration was professionalized and an energy manager was hired. Financial resources consequently matter also for the ability of projects to access additional sources of income.

Similarly, both the income generated by Garmony Hydro and the organizational resources provided to the project by MICT created a permanent representation of the project in the community in the form of financial and organizational outputs. An energy manager professionalized the local activities. MICT and GEM connected through the ACCESS project, further establishing in the community that MICT staff could be addressed when questions or concerns arose regarding energy activities. Individual investments could be rewarded financially in the organizational structure of GEM. But gifting 25.000 £ to The Waterfall Fund in its first year illustrated GEM's commitment to creating a meaningful influx of capital into the local community from Garmony Hydro. This repaid, to some extent, MICT's investment of symbolic and organizational resources into Garmony Hydro. Supporting an energy project carried a risk, as the project needed to be embedded in the local community to secure continued trust of locals in the work of the Development Trust. That Garmony Hydro was successfully commissioned and generated financial revenues strengthened the position of the Development Trust in turn.

9.2   Financial and Organizational Resources

On Samsø, project maintenance as an organizational process was defined as not only upholding technical installations and reporting on past achievements of energy system change, but reinvigorating community processes underpinning these changes. After achieving the REI target of 100% renewable energy sources, there was no immediate agenda of continued development. The Energy Academy interpreted its role as a multiplier and communicator of the REI project success to also include continued support for learning and transition processes on the island. Community learning had been neglected in the development plan of the competition. The focus was instead on the realization of technology and emissions targets: "we forget about the learning journey and what are the learning skills." (I25: 18–19) This realization resulted in a staff position in the Energy Academy focusing on community learning and development. "I have defined my role," this person continued to explain the emergence of the position in the Academy (Interview Samsø, May 2016). The job description resulted not from an administrative decision but from the realization of community needs and personal strengths. This was seen as a risk by some at the time. Yet, the institutionalization of a role reflecting learning processes back to the community and challenging the organization and its constituents to not rest on past success but extend developments into the future shaped the dynamic of project maintenance. The Energy Academy not only had the personnel, time and money to 'keep the torch lit' while the community resumed its daily routines, but also created a permanent internal challenger position.

This involved reflecting community developments back to islanders. "I said, we need to make movies because I can see that people stay on the website longer," one community member remembered convincing the organization to develop visualizations of the energy transition process on the island. They continued to explain the relevance of islanders' being able to see the changes they had undergone from an outsider's perspective, "suddenly people could **see** those new processes. They had a reflection." (Samsø Energy Academy, May 3, 2016: 189–192, original emphasis) In addition to creating permanent positions to support and extend energy services, similar to the position of energy managers in Scottish projects, a representation of community involvement was created. The REI master plan had been drafted by experts of technological system change. New targets and processes of change should now be conceptualized by the community: "We need to redefine ourselves in this moment... And we need to make a community process as we did the first time" (I25: 182–185). The team member talked about repeating a community process, but it was actually the first process, wherein the community defined targets of cooperation. This was a result of previous success as well as newly created positions in the Energy Academy. Successfully transitioning the local energy system was assumed as a collective achievement by islanders ('this **is** us'), and secured funding for further activities. Through a permanent staff position, the project ensured that community opinions were invited into and subsequently represented within the organization. "We invite them [the citizens] to look into the future. And it is also important that we take what they say about the future **seriously**," said one interviewee remembering internal debates on future project activities within the Energy Academy. Slapping the table for emphasis, they added "so many, when they do these facilitation methods, then people get a kind of disillusion after because you don't follow and integrate those pieces of good advice you actually get" (I25: 198–202).[1]

This statement points to an important pitfall to community projects. Continued community support for project efforts often relies on the perceived accountability of project leaders to their targets and values. This is exemplified by the double donation to EWS following a handwritten letter of thanks invoking shared values (see section 8.1), as well as the resonance of project activities with community values in the case of Mull, "if there's a problem ... they won't come to us." (I20: 407–409, see section 6.2). The institutionalization of organizational resources through financial support of the national government sustained a different kind of feedback loop in the case of Samsø. Organizational resources of the Energy Academy mobilized organizational resources within the islands' communities, as islanders engaged in its activities. Similarly, knowledge resources of the Energy Academy, for example in the facilitation of community participation, mobilized knowledge resources in the community which then provided input to the Energy Academy's efforts to further develop project strategies.

---

[1] Pronoun was chosen to protect anonymity.

Although important, financial resources are not the only determinants of the ability to hire staff. Said one EWS representative when asked about major company challenges: "Recruiting. We have some vacancies where we have been looking for months" (I5: 50–52). This is assigned to remote location, "[recruiting] is not easy in rural areas in the Black Forest, you have to be honest, we are not Berlin. Somehow, you have to get people to come if they are not here. On the other hand, we are not a company like [name] that can just wave an annual salary of a hundred grand and then you'll find someone" (I5: 61–66). As small market actors attempting system innovation that requires highly trained professionals, community energy projects can struggle to mobilize organizational and knowledge resources required even with financial resources.

A steady source of financial resources also affected projects' abilities to network and cooperate with other energy projects. In Schönau, the community energy project grew in numbers, output and complexity with diversification of energy services and establishment of cooperative ('EWS eG') structures, as well as the creation of a growing number of subsidiaries after winning the concessionary application in 1996 (EWS 2018b, p. 52). Successfully established as a community energy actor, EWS co-created a fourth arena it subsequently performed in as a pioneer: the German community energy arena. EWS showcased the potential of community energy engagement and was an active supporter of new entrants to this sphere, extending "solidarity… [to] many smaller projects" (I2: 202). Financial resources could be mobilized from the diverse energy services provided by the cooperative. This was supported by regulatory conditions of the national energy arena, most notably the liberalization of the electricity market in 1996 allowing EWS to expand retail services nationally, and the regulation of distribution fees which EWS could raise from customers due to the continued political mandate of the municipality. EWS shaped the emergence of a community energy arena as a project pioneer, providing organizational, knowledge, symbolic, and even financial resources to emerging actors.

All projects are members of regional and/or national associations relating to their organizational form (as a cooperative or a development trust, for example) and/or their agendas of community owned energy system change or development. Those projects with a longer track record, and many successfully implemented targets to showcase, often have prominent positions in these networks and are called upon by community energy actors, as well as other commercial actors of the energy sector and government representatives. Within the network of community energy actors, these organizations provide important anchors of knowledge and support. Said one EWS representative, "We are… sort of like the 'Big One' that you lean upon" (I2: 164–165, see also chapter 5). The assessment of EWS as a source of knowledge, support and financial resources is echoed by smaller organizations throughout the country which further supports the role of EWS as an important initiator of the German community energy arena. Olegeno, although located almost at the other end of the country, made reference to the story of the Schönau electricity rebels to explain their actions in public events and relied on consultancy from other community energy organizations. Similarly, FDT in Fintry was the first to institutionalize a cooperation of citizens and commercial wind energy developers in the form of a development trust. This pioneering role has inspired other communities to seek their help but also led to requests for consultancy from project developers, who "if they have done their homework... know that we know what we are talking about." (I19: 149–153).

Projects themselves have also begun organizing forums of interconnection. The Energy Academy hosts bi-annual conferences to develop new ideas under the slogan of 'From Best to Next Practice' (Flemming 2013). Fintry Development Trust hosts the 'FreSh Fintry Renewable energy Show', including educational, networking and awareness raising activities for the local community as well as reaching out and building networks of political support (the fourth event in 2014 counted various representatives of Scottish and UK government among its visitors, FDT 2018). EWS focuses on solidarity and awareness raising among community energy actors in its annual award of the 'Electricity Rebel of the Year' (EWS 2019). These activities present different ways of recognizing simultaneous technological and societal changes that are part of an energy system transformation.

9.3   Financial and Symbolic Resources

Across projects, financial resource mobilization was actively embedded in the symbolic resources mobilized during emergence. On Samsö, the Energy Academy referenced its role as a moderator and driver of continued successful system change and the narrative of Danish leadership in energy systems. On Mull and in Fintry, the charitable status of the organizations operating installations and/or governing financial revenues institutionalized the connection between financial and symbolic resources. In the case of EWS, governance of financial revenues was not an organizational feature but actively phrased as an expression of the project's continued ties to the underlying social movement for energy system change. Financial resources were mobilized in a feedback loop with organizational resources. But this feedback loop was framed by symbolic resources. In their own project work, this refers to their self-classification as a '2nd generation energy utility' (I3a: 24) wherein energy services are organized to actively increase the share of renewable energy generation, and increase awareness, responsibility and the decision making power of customers. Financial resources are invested in projects of energy system change in institutionalized funding mechanisms such as the 'Sonnencent', as well as through individual and ad hoc partnerships. For example, EWS and a local cooperative are offering electricity retail services to customers in Berlin wherein the 'Sonnencent' raised locally is invested directly into projects of the local cooperative (BürgerEnergie Berlin eG 2019). Said one interviewee on the willingness of the cooperative to invest in energy system change: "The mission of our company is a successful energy transition. And to achieve this we will do anything, beyond any interest in revenues. And that is why we are prepared to invest a lot of money without expecting any kind of return" (I2: 37–40). This position of the management was underscored by decisions of the General Assembly to limit members' dividends of the cooperative to 3.5%. After what one member described as "intense discussions" (I1: 199), the General Assembly decided to lower the so-called Gierbremse (greed limit, author's translation, Faltin and Schulz-Braunschmidt 2013) "The best was really that we discussed, and members who were opposed talked to those in favor," one member recounted the discussion, "And in the end, an overwhelming majority was in favor of reducing the dividend. I thought that was just wonderful, because it shows that our members really are people who want to change something and who are not seeking the highest possible payout." (I1 199–212). This quote is indicative not only of the strong embedding of financial resource mobilization in shared values of investment, but also of procedural values of discussion and consensus orientation in the project.

In 2016, the cooperative reported its best commercial performance yet. But management representatives at the annual General Assembly highlighted that the term dividend would need to be redefined. Financial dividends would not resonate with the cooperative's self-concept, said manager Sebastian Sladek, continuing, "for future discussions we want to increasingly establish a concept of a social and ecological dividend in our company actions and with our members" (EWS 2016, 7. General Assembly). These terms have not been further defined to date. Still, Sladek's statement indicates sustained leadership commitment to the symbolic resources mobilized during emergence and establishment. The credibility of EWS as an actor of the energy transition is further underscored by the symbolic resources the cooperative was able to mobilize towards the German anti-nuclear movement, culminating in political support for nuclear phase-out in the renewable energy law of 2000, which was met with a surge in customer numbers of EWS retail services (FUSS 2007). Nevertheless, one member dryly commented that while the cooperative was part of many political campaigns with large public support, this would not necessarily translate to customer numbers. "If the 175.000 people signing the… campaign had all switched to us, that would of course have been great, but that is not the case" (I1: 318–321). The person continued to underline the company commitment to advocacy work nonetheless, saying, "still, we will continue running these kinds of political campaigns, because their goal is not advertising for us but rather supporting (or opposing) the respective political issue" (I1: 321–324).

In the cases of Fintry and Mull, the orientation of financial resources towards symbolic resources of emergence and establishment is predominantly realized through the investment structure of a community trust or charitable foundation. The projects chosen for the community then make reference to the founding ideas of the respective organizations, namely the realization of climate change mitigation measures in the village of Fintry and the support of vulnerable island communities on Mull. On Samsø, it is not so much financial resources but instead organizational resources of the Energy Academy supported by financial resources of the national government that secure continued orientation of the project towards locally shared symbols.

Symbolic values also manifest in technical installations realized through the mobilization of material resources. This can refer to the visibility of installations, such as the PV panels of Schönau's 'Schöpfungsfenster' installed on the village church or the offshore wind turbines at Paludan Flak signaling energy system change along nautical routes of the Kattegat. It can also mean invisibility, such as the ability of Garmony Hydro to vanish into the hills. Financial resources increase the ability of projects to follow up with additional installations, visibly underlining their commitment to project symbols and values. Visibility (or the intentional lack thereof) can in turn be mobilized as a resource by projects, underlining legitimacy or accountability. These cases again underscore that while natural resources may give indications of what can be done, they do not by themselves define project actions. Symbolic resources, although rarely actively mobilized during project maintenance, frame and reinforce mobilization of, especially, financial and organizational resources that dominate this project stage.

9.4   Maintenance in the Absence of Financial Resources

The relevance of financial resources for project maintenance is further underscored by projects wherein these resources are scarce. Of the five cases, olegeno's position is most precarious in this respect with no financial income stream to support their activities beyond honorary engagement.[2] Membership shares enabled smaller investments but were not dynamic and growing asset bases themselves. olegeno was unable to enter the energy production market, as the phase-out of the German feed-in tariff created an investment environment too insecure for a small actor such as the cooperative and neither its material nor its organizational resources could support participation in tenders for renewable energy generation (Grashof 2019). One interviewee commented feeling as if the cooperative were "late to the party" (I9b) when speaking to other community energy projects that had realized installations prior to EEG reform. Funding opportunities through associations that are available to commercial community energy projects such as the UK's 'Community Shares Booster Programme' have not been established in Germany. Community energy associations provide knowledge and organizational resources, but no financial support to members. The lack of financial resources limits the cooperative's ability to implement projects. It is also a challenge for oversight. Mandatory audits by oversight associations for cooperatives ('Prüfverband', §54 Federal Republic of Germany 2006) will challenge the organization when company assets decline over time and without any indication of future growth potential.

olegeno struggles to mobilize organizational resources. The cooperative relies exclusively on non-paid engagement. After a change of management and the loss of many team members following the loss of the concessionary application, olegeno is now maintaining relatively low levels of fluctuation in team membership. New team members still join. Members attribute this to their commitment to regular face-to-face meetings, "attending a team meeting… for the… group that makes a huge difference. That I took the time to be present" (I9: 316–320). Still, established members of the team also reported high levels of personal commitment and frustration. One spoke of the difficulty of maintaining personal boundaries: "You may have said to yourself, I will only tell friends outside the cooperative once of the electricity tariff, but then you catch yourself nagging them again, anyway." (W2).

When new organizational resources can be mobilized this immediately creates a strong feeling of support and reinvigoration amongst team members. The cooperative was able to move into its first private office space, including an open space for team meetings in May 2018. This was used for a friendly gathering

---

[2]The cooperative commenced electricity retail but had not acquired a significant customer base within the time frame of this study (see section 7.2).

following the next general assembly, with team members reporting how this invoked a sense of belonging (WS2). Within the city of Oldenburg, this also created individual visibility for the cooperative. Having an office of its own also mattered because the cooperative had so far not succeeded in creating visibility for its actions, for example through installations. Although several installations were built with the support of knowledge resources provided by the cooperative, none of these installations were ultimately realized by the cooperative. This was interpreted differently by project members. Some voiced frustration while others saw installations as visible manifestations of their educational work. Said one workshop participant: "None of these installations would be there without us. We are the seed, the catalyst" (W2). Another member cautions that this could not sustain the cooperative, "they are glad we provide knowledge, but then somebody else reaps the benefits." (W2). The cooperative is predominantly providing knowledge resources much like the Energierat, its predecessor and origin.

olegeno repositioned itself in the local energy arena during maintenance. DSO related actors and institutions were no longer relevant. Instead, the arena was shaped by olegeno's attempts to establish alternative local energy activities, including initiatives for PV installations for apartment buildings and RE services for tenants ('Mieterstrom'). The municipality now welcomed an energy cooperative, after keeping a more formal and distanced relationship during the bidding process. Oldenburg is home to Germany's first energy cooperative run exclusively by women (Windfang 2011). But olegeno is Oldenburg's only local energy project, as Windfang's projects are remote. This is acknowledged by the city council: "the city supports that and is interested in it and that there might be projects where it is welcomed that olegeno is there and involved" (I8: 243–245). The anti-nuclear arena no longer formed a reference frame to team members. A growing community of organizations and citizens calling for an analogous phase-out of coal formed a new arena of national protest. olegeno supports coal phase-out. But the national movement so far was not strong enough to provide resources to local projects in the way the anti-nuclear movement did for EWS, nor are olegeno's activities locally immediately connected to an agenda of coal phase-out. olegeno remains engaged in the national energy and community energy arenas through the operation of retail services and membership in associations. The cooperation with Bürgerwerke provides organizational and knowledge resources as olegeno does not operate retail services itself, although the influx of financial resources is small. Symbolic resources mobilized through the association with a national network of community energy projects are insignificant to local project work. This underscores the importance of mobilizing knowledge and organizational resources within the voluntary team of the cooperative, and the subsequent role of the cooperative as a 'catalyst' providing knowledge resources locally.

Financial resources also mattered for the ability to network. Smaller community energy projects like olegeno may already struggle to pay an association's membership fees.

9.5   Summary of Resource Mobilization for Project Maintenance

Table 9.1 summarizes the mobilization of resources during project maintenance. In a notable difference to mobilization during emergence, financial resources are mobilized across all projects to maintain activities, although olegeno's success in generating financial income despite extensive attempted mobilization remains limited. The availability of financial resources matters both for the mobilization of organizational resources within the project and for the ability of projects to deliver and develop project functions and services as intended. Although almost all projects engaged in energy services related to technological installations, not all projects mobilized additional technological resources to extend energy generating or distributing capacities. GEM continues to operate Garmony Hydro and to participate in ACCESS, but no additional technologies or installations have been introduced. olegeno attempted to realize technology installations, especially energy generating capacities, but so far was unsuccessful. Concequently, olegeno also did not mobilize natural resources. The provision of energy supply services by olegeno and Bürgerwerke is balance-based and not tied to physical generation locally. Knowledge resources were mobilized in all projects in combination with organizational resources. Like during project emergence and establishment, the availaibility of natural resources may influence which actions projects choose to pursue but in itself is not constitutive to project activities.

Symbolic resources remain present as the overarching narratives developed during the emergence and establishment of projects. These narratives influence how material, especially financial and organizational, and knowledge resources are mobilized and employed. But in day-to-day activities of project maintenance, references to narratives are implicit.

Three lessons can be drawn across cases for the analysis of resource mobilization during the phase of project maintenance. First, financial resources mattered for the ability (or lack thereof) to mobilize steady organizational resources. Hired staff shaped the ability of projects to mobilize knowledge resources and maintain project activities more generally. They created a center point for project activities

Table 9.1   Mobilization of resources by type during maintenance

Project	Mobilization by resource type					
	Financial	techno-logical	natural	organizational	knowledge	symbolic
EWS	+	+	+	+	+	+
Olegeno	+	−	−	+	+	+
Samsø	+	+	+	+	+	+
FDT	+	+	+	+	+	+
GEM	+	−	+	+	+	+

recognizable to members of the project as well as outside parties. Organizational resources were also mobilized to create larger networks and symbolic alliances, connecting community energy projects to actors and institutions of other negotiating arenas. While mobilization of financial resources may include the distribution of revenues to members, this did not determine successful project maintenance as the decision of the EWS General Assembly to reduce individual dividends, or the lack of individual dividends altogether in the case of FDT in Fintry indicate.

Second, all projects strived to establish financial income independent of government funding, or to at least diversify sources in cases were government regulated funding was embedded into the core funding model of the organization. For example, both Fintry and Samsø project leaders sought to diversify sources of government regulated income (e.g. through direct funding, grant funding, project funding), as well as to establish their own funding mechanisms. EWS in Schönau did not receive government funding, but the endorsement of EWS as a concessionaire (indicating its organizational resources are provided by the municipality) installed the cooperative as a distributions systems operator and secured respective revenue streams. olegeno was unsuccessful in receiving administrative endorsement and was the only case where policies supporting the market entry of small scale investors expired when respective activities could begin. This case selection indicates that either financial or organizational government support matters to establish financial resource mobilization in community projects.

Lastly, symbolic resources frame resource mobilization during maintenance, even if they are not actively mobilized themselves. While symbolic resources mattered for project emergence and establishment across all projects, what could be mobilized as a symbol, and how, differed and affected the subsequent shape of the project. This is underscored by the analysis of project maintenance. In successful cases, deployment of material (financial, technological) and organizational resources during project maintenance corresponds to the narratives of emergence and establishment. This last aspect was even more important when projects were challenged, as will be shown in the following section.

Challenges: Returning to Shared Symbols

10

Challenges to community energy projects can arise from within the project, from project conditions, or as a result of external events. Internal challengers include leadership change or resource scarcity, or events immediately connected to but not directed by the project such as a critical audit of a cooperative. External challengers considered include external actors such as policy makers, utilities or communities; policies (often associated with (groups of) actors or communities); or external events unrelated to project actors or energy sector dynamics. Across cases, organizational resources feature most prominently when a project is challenged, combined with symbolic and knowledge resources. Sustaining a project in distress can therefore be interpreted as project renewal or re-emergence.

As indicated in the introductory illustration referring to development stages and challenges of projects (Figure IV.1), most cases under study faced challenges during emergence and establishment. These typically referred to policies or sectoral practices within the energy sector, often in combination with actively opposing actors, such as policy makers or energy utilities. EWS in Schönau, after overcoming a coalition of the local political decision-makers and the energy utility, faced a referendum initiated by the utility to fight the concessionary decision of the municipal government. Falck Renewables declined cooperation with villagers of Fintry outside of a cooperative. MICT on Mull faced new regulation which put grid access of Garmony Hydro into doubt. This led to the establishment of a much larger cooperation which could have surpassed the lack of grid connection threatened by regulation. olegeno in Oldenburg challenged the local incumbent in an administrative application despite its close ties to the local community and lost this race. The only exception was Samsø as a result of the municipal decision to participate in the REI competition to create a showcase for Kyoto protocol commitments, a policy-related impulse at the national level.

© The Author(s), under exclusive license to Springer Fachmedien Wiesbaden Gmbh, part of Springer Nature 2021
A. Colell, *Alternating Current – Social Innovation in Community Energy*, Energiepolitik und Klimaschutz. Energy Policy and Climate Protection, https://doi.org/10.1007/978-3-658-32307-3_10

Community energy projects are often framed as an opposition or challenge to the incumbent energy system (Mey and Diesendorf 2018, p. 109; Blanchet 2015).[1] But analysis of resource mobilization underscores learnings from the study of emergence and establishment. Only the German projects were conceptualized as challengers, and only one actively referred to these symbolic resources when challenged from the outside. The other three projects focused on the mobilization of organizational and knowledge resources, reinforcing the community oriented symbols of their emergence.

10.1   Reacting to External Challenges: Campaign, Community and Policy Challenges

In Schönau, the project faced the opposition of the municipality and the local incumbent, as well as parts of the local community. It was actively challenged by this coalition, when the utility initiated a referendum to revoke the concessionary decision of the municipality. This second referendum challenged the decision of the municipality, and required supporters of the cooperative to flip their response.

During the 1991 referendum, initiated by the community energy project, supporters had had to vote 'Yes'. Community organizers rallied accordingly and framed the campaign as an affirmation of locality and what was 'right': "Everybody knew what the right answer was" (EWS 2018c; see chapter 5). The images of citizen engagement and self-reliance as an affirmative stance of involvement, rather than withdrawal, finds its expression today, in the cooperative's slogan "we were always in support of this" ('wir waren schon immer dafür', EWS 2018a) on printed ads for renewable energy services.

The cooperative attempted to invoke the same symbolism of being the 'right thing to do' in response to the second referendum. But the answer in support of the cooperative was now 'No', which required additional campaigning. Also, while EWS had advanced as a symbol of a national anti-nuclear movement, the concessionary decision rested with the locals, and personal alliances and trenches ran deep within the village (SPIEGEL 1996). Members spoke of nightly meetings held during the time when they would share their stories of support or rejection and reinvigorate their commitment to resume campaigning tomorrow: "We met at night, half past ten or half past eleven, [in the Sladek's home, author's note] and everyone told their stories and some were devastated because they had met

---

[1]Such studies often draw on the approach of strategic action fields as developed by Fligstein and McAdam (2012).

someone of whom they had been sure he would vote for us in the referendum and then he revealed that this was not at all what he intended to do." (I1: 253–56). This quote points to the strong interpersonal connections between project campaigners, meeting late at night in the home of the village doctor and his wife, Michael and Ursula Sladek, who had become a center point of the organization.

Beyond the importance of sharing adverse experiences and consoling one another, members recounted how personal perceptions of alliance within the village changed through face-to-face campaigning: "But then there were those in the group who said, 'Oh, but I spoke to someone and I always thought they would be utterly opposed and they said, No, I think what you are doing is great, and I will speak to all of my friends and family and so on, and that is how we encouraged each other" (I1: 259–263). Personal commitment and contacts determined assertion of the project during this time. The challenge within the local energy arena was countered by reinvigorating the feedback loop of symbolic and organizational resources within the project arena, and building support for shared values and organizational structures within the local community.

While prioritizing community involvement and consent in project development, project leaders on Samsö also faced challenges at the community actor level. Challenges could refer to siting, but often spoke to underlying values of participation in decision-making. Søren Hermansen remembered a conflict wherein one villager opposed the location of a district heating plant because they feared the site would adversely affect the village's medieval church: "[we] produced a lot of expensive visualizations to show… that the plant wouldn't overshadow the church. We had chosen that site because there were barns and a store-house already in place, so it was the economic solution" (Hermansen in Papazu 2018, p. 13). His continued explanation shows how implicit motives could be more influential, stating that opposition arose from "fear that the steam from the plant would pollute… houses in the village. But there is no real danger of pollution from the plant; the Environmental Protection Agency checks the levels each year" (ibid.). While initial measures required mobilization of knowledge and financial resources to produce complex imaging, other resources were ultimately required to assuage the conflict. Hermansen concludes, "We moved the chimney but not the plant… In the end, [person] retired and moved to the mainland, and the dispute was forgotten" (ibid.). This example is indicative of the efforts undertaken to ensure meaningful participation of villagers (Papazu 2018, p. 14). It underscores the commitment of organizers to take community input "about the future **seriously**" (I25: 199), and frame project development as a community-led process rather than the realization of the smallest common denominators of a centrally planned transformation.

Responses to opposition regarding the project campaign in the Fintry, Mull and Oldenburg projects, by contrast, were characterized by a mobilization of organizational and knowledge resources. In none of these cases a challenger-incumbent frame was established during their emergence. Still, all of these projects endorsed an agenda of community empowerment.

In Fintry, the commitment of project volunteers to community involvement in a non-hierarchical structure sustained opposition to the cooperative model suggested by Falck Renewables. A windfall influx of knowledge resources provided by the DTAS campaigning for communities to establish development trusts independent of an agenda of energy system change ultimately determined organizational choices (see chapter 6). The offer of an alternative legal form reinforced by the credibility of an actor representing the community development agenda of the Scottish national level and providing knowledge and organizational resources was enough to surpass the dominant but still young sectoral practice of involving citizens only if organized as a cooperative.

GEM and MICT on Mull faced more substantial opposition as Garmony Hydro was challenged not by sectoral practice–Falck Renewables had had no legal reasons to dismiss FDT–but by the regulatory authority Ofgem restricting grid access. Correspondingly, MICT mobilized organizational resources itself as well as creating new structures wherein both organizational and knowledge resources would be mobilized in the form of the ACCESS project. This created new knowledge on community energy installations but also new partnerships for the project as MICT entered a cooperation with the national association of community energy projects, the national grid operator, and various specialized service providers (see section 6.2). This speaks to an affirmative understanding of self-reliance, rather than a withdrawal from cooperation. GEM was able to mobilize additional organizational resources creating new partnerships beyond the project arena. And while a funding agreement with the Scottish government qualified that the intellectual property rights of knowledge obtained within the cooperation rest with the government, the community project nevertheless established itself as a provider of specialized, original knowledge. As one project member ascertained, "Will we use it for leverage for other projects? Of course we will!" (I20: 360–361). This emphasizes the importance of proof of concept, even for future activities outside of energy system services. This person continued, "It might be for different and unrelated projects, we'll add it to our portfolio" (I20: 361–361).

ACCESS created a cooperation wherein a rather inexperienced, new community energy project (GEM) represented through an experienced actor of community development (MICT) entered a partnership with an energy sector incumbent, specialized service providers, and the association representing the community-organized energy projects. Beyond expressing GEM project leaders' commitment to realizing Garmony Hydro, the establishment of ACCESS speaks to the availability of organizational resources locally and within the Scottish energy arena. Locally, GEM could rely on MICT to coordinate the establishment and maintenance of both the network and coordinating work, as well as the added efforts of a second energy project on the island. Within the Scottish energy arena and the additional ACCESS project arena, Community Energy Scotland (CES) provided key organizational resources. Community Energy Scotland is an independent charity which developed out of the Highlands and Islands Enterprise's (HIE) subsidiary Highlands and Islands Community Energy Company (HICEC). The former is the Scottish national government's community development agency for the diverse region of Scottish islands and mainland highlands, covering more than half of the Scottish landmass while accounting for approximately 10% of the Scottish population (HIE 2018). HIE's operations are guided by the 'Economic Strategy' of the Scottish national government, and center on economic development of the region (ibid., p. 3). Community Energy Scotland was established in 2008 to *"strengthen and empower local communities by helping them to own, control and benefit from their local renewable energy resources"* with a view to building a "more localized, democratic and sustainable energy system" (Community Energy Scotland 2019). While this speaks to the economic development goals of HIE, these were complemented by advocacy for community ownership and decision-making power ('democratic') in energy system change.

An office was established by CES in Glasgow, relatively close to Mull on the Scottish mainland. A full-time position was created, supporting ACCESS project in its first year and continuing part-time support throughout project completion. CES is performing project management duties and delivering technical expertise. "We worked with the community really to set up a project that would assist them, working with the other project partners," a CES representative explained the role of the organization in ACCESS (interview MICT, April 2016).They continued: "We've got that technical expertise and know-how... helping to design the network system" (ibid.).[2] While CES often provides project planning and management services to communities seeking to engage in energy services, the scale of the ACCESS project is unique to the organization, "it's quite a new way of working. So very different organizations, very different cultures... it seems like a slightly new niche that is opening up" (interview MICT, April 2016). While this refers to cultures of cooperation and operations, ranging from the quality management procedures of a national service provider to the often informal modes of

---

[2]Pronoun was chosen to protect anonymity.

communication in a community setting, ACCESS project partners stressed the positive experience of this new form of cooperation in the energy sector (see 6.2). While CES supported MICT by providing project management and oversight, the local partner coordinated all relations with community members; this far exceeded community relations for Garmony Hydro. Co-operations beyond the community level were very resource intensive for community projects, further underscoring the importance of the organizational resources provided by CES (see section 6.2). Regulatory grid constraints on community level installations were a challenge which was answered by an increase in community energy project work, supported by organizational and knowledge resources provided by a charity which was politically mandated within the energy landscape.

olegeno had framed its competitive position to the local incumbent as providing an informed and citizen owned alternative. This centered on knowledge resources and more cautious symbols of community empowerment. When challenged on this position during the application process, the cooperative reacted by providing additional knowledge resources, including the mobilization of external actors with specific expertise (see Sections 7.2 and 8.1). Upon losing the bid, the cooperative invoked organizational and knowledge resources to sustain project activities within the symbolic reference frame of a citizen owned local energy alternative. An important difference to the other projects lay in the timing of this challenge. While other projects were still striving to achieve their initial goal, olegeno was struggling to reinvigorate and redirect community engagement after failing to achieve their initial target.

This difference shows in its patterns of resource mobilization. Existing organizational resources could not be mobilized and extended. Instead, many who had carried the project until then were no longer interested or able to sustain their engagement. One management member immediately stated their resignation, a second followed shortly after. Several team members also withdrew from active roles. The board was urged to renew the management team, resulting in a team of three newly recruited but previously connected members, and initiated a process of strategy development including a survey among members and open workshops. Two workshops were hosted within one year, inviting all members of the cooperative. "The first... had people that were still with us from campaigning in the concessionary application... some of them... then left the more active circle... and of course management and board... around ten or twelve people. And then, a year later, there was another workshop" (I7a: 64–72). The purpose of survey and workshops was to reorient the cooperative towards new project activities that would continue the work of the Energy Council, which had been engulfed by the cooperative, and work on energy projects beyond investing in generating capacities as the original idea of grid management had implied. As one remembered, "we were careful [with energy generating installations], because we had other ideas. Because we believe that efficiency and sufficiency and reducing consumption are also part of the energy system" (I9: 121–124). Also, the 2014 reform of the German renewable energy law had rendered investments less attractive, requiring installations to be oriented more strongly towards personal consumption. This member referenced policy reform, indicating that the cooperative so far had not managed to establish a respective project because of the organizational difficulties, "it is immensely difficult" (I9: 127). A second course of activities therefore centered on supportive actions, "to help establish citizen owned energy infrastructures. We are a member of the 'Bürgerwerke' [network of cooperatives] and are offering an electricity product with them... we try to be an alternative in the energy industry" (I9: 128–131). Members felt the responsibility of continuing active engagement in the energy sector to honor the activities of past associations that had been absorbed by olegeno.

Beyond establishing partnerships with other networks of cooperatives, olegeno also worked with the local university. In a co-operation with a seminar of business students, various groups were formed to develop diverse suggestions of services that the cooperative could provide. These were then presented to the cooperative in a workshop at the University of Oldenburg (March 7, 2016). Team members in support saw this as an urban lab of sorts, wherein students would work on actual social situations, while the cooperative would benefit from an influx of knowledge which could not have been obtained through voluntary structures. But the attempt to form new partnerships within the project arena to infuse new knowledge resources was not successful, in large part because it had not been commissioned based on consensus of the team (see 7.2). The concept for strategy workshops was successfully shared with another cooperative facing a similar challenge. But this cooperative purposefully decided not to open this process of reorientation to non-members to avoid conflicts of interest (Interview Dec 2015). Nevertheless, olegeno is sustaining a process of reorientation through continued voluntary engagement. Core symbols of providing alternative energy system services and civic responsibility have prevailed in mobilizing organizational and knowledge resources to support local projects.

Lastly, across projects, challenges to project work originating in the respective policy environments have become more frequent across cases since 2014 (Figure III.1). These challenges relate to policy changes that include the rescinding of financial, and often also organizational, support for community energy actors and the introduction of instruments with investor-specific (adverse) consequences at the community level (Grashof 2019). Across cases, this has reinvigorated the mobilization of organizational and knowledge resources, corresponding to key shared symbols of the respective community energy projects.

The establishment of ACCESS on Mull is exemplary of a successful project reaction to a regulatory challenge. By mobilizing additional organizational and knowledge resources, embedded in the narrative previously driving project emergence, socio-technical solutions were developed to overcome the regulatory obstacle. In this case, existing services of demand-side management were reconfigured in cooperation with energy generation and distribution to create a new service which answered to the requirements of the local community. Similar project reactions were the establishment of SMART Fintry by FDT, a two year project developing electricity trading and charging services that enabled customers to buy power directly from local renewable generating installations in cooperation with commercial and academic partners (Smith 2018; SMART Fintry 2018), or the establishment of a subsidiary concentrating on direct sales of renewable energy to commercial customers as well as developing models of production and retail for apartment installations ('Mieterstrom') by EWS (EWS Direkt GmbH, EWS 2017, pp. 43–44). Additional organizational and knowledge resources have been mobilized by EWS through the creation of an office in Berlin to create a permanent representation of the organization, improve communications and policy monitoring, and increase its political influence (EWS 2018b, p. 77). On Samsø, reactions included the extension of energy services, including the mobilization of financial, material and non-material organizational, and knowledge resources, and the maintenance of political support for the Energy Academy. The latter could be achieved by mobilizing non-material organizational and symbolic resources by building on established networks of political communication. While the island has pursued diverse projects extending energy services, such as the establishment of ferry services fueled by liquid natural gas (LNG) to replace the former diesel powered connection to the mainland, none of these projects corresponded to regulatory constraints in the respective areas of action.

olegeno has mobilized organizational and knowledge resources extensively to establish alternative operations under adverse regulatory conditions, as discussed under maintenance. This has so far not resulted in successful operations. As indicated by the quote of feeling 'late to a party', the cooperative is feeling the lack of either existing larger organizational structures to mobilize (such as MICT or Community Energy Scotland), or financial means to support mobilization of organizational resources (such as the revenues of energy services that FDT or EWS can work with). Project maintenance, in the case of olegeno, could therefore also be categorized as a permanent process of responding to project challenges.

10.2   Reacting to Internal Challenges: Leadership and Volunteerism

Internal challenges to community energy projects referred most importantly to developing leadership, and reconciling paid and non-paid engagement. Three projects stand out in this respect: olegeno, EWS and the Energy Academy. Oldenburg's olegeno faced both the challenge of leadership change in combination with the external challenge of forced reorientation of project work, and the challenge of sustaining non-paid engagement. EWS in Schönau as well as the Energy Academy on Samsø face a different situation. In both, leadership has rested with the same small group of people for many years, creating personalized representation and networks but also the challenge of sustaining the project in the future.

In Schönau and on Samsø, leadership rested with a small number of people since the very beginning, personified in many ways in Søren Hermansen on Samsø, Denmark's self-proclaimed most famous energy person, and his partner Malene Lundén, and Ursula and Michael Sladek in Schönau, co-founders and project leaders for over 30 years. Hermansen referred to the benefits of sticking with the same leadership team, pointing out that continuity in leadership creates networks of reciprocity over time. In interactions with political decision-makers as well as energy utilities and other projects or partners in the energy sector, this created a recognition factor for projects. Over the course of their work, both Hermansen and the Sladeks received a number of awards and honors. For the Energy Academy, this included the Göteborg Award for Sustainable Development for Sören Hermansen in 2009, and the Danish Svend Auken prize in 2011 (Energy Academy 2019). TIME Magazine named Hermansen a "Hero of the Environment" in 2008 (Bryan Walsh 2008)). For EWS, this included the Goldman Environmental Prize for Ursula Sladek in 2011 (Wolk 2016), the German Founders' Prize (Deutscher Gründerpreis) for EWS and its founders in 2007 (Deutscher Gründerpreis 2007), and the German Environmental Prize (Deutscher Umweltpreis), Europe's most highly endowed environmental prize awarded by the Deutsche Bundesstiftung Umwelt for Ursula Sladek in 2013 (EWS 8/9/2013).

External recognition, for example through awards, created public awareness for the projects in diverse ways beyond generating media interest. Ursula Sladek's receipt of the Gründerpreis was an expression of the German entrepreneurial community's recognition of the community energy project as "a serious and professional business" (Ursula Sladek in Deutscher Gründerpreis 2007). This mattered also because the local energy utility had repeatedly attempted to belittle citizens' initiatives, denouncing their application as idealistic and unprofessional—an experience shared by other community energy projects across the country

(Bernd Hirschl 2016; Joel Stonington 2013). Beyond the individual project, this kind of recognition could therefore also augment the reputation of a growing community energy sector. The high visibility and increased legitimacy of projects through awards and honors could improve accessibility of decision-making structures. One project member on Samsö reckoned that representatives of the Energy Academy could probably simply "knock on the door to the ministers and talk to them" (I24: 392). The opportunity to place items on the agenda could also arise in unsolicited ways. Ursula Sladek, infamously, smuggled a copy of the cooperative's manual "100 good reasons against nuclear power" into the Oval Office, presenting it to President Barack Obama when he welcomed the winners of the Goldman Environmental Prize in 2011 (Wolk 2016). Sladek had had to be wheeled in in a wheelchair due to a broken leg and had sat on a copy of the pamphlet, surpassing all security and protocol controls (ibid).

Still, project members of EWS caution that while the project's effect on other community projects and citizens may be considerable, influence on decision-makers remains limited. "The societal effect of EWS is the largest effect we have," one member said, continuing to recall how "before [reform of, author's note] the EEG, [EWS] wrote to all members of parliament and if I count the responses, then that was…a handful? Not more." (I1: 53–57). In Schönau, the continuity of personal leadership was also recognized on the local level. In 2017, Ursula and Michael Sladek were honored with the civil medal of their hometown Schönau, recognizing citizens for extraordinary local engagement (EWS 11/21/2017). The laudation given on their behalf remarked on the strong connection both had to their town, the city mayor remarking on how the Sladeks were "not distant images, working wonders somewhere faraway" (ibid.). This echoes the assessment of Hermansen as a person of 'street cred' (see section 6.1). The continued leadership of Hermansen and the Sladeks also mattered to mobilization within the project arena. This assessment echoes findings in the literature which positively link the satisfaction of volunteers within an organization to their recognition of project leadership (Dwyer et al. 2013).

Accounts of what grounded members' support for long-term leaders were often unspecific. Members reported a 'feeling of getting good answers when one needed them' or remarked more generally on their appreciation of leaders (see 6.1.2). The literature on honorary organizations argues that leaders in voluntary project environments matter to those engaged in honorary work, but what constitutes effective leadership often remains unspecified (Posner 2015, p. 895). The personalization of leadership also creates challenges, when leadership changes. In 2015, Ursula and Michael Sladek retired from active management of the cooperative (EWS 6/29/2015). They were succeeded in the management team by, among others, two of their five children. While both had worked in other contexts previously, acquiring sector related skills that underscored their leadership position this nevertheless also indicates family business dynamics underpinning community engagement. The Energy Academy had not actively commenced the search for a new leadership team when field work for this study was closed.

Team members of Samsø reported on the exhaustion that comes with being the one who 'keeps the torch lit' (see section 6.1). "I could see some shadows coming in," one member reported, saying that after completing the 10 year transition, "people were tired" (I25: 209–210). Similar to the benefit of having a paid moderator for project and community development during emergence, the Energy Academy could again benefit from institutionalizing positions otherwise often neglected in community energy projects. The organization had hired a project leader who not only sustained community involvement with the Energy Academy but also challenged prior successes with a view of redirecting resources that were invested to tell the story of past transitions towards future projects. One response lay in the mobilization of additional organizational and knowledge resources through the development of partnerships and networks. "We need[ed] to renew this concept," one community member said while remembering the development of new activities in the Energy Academy. They continued that this gave rise to the recognition that, "maybe we need to hand it over and do it together with two or three organizations" (I25: 208–209).[3] The Academy created a biannual forum of exchange wherein the focus was "how do we go from best to next [practice]" (I25: 206; see also Flemming 2013). The second response centered on the reorganization of existing knowledge resources, by establishing the Energy Academy as a provider of consultancy and support to academic, political and community organized work (Energy Academy 2016). Within the organization, the position of an internal challenger questioning methods and targets created a built-in defibrillator. When the heartbeat slowed down, during implementation or after an achievement was completed, this position legitimized a new impulse and ensured that resources were dedicated to developing these stimuli. This institutionalization suggests that the Energy Academy may have established organizational resources that also benefit the search for new leadership.

Beyond management and leadership, community engagement and the consolidation of different levels of involvement within the project are key internal challenges. On Samsø, the account of project emergence points to the involvement of diverse groups on the island. But representation in advisory boards or committees was frequently shared among a relatively small number of people, as

---

[3]Pronoun was chosen to protect anonymity.

exemplified by the recruiting process described above (see 8.1). EWS faced the same challenge in a different setting, as the dynamics of a civil society initiative connected to those of a family business in management changed. Organizationally, this points to the intersection of personal and professional involvement.

Another example of the connection of personal and professional spheres is the coordination of paid and non-paid labor. Shared values (such as grandkid suitability, responsibility, localism, etc.) connected EWS to the village, and the national anti-nuclear movement. Yet, organizationally the project relied on a relatively small circle of founders and team members who devoted their time and energy to the cooperative—many of them without payment. This ensured a continued accumulation of knowledge and networking resources among these people and within the organization. However, it also resulted in a group gradually growing older together. As one says, "many older people, who were involved, are already dead or have grown more fragile... The mail, for years until she moved away, was done by a lady; she was well over 85 in the end. She came every day and sorted the mail, opened letters, helped out. And it would have never occurred to her to ask for payment, because it was just important to her" (I3a 273–278). This refers to commitment to the goals of an organization. But it is also indicative of a community tightly knit by joint action. This speaks from quotes describing community functions of the cooperative in early years ("somebody was always on the brink of resignation", I2: 160). It is also apparent in the organization's sense of self today, as this member described, "These projects bring people together. You share a table, you chat, you laugh and feel human companionship again" (I2: 156–158).

As paid employment grows (EWS reported 110 employees by Dec 31, 2016, EWS 2017: 66), paid and non-paid engagement must be reconciled. Members were critical, as one pointed out, "this company has many employees that must be paid. But do they keep the others on track, also? This hasn't worked out as well as I had imagined it" (I3a: 272–273). Paid employees show strong personal commitment to sectoral and organizational goals, as one member described, "People that are passionate about energy politics... EWS has heaps of them... this clear desire for change is embedded [in EWS] and that shows constantly among colleagues" (I5: 25–29). Yet, as the 'table' grows larger and the challenges grow more diverse, companionship becomes fragmented and the joint overcoming of obstacles is no longer felt throughout the whole group.

On Samsø, the continued engagement of communities is not expected by all in project leadership. The Energy Academy is the institution charged with 'keeping the torch lit' while the community is 'at work' with other things (see section 6.1). As the first transition was completed and more complex targets were developed in community consultation processes, the municipality also began assuming a more active position. One interviewee explained how the municipality had taken over project leadership from the Energy Academy who is now predominantly a story telling agency building national and international networks. "The municipality kind of took over. And now it's the municipality that is taking care of building biogas infrastructure, building a ferry that is fueled by LNG [liquid natural gas, author's note] and other projects, biodynamical farming also. We take over from these more concrete projects from the Energy Academy" (Interview Samsø, May 2016). This view is not shared by all islanders, "the municipality does a lot of the writing. But I am absolutely positive that the Energy Academy has the contact. Or they initiated the contact, and they are still the ones who are the **obvious** people to ask from the outside" (I27: 118–120). Despite managerial involvement of other partners, the track record of the organization–and its storytelling skills– have retained the organizational and symbolic relevance of the Energy Academy as the center point of project work.

The storytelling power of the Energy Academy is exemplified in the only reaction (among selected cases) to an external event outside energy infrastructures. In 1999, two years after Samsø had begun its REI transition the local slaughterhouse closed and moved operations to the mainland. This resulted in the loss of the island's single largest employer, toppling the community into crisis. The project reaction was an intriguing indication of narrative power in resource mobilization. The image of crisis is central to the transition story told by the Energy Academy. It refers to the economic potential in energy system change on the island, as one said "those people were **very** frustrated about the possibilities of the future. And then this project could potentially produce some jobs. So, I think that was the main reason for a high percentage of the people to look into this as a possibility." (I24: 57–59) But it also refers to the community building process involved. "We have an island that is known for its potential that it helps and shares and a lot of people actually come back into jobs," as another remembered, continuing that it was important to "meet the slaughterhouse people where they were–in crisis–they were thinking they were going to need to leave the island." (I25: 119–122) The account of energy transition on Samsø moves closure of the slaughterhouse forwards in time, framing the start of the project as an immediate opportunity responding to crisis and "creating a narrative so well-crafted that the original chronology of events has become irrelevant" (Papazu 2018, p. 16). The actual time line of events neither lessens the effort of system change undergone to achieve local energy system change, nor does it soften the critical blow of losing a large employer in a rural community in economic distress. Yet, the symbolic account of events in the story of Samsø's transition points to the importance of shared images of a process that are universally applicable, as well as the messiness of resource mobilization without such a narrative at the time of project emergence (ibid.). Energy system change on Samsø was a top-down initiative which needed to be embedded in bottom up processes (see 6.1). Implementation relied on public acceptance, as a minimum category, and in many ways on public participation, both financially and in decision making processes. The narrative of crisis and opportunity reinforces the image of Samsø as a model of transition which could be tailored to other local contexts, as Papazu (2018, pp. 15–19) argues.

Classic storytelling images were employed: A crisis, the closing of the slaughterhouse, a local hero, the long-time leader of the Energy Academy, and the golden opportunity, the REI competition. But this narrative was not available at the time, meaning that project organizers could not rely on its momentum in mobilizing resources in the project arena. The narrative serves as a symbolic resource today, when mobilizing resources beyond the local arena, for example based on the transferability of this process to other communities which can lend argumentative power in the national energy arena. But while not available to mobilize community support at the time, today these symbols of perceived community participation may also reinforce actual community involvement. The transition narrative retold by the Energy Academy augments community roles beyond actual influence and questions the empirical salience of the image of "energy democracy" invoked by project leaders (Papazu 2018, pp. 14–15; Sovacool and Brossmann 2014). Other communities might learn more from the tools of organizing underpinning the creation of this narrative than its notion of replicability (Papazu 2018, p. 14). The power of these symbols may prove important for the continued mobilization of organizational and financial resources on the island itself. Independent of actual ownership, the narrative of an island (re)defining its own development has created a sense of assumed ownership among islanders. Investment conditions have changed with feed-in tariffs expiring on wind energy. Investments of the first round may not be renewed under these conditions, not only reducing production output but also dismantling material manifestations of processes of energy system change. As the first period of operations expires on various installations, the lack of awareness of a 'masterplan' for transition, superimposed by ideas of community ownership, may prove a key resource to continued mobilization of organizational (and financial) resources at the community level in a changed funding environment.

The analysis of patterns of resource mobilization for project maintenance in the case of olegeno and its response to external challenges suggested that project maintenance could also be interpreted as a permanent response to challenges.

This corresponds to the analysis of olegeno structuring leadership and volunteerism. Two observations stand out in relation to the response to internal challenges: the maintenance of leadership structures and personal boundaries, and the ability to integrate voluntary engagement of members and efforts to professionalize operations.

Leadership was sustained throughout the resource intensive application process of olegeno, but the cooperative has since experienced several changes in the leadership team. This corresponded to changes in the cooperative's overarching targets which lacked a comprehensive external framework, such as the DSO application. Members taking on leadership roles often did so in response to certain strategic decisions. An example of this was the project member who was professionally connected to electricity retail and project development beyond the cooperative and became part of the management team at a time when renewable energy generating installations seemed a more promising avenue of commercial operations. Leadership could change again as strategies were reviewed and altered.[4] As a clear, overarching framework for activities was lacking, increased fluctuation within the management team suggests that leadership roles became harder to sustain (North Data GmbH 2017). At the same time, the management team is typically recruited from within the project arena (ibid.). This indicates that the sense of responsibility for the co-operative, described also in reference to its history with the local Energierat initiative, still serves as a symbolic resource to mobilize organizational capacities. Still, members reportedly feel challenged by the perceived lack of traction in the cooperative's projects, feeling compelled to surpass personal boundaries in mobilizing for the cooperative (see chapter 9). One management member working at the local university established the university cooperation, which while supported among members at first was called into question when the diverging agendas of the partners were revealed in strategy workshops (see 7.2).

While a challenge to the core values or activities of a project can serve as a reinforcement of shared symbols and organizational processes when members still feel the initial goal is attainable, this pattern of resource mobilization could not be repeated when a challenge coincided with the failure to do so or with a lack of an overarching strategy framing actions of the community energy project. This explains the attraction of the slaughterhouse narrative, and underscores the importance of closely examining actual resource mobilization processes which indeed do not follow these storytelling rules.

---

[4] An overview of management members of the cooperative over time based on public records is available online (North Data GmbH 2017.)

10.3   Summary of Resource Mobilization for Project Challenges

Table 10.1 summarizes the mobilization of resources by type. The analysis of projects' assertiveness in the face of external or internal challenges points to the relevance of organizational resources. This is true across project stages, as well as for projects operating on the predominantly paid or un-paid engagement of members. Knowledge resources, both to foresee potential challenges as well as to overcome barriers, were similarly important. Their mobilization again relied on the availability of organizational resources. Symbolic resources could be mobilized directly in combination with organizational resources, and could substitute knowledge resources entirely depending on the external challenge. Examples of this are the mobilization of village support in the second referendum EWS faced, or the reminder of the symbolic power Samsø held vis-à-vis the government threatening to cut funding because of its high international visibility. In other cases, symbolic resources directed the mobilization of organizational and knowledge resources. This occurred in the cases of FDT and GEM when energy services were framed as community building services and mobilization of organizational and knowledge resources in moments of assertion re-invoked community centered values, such as in the mobilization of knowledge resources from DTAS to create a community development fund instead of a cooperative during the project establishment phase in Fintry.

Material resources could matter for the mobilization of additional organizational resources, such as in the cases of EWS or Fintry. The Energy Academy on Samsø presents a unique case in this respect by permanently allocating financial resources to a staff member designated to challenge the project from within. But financial resources did not determine successful assertion. This is exemplified by the case of olegeno, an organization prevailing despite an abundance of challenges and a distinct lack of funding. Natural resources were not mobilized distinctly in these instances of project work.

Table 10.1   Mobilization of resources by type when challenged

Project	Mobilization by resource type					
	Financial	techno-logical	Natural	Organizational	knowledge	symbolic
EWS	(+)	–	–	+	+	+
Olegeno	–	–	–	+	+	+
Samsø	+	–	–	+	+	+
FDT	(+)	–	–	+	+	+
GEM	–	–	–	+	+	+

The image of resource mobilization by type when projects were challenged is therefore quite similar to that of resource mobilization during emergence. This suggests at least one similar lesson to be drawn. Storytelling, the symbols and images attached to community energy services and the organizational resources to tell these stories and build respective networks, is as important to project assertion as it is to project emergence. This is true for external challenges, if a larger audience can create pressure in favor of the project's position. It is also true for internal challenges, when the mobilization of organizational resources itself is under pressure. Shared symbols can revive the feedback loop of organizational and symbolic resources. The combination of symbolic and organizational resources is less powerful if public debate cannot be re-invoked to reinforce shared symbols or narratives. In these cases, the combination of organizational and knowledge resources proves more powerful.

Structural resources providing organizational and knowledge resources can determine the ability of a project to overcome challenges. While financial resources derived from structural resources, such as the revenues of the wind turbine in Fintry, may support the mobilization of organizational resources which in turn can aid project assertion, this path of mobilization is indirect and not necessarily successful. Organizational and knowledge resources provided to projects could, for example, take over coordinating roles for energy projects with community participation. This was the case when Community Energy Scotland and ACCESS bridged gaps in MICT's abilities to overcome impediments to the realization of Garmony Hydro. olegeno's establishment of a retail product with the help of a cooperative network (Bürgerwerke) could also be classified as a form of assistance. Without the structural resources this cooperation provided to olegeno in the form of organizational and knowledge resources, the cooperative would not have been able to establish retail services and the steady stream of revenues these should secure to the community energy project.

Part V
Social Innovation in Community Energy

Key Findings　11

This chapter synthesizes findings across the five cases of community energy. It points out three key aspects the case studies reveal. First, projects have distinct innovation biographies that relate to frames of energy system change and community development, and create corresponding power relations within which energy system alternatives can (or cannot) be implemented (11.1). Second, patterns of resource mobilization point to distinct development stages of social innovations across projects, while resource mobilization within projects relates to individual biographies, frames and arenas of negotiation (11.2). Lastly, innovation biographies and resource mobilization point to how social innovations in community energy create new roles within the energy system, and to the challenges to energy system governance these entail (11.3). This chapter outlines what kind of energy system change occurs in community energy projects, before turning to theoretical implications of empirical findings.

11.1　Innovation Biographies of Community Energy

In analyses of the energy sector, innovation biographies so far predominantly referred to technological innovations, explaining how constellations of actors, technologies, symbolic and natural elements developed as implementation unfolded (Bruns et al. 2011). Innovation biographies of community energy projects, by contrast, focus on the development of an organizational paradigm over time. An organizational form, community energy projects, is considered within its specific setting, including how power is administered throughout the innovation process (Ohlhorst and Schön 2015, p. 267). Comparison of innovation biographies indicates similarities and differences in the conditions of social, rather than technological, innovations. This includes the role of state and regulatory authorities, and the "strategic use of policy planning and control instruments in innovation processes" (ibid.). The advantage of focusing on the innovation biographies of the organizational rather than the technological phenomenon lies in the direct assessment of socio-cultural parameters, which are frequently highlighted in technology-centered studies but only as a backdrop to technology development (Bruns et al. 2011, p. 40). Key findings on the innovation biographies of community energy projects include the continued relevance of shared frames developed upon initiation, and the ability of projects to create frame resonance corresponding to diverse actor constellations and power dynamics.

11.1.1 Community Project Frames for Energy System Change

Projects of this case study developed distinct and salient interpretive frames that remained consistent over time and characterized and structured their argumentative positions and actions. These characteristics shaped actor constellations within the project as well as relations to other actors, such as regulatory authorities or energy utilities. Frames affected the distribution of power within the research situation. Projects' innovation biographies were therefore named after these characterizing features: electricity rebel, moderator, reluctant subversive, pioneer, and catalyst. This does not suggest a typology of community energy projects. Instead, it points to the relevance of dominant interpretive frames for project development over time.

   These characteristic, biographical frames were not results of projects' organizational form or regulatory conditions. Projects of the same organizational form or within similar regulatory contexts developed different biographical traits. For example, EWS and olegeno were both cooperatives but developed distinctly different biographical traits. Similarly, projects in Fintry and on Samsø were both established in regulatory conditions including feed-in tariffs for renewable energy generation and institutionalized regulatory and organizational support for community participation, but did not resort to the same biographical traits. Instead, characteristics of the innovation biographies of projects corresponded to the conditions of project formation. Frames could relate to specific targets of energy system change, such as nuclear independence or financial benefits for the community, or to procedural qualities of energy system change, such as embedding transition in existing community organizations and narratives.

In Schönau, project initiation occurred in response to the nuclear incident of Chernobyl and was shaped by the subsequent inaction of policy makers and the energy utility, and their disregard for community wishes in subsequent regulatory and commercial decisions. This resulted in a reactionary, 'rebellious' stance of community organizing and a move to provide energy system services in accordance with project initiators' values. It created an actor constellation whereın project supporters opposed the energy incumbent and municipal authorities despite their considerable power imbalance. Consequently, the project narrative formed based on shared frames of what project founders believed to be 'right' for the community against the frame of a Goliath-like coalition of economic and political decision-makers who until then had determined the shape of the local energy system. This narrative remains central to project communication and strategy development today, despite the cooperative's successful establishment at local and national levels. The project arena on Samsø, by contrast, was established not by the community but for it; the municipal administration leveraged national level policy instruments to initiate community energy action. This resulted in a distinct role for the Energy Academy and corresponding interpretive frames focusing on community integration and moderation, while attempting implementation of ambitious overarching targets. Like in Schönau, these frames stuck even after establishment of community energy structures within the project arena.

Characteristic traits of projects' innovation biographies stuck even when the conditions of project formation had since been resolved or no longer applied. Oldenburg's olegeno was founded as a commercial organization of the local Energierat, which had previously been confined to monitoring and consulting for the municipal authorities, and public awareness campaigning. The cooperative's campaign was based on the knowledge obtained in previous monitoring functions, a narrative wherein citizens had contrasted the performance of local economic and political decision makers to their knowledge of system potentials. This resulted in a double role for the community energy project. The cooperative was simultaneously applying for grid operations, and monitoring conditions for application. This ambiguity continued after the application process failed, as the community energy project struggled to find new areas of occupation. The cooperative still attempted to both realize its own projects and consult on energy system change locally. This created a catalyst-like setting, wherein the project was involved in the initiation of various local energy installations but ultimately not part of their realization.

Across projects, energy installations were linked to specific interpretive frames in recurring linguistic images. Although the images varied in accordance with specific biographical characteristics, the functions of these images to projects were similar. For EWS, these images include energy rebellion, grandkid-suitability, and empowerment of citizens. These images immediately connect the energy system alternative presented by the cooperative to values of responsibility and accountability. At the same time, the project's multiple service options enabled the individual to immediately act upon these beliefs. In the early years of mobilization, a strong connection to the locality was another recurring image of the projects' language ("Yes to Schönau"), combining responsibility and localism. This emphasized the ability of the individual to act upon energy system values within their local context, despite facing challenges seemingly beyond local influence. The project also created a forum wherein those previously not invited to energy system decision-making had power and voice. Energy system services were 'not rocket science', but open to political and cultural debate beyond the previously dominant circle of technology experts and political decision-makers. In Oldenburg, similarly, the cooperative publically presented an energy system alternative, competing 'at eye level' with established actors of the energy sector. This focus on individual and community ability to act upon shared frames is echoed by projects in Denmark and Scotland. Samsø islanders were not presented with a master plan to enact. Instead, the master plan (key element of a successful application) by including more options than required by the overall emissions reduction target was designed in a way that would conserve the agency of islanders during implementation. Images invoked, consequently, emphasized agency and responsibility in connection to the locality: 'good housekeeping', 'Viking leadership', or 'skafning'. In Fintry, project initiators rejected the suggested organizational structures of the commercial developer and insisted on institutionalization that would speak to community values. On Mull, the energy potentials expressed in the national energy strategy did not determine local choices; rather, local values that related to the utility and relative negligibility of energy system services were key. Energy systems should serve the community, but not determine its actions: 'all you're doing is getting clean renewable energy for the next 100 years' (section 5.2). Beyond linguistic sophistry, these frames indicate the development of shared cognitive models of understanding energy system services (Ziegler 2017a, p. 103). These models forge connections between energy system services and other aspects of community life in accordance with the respective dominant project narratives, and are characteristic of community energy projects across this study.[1]

The development of shared interpretive frames served similar functions within projects, although frames differed. Some interpretive frames were also shared

---

[1]Rafael Ziegler's analysis of community led water activism in Bavaria includes similar observations on community engagement (2017a, pp. 103–104).

across projects. This indicates their specific relevance for the organizational model of community energy. These frames are:

- **Community**: Community is closely connected to the locality in project terms ("Think local. Act local.", Søren Hermansen 12/20/2013). It is a central organizational platform of realizing change jointly rather than individually. This platform goes beyond the local in project attempts to create networks of empowerment and self-efficacy in larger contexts, such as ACCESS, the 'From Best to Next'-communities, or the consumption decrease of EWS electricity customers. But it also transcends the local conceptually as community definitions are invoked in reference to public utilities and welfare ('Daseinsvorsorge').
- **Ownership:** While emphasized across projects, ownership does not necessarily refer to actual ownership of energy assets, but may also be realized in assumed ownership of processes of energy system change ('I own this project; this is us', Samsø). The loss of actual ownership is felt acutely, however, as a loss of community control ('it is a bit our own, but not in quite the same way', Samsø).
- **Responsibility:** Responsibility relates to the local, community level ('skafning') and to a more general understanding of intersubjective responsibility ('grandkid-suitability'). It is also phrased in reference to positions of privilege ('we can afford to think about energy transition', Fintry) and power acquired through previous project actions. This connects closely to the responsibility projects feel for their locality, as they have developed organizational structures that serve functions beyond energy system services, which might be substituted for by utilities. These can refer to the economic or social vulnerabilities of communities, but also to the establishment of organizational structures providing associative benefits to communities.
- **Empowerment and Self-efficacy:** Across projects, members report a sense of empowerment and self-efficacy derived from project work that often transcends membership, extending to community members not themselves engaged in the project. Members felt empowered by associative structures during project work and within the membership structure ('you just feel human companionship', EWS). Locals express ownership and pride of the project independent of their personal involvement in a way that speaks to a sense of power derived from the achievement of the community ('we are from Samsø, you know, **that** Samsø').

These frames form CORE (Community, Onwership, Responsibility, Empowerment) frames for the various projects' conceptualization and performance, similar both in the function they fulfill within projects and the interpretations they include.

Frames developed or assumed upon project initiation shaped project actions and interactions among local constellations of actors, policies and regulation, and economic structures, for a very long time. This is captured by the concept of frame resonance, the connection of a frame to established narratives and its ability to mobilize support of a larger community through credibility and salience (Benford and Snow 2000, p. 619; see section 2.2). Credibility refers to the degree of consistency between frames and actions, the ability to connect empirical evidence to the frame's claims,[2] and the credibility of those articulating and promoting the frame (Benford and Snow 2000, pp. 619–620; Snow and Benford 1988). Salience refers to how important underlying values are to a frame's intended audience, and the embedding of the frame in the everyday lives of said audience as well as shared narratives (Benford and Snow 2000, pp. 620–622). Although developed for the study of social movements, these processes could also be observed in community energy projects. Innovation biographies are characterized by a stickiness of established frames, and longevity of project leadership. One example of the importance of a consistent message and returning patterns of engagement is the second referendum faced by EWS. The project had successfully established support of the electorate and the municipal authorities for its core frames of a nuclear-free, community driven orientation of energy system services, but had to flip its communication compared to the first referendum. Project supporters had to now vote 'no' to remain supportive, creating uncertainty and endangering success. The project's target and its communicated values had not changed. Still, this relatively small diversion from the established message jeopardized success.

Established narratives could also result in the long-term redistribution of power in local actor constellations. Frames for project mobilization on Samsø referred to community self-help and individual and community leadership. This built on individual participation and involvement of established organizational structures on the island that had had no prior connections to energy services. The rallying point was the Energy Academy, although municipal structures provided the backdrop of project activities through the development of the master plan and the signing of financial guarantees to secure implementation. Islanders on Samsø still turn to the Energy Academy with any energy system related issue, although the municipality has since attempted to establish itself alongside the project moderator in implementing and planning local projects and has successfully realized its

---

[2]This does not refer to the factual credibility of the frame as such, but to the ability to empirically verify the claim in the eyes of the intended constituency (Benford and Snow 2000: 620, Snow and Benford 1988, Gamson, William A., David Croteau, William Hoynes, and Theodore Sasson 1992): "Empirical credibility is in the eyes of the beholder" (Jasper and Poulsen 1995, p. 496).

own projects. Similarly, EWS replaced KWR as the dominant local partner for energy-related projects in the village, while adding connections to parts of the village previously not linked to energy system services, as well as to organizations at regional and national levels. This aspect will be further developed in the following section discussing frame resonance.

The continued adherence to certain interpretive frames can also imperil project operations, as the catalyst example of Oldenburg exemplifies. Frames adopted in community energy system change affected the kind of energy system change these projects entailed. This refers to the physical configuration of energy systems, as well as to frames' resonance with values and emotions of those providing and/or receiving system services, and their organizational embedding.

11.1.2   Frame Resonance and Arenas of Negotiation

Frame resonance refers not only to the ability of a frame to resound with values of an intended audience of its own accord, but to the ability of those advocating in its favor to "connect [...] to established ideas" (Payne 2001, p. 39). Empirical credibility is not an objective marker of factual credibility or immediate visibility (Benford and Snow 2000, p. 620). It requires the intended audience to integrate empirical observations and new ideas with established values or cultural meanings. To forge these connections and mobilize support beyond those personally connected to project initiators, which would refer largely to the credibility of the initiating actors, frames must be connected to "cultural images [distributed] in society at large" (Jasper and Poulsen 1995, p. 496).

All projects proactively attempted to connect frames of project actions to locally established ideas. Energy project frames typically related either to values of energy system design or to values of community welfare and relief. Energy system design included references to energy infrastructure services, such as 'Daseinsvorsorge' or public utilities in the case of Oldenburg, or energy security, such as nuclear safety in the case of Schönau. Community welfare and relief included frames such as the creation of financial income in the cases of Fintry and Mull, or neighborhood support in the case of Samsø, as well as more general frames of responsibility, accountability and solidarity within the community. Energy technology specific reference points were made only in those projects where either an external event (such as Chernobyl in the case of Schönau) or a pre-existing organizational focus point (such as the Energierat in the case of Oldenburg) opened the debate. Although individuals and smaller groups within the local community had been exposed to general debates of energy system design in the other three projects, an existing discourse could not be invoked nor did an external rupture support the initiation of broad debate on the local level. Project initiators that had commenced mobilization on frames of energy system preferences—in Schönau, Oldenburg and Fintry—quickly branched out their argumentative positions to include values of community welfare and relief, connecting to the CORE frames. These values were invoked by initiators across projects.

On Samsø, argumentative connections between project goals and the local community were not made by initiators seeking community consensus to begin project preparations, but by organizers charged with delivering a societal consensus upon which to build a transition process mandated top-down. Energy project aspirations did not first grow locally and then become connected to local values, rather, they were thrust upon the community. Even more than on Mull and in Fintry, where energy project ideas had developed out of their potential to increase community welfare rather than a passion for energy system change, project organizers on Samsø depended on local discourses of community development and the quality of neighborhood relations independent of energy system design. Frame resonance on Samsø, from the perspective of the larger community, could be interpreted as being established in reverse, as existing values were opened to include energy references. More generally, the creation of frame resonance within communities suggests that community oriented values served as more potent resonators than energy system frames.

Analysis of arenas of negotiation indicates how frame resonance related to power dynamics. The diverse constructions of arenas underscore the argument that frame resonance was constructed individually in reference to interpretive frames that had been negotiated locally. Oldenburg and Schönau projects closely relate to local energy arenas. Projects on Mull and in Fintry relate to energy arenas on the national level, but emphasize connections to arenas of community development. The project arena on Samsø was embedded in national energy and community development arenas, but drew on neither for local mobilization.

Power distribution within the arena further influenced how frame resonance could be constructed. EWS faced an opposing coalition of the municipal authority and the incumbent utility in the local energy arena, which held decision-making power over questions of local energy system design posed by the community energy project. This resulted in project initiators branching out argumentatively to strengthen their position. Shared values beyond energy political beliefs (grandkid-suitability, responsibility for community and environment, accountability of utilities and policy makers) were invoked to build support within the project arena. Also, the project's strong anti-nuclear agenda resonated with the national anti-nuclear movement. Frame resonance of community values, as well as energy system values within the project arena and the anti-nuclear arena served to counter opposition to project frames in the local energy arena. By contrast, project organizers struggling to establish frame resonance for their values in Oldenburg faced a more ambiguous coalition of public hand and incumbent utility, offering no clear counter-constellation to mobilize upon. And although the national anti-nuclear movement had been reinvigorated by the nuclear incident of Fukushima in early 2011, it had also been placated by the government's reinvigoration of nuclear phase-out plans. The growing German sector of community energy projects, on the other hand, focused predominantly on renewable energy generation and offered no overarching narrative to connect to. Olegeno could therefore not forge connections to arenas of negotiation beyond the local. The project could, however, build on the established narrative that energy infrastructure design should follow expert opinions, shared by citizens, municipal authorities and representatives of the energy sector alike. The cooperative could create frame resonance for a campaign based on extended and improved knowledge of infrastructure alternatives.

The connection to arenas of negotiation beyond the local also points to the potential and limitations of regulatory conditions and policy instruments to shape frame resonance and innovation processes. In three out of five cases, the regulatory context provided distinct parameters of energy system design that immediately framed community energy aspirations.

In Schönau in 1986, there was a vertically integrated market resulting in regional monopolies for energy incumbents; energy system services were governed by concessions awarding the political right to own and operate infrastructure for a fixed amount of time. Regulation was primarily concerned with energy security and affordability rather than ecological performance. There was, moreover, limited access for communities or individual citizens to decision-makers or commercial structures beyond electing public actors. Renewable energies were supported only in research and development contexts and civil society participation was not mandated. The sector was characterized by close argumentative proximity between political decision-makers and incumbent utilities at all levels of government. For community organization behind energy system change, this created a clear frame of opposition. In Oldenburg in 2011, the energy market was liberalized, services of generation, distribution and retail were unbundled and customers could choose from an array of, at the time, approximately 1.000 energy suppliers. A nuclear phase-out had been decided, and renewable energy generation was supported through feed-in tariffs leading to rapid growth in renewable

energy generation and corresponding diversification of the sector's actor landscape. While security and affordability of supply still dominated energy industry
regulation, ecological performance of the sector had also been addressed in new
regulations. Olegeno did not frame its actions as an opposition push for necessary
changes that would bring distribution system operations in line with its values.
Instead, it based its campaign on arguments for alternative energy system design
without attacking the incumbent.

In Denmark, national energy policies have endorsed renewable energy generation and community involvement in various forms since the 1980s. Wind energy
technologies became an expression of a Danish culture of innovation, participation
and sustainability. This contributed to the energy political frame which suggested
that Samsø's renewable energy project could not fail. Projects in Fintry and Mull,
by contrast, were established in a regulatory environment which at the Scottish
national level did not include energy political authority. Instead, prominent regulation and policy debate made reference to community development and municipal
authority over local resource development, especially in reference to projects of
sustainable development. UK level strategies for energy system design included
the development of renewable energy resources in Scotland but did not address
community level actors and were largely ignored by Scottish communities. This
framed local community engagement, independent of its thematic focus, in terms
of community welfare, and supported community actors in its pursuit. The regulatory context therefore influenced both arenas of negotiation open to community
energy project representatives seeking to create frame resonance, and the relative
positions of power(lessness) actors held within these arenas. It also prompted the
question of how positions of power within one arena could be transferred to other
arenas within which projects were engaged. This is a point which will be analyzed in more detail when the patterns and trajectories of resource mobilization are
examined.

Within arenas of negotiation, the creation of frame resonance was often an
interactive process. Joint meanings were typically negotiated in the project arena,
while frame resonance could provide supportive or adverse connections to arenas
beyond the local level. This is most apparent in the case of Samsø, where project
organizers deliberately and repeatedly initiated such processes of 'talking your-
selves into consensus' within the project arena to determine how energy system
change should occur on the island, and how future projects would be selected
and implemented. The discussion of technologies, followed by the commissioning of a feasibility study for siting and a survey to secure public support on Mull
indicates similar commitment to local negotiations of shared values underpinning
measures of energy system change, as does the branching out of the anti-nuclear

narrative into a narrative of family and community support in Schönau. Findings across cases also suggest a geographical component to frame resonance within the project arena. Projects characterized by geographically confined arenas of negotiation, such as on the islands of Mull and Samsø, emphasized frame resonance and cooperation in numerous ways, highlighting the need to work together and find consensus in shared projects. Further research into geographical parameters of frame resonance in community settings could assess these tendencies.

This leads to a controversial question on the role of those initiating projects: How much does the ability to connect new and established values depend on the individual capacity of project leaders? Aspects of personality or hierarchy were not specifically researched within this project; however, sources indicate that project members value individual traits of their leaders. On Samsø, this is referenced most directly as interviewees speak of the importance of a person locally established (someone that 'people know the parents of', see section 6.1) and individually talented ('if you don't have this ability to speak and create trust, then forget it', ibid.). In Schönau and on Mull, the group of project initiators included members known to the community in positions of trust, such as the director of the local Development Trust on Mull or the village doctor in Schönau.

This study does not provide data to elaborate on the personal qualities and traits of norm entrepreneurs seeking ideas upon which to saddle new projects (Payne 2001, p. 47). Rather, it focuses on the procedural dimension of frame resonance as indicated across cases by referencing the credibility of actors' engagement and their sustained commitment over time. Samsø islanders not only referenced the ability of Søren Hermansen to speak publically, but also pointed out that "everyone feels they get good answers when they need it" (Lundén 2003, p. 39). EWS project members underscored the importance of members' persistence when talking about the transferability of community energy concepts. This refers to the importance of sustained procedural commitment within the group—a position supported by data from all projects. It was also referenced when cautioning against transferability: "that those inspiring people acting on conviction, those Schönauers, are not there... that is why it is difficult" (I3: 163–164).

Frame resonance was therefore analyzed within project arenas as well as in interactions of the project arena and arenas at the national or subnational levels. The regulatory context mattered as it opened (or closed) arenas of negotiation for community energy projects, and could influence their relative positions of power(lessness) within them. External events mattered only if members of the project arena could mobilize others corresponding to the event. Frame resonance within the project arena, however, dominated project decisions. For example, the

local energy arena offered a position of relative powerlessness to community organizers in Schönau reinforced by the national energy arena, prompting frames of
defiance and rebellion among project members. Similarly, despite UK support for
wind energy installations and Scottish support for community organized projects,
project leaders on Mull decided for a considerably smaller installation in accordance with local values. But project leaders countered regulatory constraints by
setting up a collaboration with sectoral partners wherein the project's relative sectoral powerlessness within the national energy arena was counterbalanced by the
support of organizations in the Scottish energy arena. Beyond general insights
into the diverse factors that shaped frame resonance, analysis through arenas of
negotiation offers a way to organize and understand the interactions determining
which frames become dominant and advance to characteristics influencing the
overall trajectory of community energy project development over time.

11.2   Resource Mobilization in Community Energy

Resource mobilization features in many studies of community energy projects,
because of its connections to concepts of powerlessness and empowerment. This
study expands on this research, by specifically focusing on differences in the
mobilization of material and non-material resource types over time. This adds
depth to the study of changed power relations and improved capabilities in concepts of social innovation, while also providing a more nuanced understanding
of the conditions of social innovations and their biographies. Patterns of resource
mobilization by type point to shared conditions and characteristics of project development, and qualify the relative importance of material and non-material power
dimensions. Resource configurations within projects tie these types to specific
innovation biographies.

11.2.1   Patterns of Resource Mobilization—Innovation
           Biographies and Resource Types

Across projects, patterns of resource mobilization indicate similar stages of project development. These stages varied regarding who or what (e.g. regulations,
technologies, etc.) was involved within the research situation, as well as relevant
arenas and argumentative positions. Development stages could also last for different periods of time. But across cases they were similar with respect to the *types*
of resources mobilized by community energy projects. Changes in the patterns of

resource types therefore indicated changes in the development stage of a project. Overall, three consecutive stages of development could be identified. (1) Project emergence lasted from the initiation of project activities until the resolution of the first planned target, (2) project establishment, followed by (3) project maintenance, the sustaining of project operations after establishment. A fourth resource pattern consistently indicated response to an internal or external challenge to the project. This could occur throughout all three consecutive development stages.

This study considered material and non-material resources based on the understanding that a resource may be "any social, political, economic asset or capacity that can contribute to collective action" (Jenkins 2001, p. 14368). Material resources included financial and artifactual material resources, such as money, equipment, or technological installations; as well as land or natural conditions such as climatic or topographic conditions. Non-material resources included organizational resources such as time, labor and social networks; knowledge and information; and symbolic resources or shared understandings and interpretations of what is meaningful within the organization as well as its cultural context. In the analysis, the overarching types of material and non-material resources were distinguished in six sub-types, namely financial, technological, and natural resources as sub-types of material resources, and organizational, knowledge and symbolic resources as non-material subtypes. In addition, structural resources were considered referring to "formalized resource allocation mechanisms" (Schreuer 2015, p. 64). While structural resources cannot be owned by community energy projects, different kinds of resources could nevertheless be mobilized on their basis. Financial support mechanisms used by projects appeared as material resources. Support for community development in the agenda setting process of national governments resulted in the mobilization of interpretive frames linking energy system change and community support, providing non-material, symbolic resources.

Project emergence is characterized by the mobilization of non-material resources: organizational resources, as well as knowledge and symbolic resources. These resources were mobilized by all projects during emergence. Although only one project mobilized financial resources directly (paid staff hired on Samsø), two other projects indirectly benefitted from the availability of natural resources. Project establishment was characterized by a surge in resource mobilization, wherein all of the above non-material resources continued to be mobilized, while financial resources were also mobilized. All but one project mobilized technology resources upon establishment; natural resources were mobilized in three projects upon their establishment.

Three lessons can be drawn across cases regarding resource mobilization for emergence and establishment. For one, symbolic resources and organizational resources, the stories that were told and those bringing them to life, were key to project emergence. These resources could either originate in the project itself, or be mobilized on the basis of structural resources that were appropriated and reconfigured to become symbolic or organizational resources. Variation refers to the order of appearance of symbolic and organizational resource types as well as their local specification, but not to the types as such. The power of story-telling as a resource is summed up by a brief conversation between a researcher and a film maker for Samsø's Energy Academy wherein the latter explains the Energy Academy to be, "first and foremost, a story-telling house. They could have chosen to tell a story about how three–only three in twenty-one! [sic!]–of the windmills are owned by cooperatives of local citizens! But that's not how they tell it, is it?" (Papazu 2018, p. 15). Secondly, while other resource types may have also been mobilized, they alone were insufficient to instigate actions. Financial resources, although important for establishment, did not provide the initiating impulse for emergence. Similarly, knowledge resources such as awareness of policy support mechanisms for community engagement in energy systems or expertise in renewable energy technology, as well as natural resources, such as climatic conditions favorable to wind energy generation or space to establish installations, were insufficient to create enough momentum for project establishment unless they became embedded in symbolic and organizational resources. Lastly, the mobilization of all types of resources for establishment underscores the broad range of activities necessary to anchor an energy project within a community, and the diversity of activities and discourses projects must engage in within the community to this end.

Resource mobilization was especially successfully when feedback loops of resource types could be established. These were observed predominantly for shared interpretive frames and organizational resources (time, labor and social networks), most notably in the cases of EWS and Samsø. Feedback-loops sustained project activities during emergence, which could take an extended period of time (both of the above projects took almost ten years to establish). In Oldenburg, the project notably managed to establish a feedback loop between knowledge resources and labor, mobilizing voluntary engagement based on the credibility of the suggested system alternative. This corresponds to the observation that knowledge resources served not only as an influx of expertise but were connected to a larger narrative of the credibility of experts in energy system design. The observed feedback-loop could be indicative of an implicit feedback-loop of symbolic and organizational resources similar to that of EWS and Samsø. Certainly, it underscores the importance of creating sustained cycles of resource mobilization reinforcing one another.

The importance of financial resources and the relative neglect of symbolic resources distinguish project maintenance from emergence and establishment. Financial resources importantly determined the sustained mobilization of organizational resources within the project during maintenance, and the ability of projects to deliver project services as intended. Technological resources mobilized upon establishment remained active throughout project maintenance, and three of five cases extended technological capacity by adding new kinds of energy system services. Symbolic resources were not mobilized explicitly. Still, they remained present as overarching narratives influencing how financial, organizational and knowledge resources were mobilized and employed. This is most prominently the case in Fintry, Mull, and Schönau. These three projects show the strongest institutionalization of financial revenue allocation to community and project-oriented services in their organizational statutes. In all projects, however, reference to narratives is indirect, or at least routinized as in the case of Schönau's 'Sonnencent', during the stage of project maintenance.

Challenges to the community energy projects had their own pattern of resource mobilization. This pattern strongly relied on organizational resources, regardless of whether challenges occurred during project emergence, establishment or maintenance. Knowledge resources were mobilized depending on the availability of organizational resources. Symbolic resources could be mobilized directly in combination with organizational resources, and could substitute knowledge resources entirely. In most cases, symbolic resources directed the mobilization of organizational and knowledge resources, therefore re-invoking community centered values. Shared symbols hereby revived the feedback loop of organizational symbolic resources, and ascertained the project's ties to initially negotiated values. Resource patterns when challenged therefore underscore the importance of projects establishing mutually reinforcing cycles of resource mobilization. This was true for external challenges in situations where a larger audience could create pressure in favor of the project's position. It was also true for internal challenges when the mobilization of organizational resources itself was under pressure. The combination of symbolic and organizational resources was less powerful if external challenges could not be addressed by building public awareness and political pressure. In these cases, the combination of organizational and knowledge resources proved more powerful. Material resources could matter for the mobilization of additional organizational resources, for example by paying project members for their time and thus enabling them to devote more time to overcoming project challenges. Material resources, however, did not determine whether challenges could be overcome successfully.

Structural resources provided by policy and regulation featured strongly during project establishment, most notably by enabling projects to mobilize financial resources (for example based on investment certainty created by feed-in tariffs) or organizational resources (for example through mechanisms for funding staff). Policy environments mattered again when projects were challenged, but less for their financial prowess. Although project challenges could also arise from regulatory measures, such as grid constraints, the policy environment could provide organizational and knowledge resources at the same time. Financial income based on structural resources could support the mobilization of organizational resources which in turn could support projects under challenge. Yet, this path of mobilization is indirect and was not always successful. Organizational and knowledge resources provided to projects could importantly bridge gaps in a project's individual abilities to overcome impediments. Structural resources provided by the socio-cultural environments of the projects mattered, especially during emergence and establishment, and again when projects were challenged. They provided larger narratives to mobilize symbolic resources in favor of the project. This was the case with projects' individual and CORE frames and the resonance of those frames.

11.2.2 Alternating Current: Resource Configuration and the Transfer of Resources and Power

Innovation biographies are characterized by similarities in resource types mobilized across development stages. One way in which they differ is with respect to what can be mobilized as a resource within certain types of material or non-material resources. Projects differ predominantly with respect to the symbolic resources they can mobilize. This has important repercussions for project development because of the role symbolic resources play in feedback-loops, for example with organizational resources during project emergence and establishment. It also refers to their role in framing the deployment of organizational and financial resources during maintenance, and the insurance they provide when projects are challenged. Beyond the simple conclusion that project initiators must be able to sell a project to the community by providing a good story, this is relevant because the frame resonance of symbolic resources affected negotiating positions and power relations by connecting the project to specific arenas.

The stories projects tell of energy system change, as indicated by CORE frames, connect to organizational structures and cultural values previously not associated with energy system services. By creating these connections, energy system change in community energy settings moves into a negotiating space where more fundamental questions of public utilities are debated. This can provide important material and non-material resources, as organizational, financial and decision-making structures are activated on behalf of energy system change that otherwise may not have been accessible. The framing of energy system services beyond technological or infrastructural parameters also subjects system design to values and mechanisms of association and negotiation forms and financing sources that previously did not apply. For example, private energy utilities typically follow profit maximizing rationales in their economic actions bound by organizational forms of private ownership. The profits of public energy utilities (if any) become part of the public budget. By contrast, community energy projects chose organizational forms in accordance with members' shared frames of community orientation (e.g. cooperative, development trust, community benefit society, etc.). This had specific repercussions for the reinvestment of profits. Projects specified investment conditions to secure adherence to shared interpretive frames (e.g. the Schönau Gierbremse, or the Mull Waterfall Foundation).

The resonance of CORE frames can also result in projects assuming more powerful positions vis-à-vis established actors of the energy sector. The large frames offer the projects connections to arenas of negotiation previously not associated with energy system change. One example of this is the strong connection of Scottish community energy projects to the Scottish community development arena. Another is the connection of EWS founders to the anti-nuclear arena. Although anti-nuclear protest addressed political debates on energy system services, the anti-nuclear arena rarely engaged in alternative models to provide energy system services itself prior to EWS. Activities were predominantly based on protest and political debate. Implementations of alternative energy systems were limited to individual or household levels. In the Scottish cases, the framing of energy system change in community welfare terms connected to an arena of community development wherein structural resources provided access to regulatory, symbolic and even financial support. This importantly strengthened the negotiating positions of community-based actors. In the German case, the benefit was mutual. The community energy project provided an opportunity to physically push nuclear power generation and associated actors of the energy sector out of the Schönau distribution grid. Beyond blocking construction of further sites, this provided the first opportunity of exnovation—the removal of an outdated innovation—to the movement. For the community energy project, in turn, the backdrop of the national anti-nuclear movement sustained symbolic and organizational resources that had become available upon the nuclear incident of Chernobyl, as the movement locked in awareness of and activism against the risk of nuclear energy generation that had sparked locally. This leads to the transfer of power between arenas of negotiation and the ability of projects to mobilize resources within one arena and relate these to other arenas.

A rather straightforward form of transfer occurs between arenas bound by the topic of negotiation and project arenas. The above example of the anti-nuclear arena in the case of Schönau is such a case, wherein the project arena sought to mobilize time, labor and later also financial resources based on a shared interpretive frame of anti-nuclear protest. The anti-nuclear impetus connected the project to the national anti-nuclear arena that shared its rejection of nuclear technologies and desire for community based action in the face of government and energy utilities' inaction. Schönau villagers mobilizing for anti-nuclear activities could see their engagement reflected on the national level, which in turn lent credibility and legitimacy to their narrative on the local level and supported mobilization of time and labor. The reverse case occurred when the anti-nuclear arena became aware of EWS' struggles to obtain grid operations locally, which presented the first case wherein the movement could claim to have achieved nuclear independence. Consequently, the project provided proof of concept to the movement, linking material and public participation in energy system design through the realization of energy political claims in infrastructure configuration.

More complex processes of transferring power occur when projects seek to mobilize resources provided by one thematic arena wherein the project holds a relatively powerful position towards a negotiating arena wherein the community energy actor holds a relatively powerless position. Empirical evidence suggests that this is more successful in cases were structural resources provide material or non-material resources that can be mobilized within one arena and strengthen the negotiating position in the other. Examples are the mobilization of organizational resources (namely labor and networks) as well as knowledge resources within the Scottish community development arena by projects in Fintry and on Mull, which countered a relatively weak negotiating position for community energy actors in the UK energy arena, wherein structural resources provided access to financial funds for renewable energy installations but otherwise adverse conditions for new market entrants. Scottish national parliament had passed legislation strengthening the position of municipalities in plans increasing community welfare, which extended to community led initiatives if these could prove charitable status and community orientation. This provided FREE organizers in Fintry with an argumentative position to counter the sectoral practice of involving citizens financially on the basis of cooperative models; they established the Development Trust with the support of policies and public institutions for community development. This model of community support had been extended to the energy sector by the time MICT explored energy opportunities. The Scottish parliament had established Community Energy Scotland (CES) as an umbrella organization for communities venturing into the energy sector providing knowledge and organizational resources. When grid constraints threatened realization of Garmony Hydro, CES aided GEM in the creation of ACCESS with its provision of organizational and knowledge resources. CES became the consortium leader of ACCESS, which also ensured that interests of the community energy project would be represented in leadership vis-à-vis the energy sector incumbents involved in the consortium.

EWS' introduction of a municipal referendum to counter premature contracting between the energy incumbent and the municipality in Schönau in the early 1990s is also an example of a community energy project mobilizing structural resources. The constitution and the municipal code of Baden-Wurttemberg offer the instrument of a municipal referendum as a regional configuration of direct democratic participation. Structural resources therefore strengthened the negotiating position of EWS as they provided legal tools to defend project interests.

Resource transfer between arenas could also fail. Following establishment as the local distribution systems operator, EWS advanced to a powerful position in the local energy arena. In the course of market liberalization and a growing community energy sector, it became of the leading actors within the national community energy arena. This did not, however, alter its position in the national energy arena. In this arena, a single community energy project, notwithstanding its local track record or relevance for the community energy sector holds a relatively powerless position vis-à-vis the four market dominating energy utilities and the federal regulatory and decision-making authorities. EWS' attempts to alter political decision-making processes, or even raise awareness for its position, remained unsuccessful. The contrasting experience of the Samsø energy project, successfully altering national policy decision-making, was not a function of its strong negotiating position in the local arena transferred to the national energy arena, but rather a result of the symbolic resources the project could wield because it had delivered on a government-induced transformation target. Transfer of non-material resources seems more successful between arenas closely interconnected, as exemplified by the position of olegeno in the local DSO applications and the local energy arena. Although marginalized by the energy incumbent that dominated services, the Energierat had gained credibility over the course of its engagement as a knowledgeable monitor of energy decision-making at the local level. In the DSO application process, as the challenger to the established concessionary, the cooperative again faced an uneven playing field as regulation impedes applications of new concessionaries. By mobilizing knowledge resources obtained within the

local energy arena as the Energierat, the cooperative could strengthen its position
by critically monitoring the formal execution of the application process.

11.3   Power Grids: Infrastructures of Energy and Decision-Making

Based on findings referring to the role of frames in the projects' innovation biographies, it was argued in the first section of this chapter that the relative position
of power within an arena can influence how frame resonance can be achieved by
community energy actors. Patterns of resource mobilization were then referred to
the frames and arenas prominent in a project's innovation biographies. This final
section on key findings turns to changes of the power relations underpinning the
development of innovation biographies as well as the mobilization of resources,
and capabilities improved within communities in the process. This section ties
the study of innovation biographies and resource mobilization to the overarching
theoretical framework of social innovation underlying this dissertation.

Powerlessness of community energy actors is frequently associated with limited material resources of community energy projects in the literature, especially in
comparison to other actors of the energy sector, and limited abilities to exert material (e.g. financial) or non-material (e.g. decision-making) power over other actors
in the energy system (see section 2.1). Empowerment, consequently, refers to
increased access to material and non-material infrastructures of the energy system.
Changes to power relations within the energy sector take diverse forms, depending on whether material infrastructures of energy system services or non-material
infrastructures of decision-making for energy system services are affected. All
projects included alterations to the material dimension of local energy infrastructures. Empirical analysis within this study also points out that successful creation
of frame resonance for ideas of the community energy projects can indeed alter
how energy systems are thought or felt about within larger communities (see
section 2.2.3). Projects gained power.

The extent of material changes to local power grids varied across projects.
The Schönau and Samsø projects included the most far reaching alterations to
energy system services (distribution, generation, retail and related services), followed by Fintry and Mull (generation, distribution and related services), and lastly
Oldenburg (retail and related services). Physical changes to power grids included
changes to existing grid structures, such as the installation of additional meters
to increase awareness on consumption of individual appliances in the Schönau
competitions to reduce consumption, or retrofitting of insulation in Fintry; the

creation of new capacities for energy generation; and the creation of new grid structures, for example in the form of district heating systems. Physical changes to the energy system also included changes in consumption patterns, such as increased efficiency through behavioral changes, and increased demand for renewable capacity based on the design of retail products. Through ownership of energy assets or related institutional structures that created forums of joint decision-making and collectivized financial and organizational benefits, all projects successfully implemented energy system services that improved community access to and control over service provision. Across projects, however, members reported that corresponding changes to the ideational infrastructures underpinning energy system services were more important and extensive than physical changes to the energy system.

The analysis of frames and frame resonance in the first section of this chapter showed that there were extensive changes to the interpretive frames guiding energy system services' community energy projects. These changes created interventions to the social grid (Beckert 2010, see section 3.1), formed by structures and patterns of social relationships, rules and norms, and shared meanings or frames. Community energy projects created new relationships between interpretive frames and arenas of negotiation, or changed existing relationships. These interventions are interpreted as evidence of social innovations in community energy within this dissertation. These changes also reflect the new roles created for individuals within the energy system. There was an alteration in the understanding of consumer and citizen roles, and new roles for energy system actors. Analysis of developments in the regulatory contexts indicates that community energy projects successfully changed norms of co-operation among public, commercial and citizen-owned organizations of the energy system. Their influence on regulatory changes, however, was limited.

11.3.1 New Roles in the Energy System

The community energy literature acknowledges new roles for citizen-owned organizations in energy systems, especially in reference to a growing sector of community-owned energy generating installations.[3] The literature highlights that community or citizen-owned organizations entered the energy sector in roles previously held only by publicly or privately-owned utilities, thereby creating new

---

[3]See for example Schreuer 2015 for an analysis of Germany, Maegaard 2013 for Denmark, and Walker and Devine-Wright 2008 for Great Britain. See also above at section 2.1.

actors in the system. This included a change in power relations, as services were assumed that previously citizens had used but not provided, and increased capacities as financial, organizational, knowledge, and symbolic resources were mobilized within citizen-owned organizations (see exemplary trend: research 2013). The empirical analysis emphasized the extent of these changes in power relations and capacities; citizens assumed power not only through organizational association, but also as individuals. Examples include households assuming control of consumption levels through increased knowledge of load distribution (Schönau), improving insulation (Fintry), and implementing solar energy generating units on balconies (rented apartments in Oldenburg). This was not limited to project members, as is indicated by the extensive reduction of energy consumption of EWS customers (not members) on a national level, or by the identification of Samsø islanders with energy system change independent of asset ownership. Community energy projects offered forums wherein the market role of the individual as consumer and his or her political role as a voter in a local referendum or as a participant in public debate on changes to the local energy system and related economies, merged. Each action sphere could provide an expression of the values of the other.

In addition, projects recruited personnel who implemented changes to the energy system within their communities. Leadership of projects remains predominantly local, pointing to the extensive increase in individual capacities as members assumed responsibilities beyond their professional training or previous backgrounds, and took on leadership within community settings. Schreuer calls this "empowering the empowered", the accumulation of resources by actors already privileged by comparison (Schreuer 2015, pp. 186–187). This can be observed within communities, as well as between communities. Negotiations on Samsø, for example, indicated how members of existing associations were invited into decision-making procedures, and sometimes leadership positions in newly established energy projects (see above at section 7.1). While this secured ties to the local community and arguably increased legitimacy of decisions and decision-making structures, it also accumulated decision-making power in diverse community contexts in the hands of comparatively few islanders. Similarly, while the continued engagement of project leaders over time can benefit project stability, especially in organizational structures that rely on volunteering, this further serves to concentrate power and resources. It has also been argued that community energy projects are predominantly realized within comparatively affluent regions, whereas socio-economically vulnerable communities benefit less from the increased powers and capacities such projects entail (for analyses of Germany see Kahla et al. 2017; Schreuer 2015). Case selection in this study does not confirm this assessment, as indeed three of five communities represent communities characterized by multiple socio-economic vulnerabilities. At the same time, the qualitative case studies of this dissertation do not provide representative data for Scotland and Denmark. This study, however, suggests that extensive and diverse resources are required to successfully initiate, establish and maintain project structures, including the wide-ranging activities their mobilization entails.

The community energy projects introduced in this study created new roles within the energy system more generally. The projects in Schönau, Fintry and Oldenburg present examples of energy system actors investing in generation and/or retail of electricity and at the same time engaging in energy efficiency and reducing consumption in extensive and meaningful ways. Three projects, in Schönau, Fintry and Mull, institutionalized structures wherein financial revenues were allocated to community-oriented initiatives in accordance with projects' respective value orientations. Such features underscore projects' commitments to the above identified key frames across projects (community, ownership, responsibility and empowerment), all of which show close ties to community welfare. In combination with the increased power and capacities at the individual levels of members and associates of the projects, this supports the assessment of community energy projects contributing to a more participatory, value oriented design of the energy sector (Klemisch and Boddenberg 2016).

Similarly, the moderator's role created in the project on Samsø constitutes a new actor in energy system design. The Energy Academy provides ideational impulses as well as organizational embedding to continued project development, and creates an overarching bracket for local actions in energy system change. But for the most part, it was not itself invested in local energy system services directly. This created a quasi-municipal actor on the island with strong and transparent value orientations alongside the municipal administration. The strong value orientation and focus on procedural performance of institutions of both the energy system and the municipal administration more generally suggests similar interpretations for all other projects of this study. Community energy projects can develop into quasi-public actors alongside municipal authorities as they provide infrastructure services as well as decision-making structures. This leads directly to questions of governance for energy system services.

11.3.2   Governance of Energy System Services

Governance of energy system services, in this context refers to administrative and regulatory handling of the new actors and roles in economic and political contexts emerging with community energy projects, as well as interventions to administration or regulation in the form of social innovations. Community energy projects include diverse governance challenges, as their organizational form and value orientations frequently link arenas of negotiation and respective administrative and regulatory institutions, which previously were not connected. This refers especially to negotiating arenas of energy and community development. Schönau and Fintry are exemplary cases of this, wherein especially actors of the respective energy arenas were surprised by a community agenda connecting energy system design and community development, as summarized neatly by one project member in Fintry: "why would a community want to have a stake in it, nobody had ever done it before, what a bizarre idea, what's that for?" (I19: 20–21). At the time the projects were forming, the British and German energy sectors' regulators were predominantly charged with technological considerations of energy security and reliability, with little to no consideration given to community participation or even consultation in the design and provision of system services (see above at section 1.3).

By contrast, energy system regulation in Denmark had been confronted with a strong and vocal community energy sector since the 1970s, which upon introduction of renewable energy technologies had played an integral part in roll-out and development of, especially wind energy installations. Danish regulation in support of community based renewable energy infrastructures was developed incrementally at first but more proactively since the late 1980s and early 1990s. Energy system, land use, and community development regulation were much more closely interconnected. Consequently, the involvement of the local community was not an innovative or radical idea when in 1997 the Danish Renewable Energy Island was selected, but rather the continuation of an established practice. Similarly, by the time MICT was exploring renewable energy installations on Mull, the involvement of community development trusts in energy developments had advanced to a more established practice following community development policies introduced by the Scottish national government.

Administrative support for community energy actors could form a proxy in the absence of a legal mandate for energy system governance. For the Scottish national government, community development policies superseded opportunities in energy regulation. Energy policies had not been devolved. The Scottish executive could, however, introduce instruments indirectly supporting renewable energy developments without assuming political responsibility for other sectoral challenges, such as development of the national grid. While the Fintry development trust therefore needed to convince project developers to enter into a co-operation with the community, the Mull development trust could even build on a national association to support its position in a project with energy sector incumbents when grid constraints threatened realization of planned installations. The Oldenburg example, nevertheless, indicates that the innovative interpretation of energy system roles by community energy actors could challenge regulatory environments of the energy system even if community actors had been established within them. Community ownership of renewable energy installations had grown significantly, following policy changes creating financial security for investments as well as non-discrimination policies in grid regulation. This did not extend to distribution grids, however, where community involvement beyond indirect representation through publically owned local utilities (so-called Stadtwerke) was still a rare exception. Although the cooperative was not suggesting regulatory reform, albeit monitoring compliance carefully, municipal authorities were cautious about transferring established practices of energy generation to energy distribution.

Governance therefore also refers to co-operations, including new co-operations between commercial and citizen-owned actors in the energy system, as well as new modes of co-operation between political and administrative bodies, and citizens and their associations. In co-operations between commercial and community actors, administrative intervention determines the playing field, as well as the moderation of uneven power relations. Uneven power relations in the case of the energy incumbent and citizens in Schönau, for example, were exacerbated by formal and informal alliances between the municipal authorities and the incumbent. Uneven power relations in the case of MICT and the national energy incumbent SSE within the ACCESS project, however, were bridged by Community Energy Scotland providing organizational and knowledge resources to embed co-operation. Cases of this study also indicate new roles for municipalities alongside community energy organizations or in partnership with them. Both in Samsø and in Schönau, community energy projects developed into dominant local figures of energy system design.

Because of their strong connection to interpretive frames beyond energy provision, community energy actors also connected to topics of community development that had previously been handled by the municipality without 'citizen intervention'. One example of this is the strong involvement of the Energy Academy, two years into the energy project, in helping those who had lost their jobs as a result of the closure of the local slaughterhouse. This even became part of the project's founding narrative. This has led to frustration at times. In Schönau, municipal representatives expressed annoyance at the prominence of EWS in local initiatives. A similar sentiment was echoed by the Samsø municipality which was outshined by the Energy Academy with its expertise in both energy system design and community organizing. In addition, such roles within the local municipality can face legitimacy challenges depending on the organizational form of the community energy project. While membership in a Scottish Development Trust is highly affordable at a 1£ rate for lifetime membership and voting rights, participatory mechanisms on Samsø are more informal and rely on the procedural diligence of Academy members. In Schönau decision-making in the cooperative is not only tied to a much higher financial investment per member, but also open to non-locals.

The assumption of new roles within the energy system as well as projects' track records have resulted in new forms of interaction between community energy projects and regulatory bodies at the local or national levels. As a distribution systems operator, EWS for example interacts directly with the federal regulator BNetzA, as well as transmission grid operators and other market roles previously unconnected to community organizations. Similarly, MICT interacts with SSE and Ofgem, sector incumbent and national regulator, within the ACCESS project. This led to considerable capacity increases within community energy organizations, as well as corresponding recognition of commercial partners in the energy system. Project members in Fintry echo this experience when reporting that their track record had since resulted in advisory functions to commercial developers, provided that these had 'done their homework' with respect to whom to consult. This is not to say, however, that inequalities in powers and resources have been resolved between community energy actors and established commercial actors of the energy sector. This is most visible with respect to projects' financial dependency on feed-in tariffs, in those cases were energy generating installations dominate the financial project portfolio. This has led to the "ironic" situation of projects replacing their dependency on local energy incumbents with a dependency on frequently unstable policy decisions (Schreuer 2015, p. 171). While projects may be able to counter policy dependencies if their symbolic relevance and visibility is sufficient, as for example in the case of Samsø, the reduction of these dependencies is a central challenge to all projects of this study.

Discussion: Who gets what, when, and how?

12

Key findings of this study point to three underlying arguments to be made in reference to the current community energy literature, and the study of civil society-led processes of transition more generally. First, the importance of frames and frame resonance points to a distinct understanding of community to be discussed in reference to concepts of locality, affectedness, and relevance or intention. Second, social innovations imply a change in power relations while the literature highlights the specific qualities of such changes when initiated from the bottom up. And, third, the comparison of biographies of social innovations indicates shared conditions for such innovations within community energy projects, and within their regulatory and policy environments. The discussion of these arguments will relate results of the empirical analysis to the literature, corresponding to the fundamental gaps found in the community energy literature and the questions of: Who gets what, when, and how?

12.1 'Three is company'[1]: Communities of Place, Affectedness, and Intention

In argumentative positions across projects, community references abound as narratives of locality ('Yes to Schönau'), action-oriented frames ('skafning'), or overarching values of energy system design ('Daseinsvorsorge'). Connotations of 'community' within projects are diverse, echoing the contested academic debate on communities of place, affectedness and relevance (Rudolph et al. 2018; Walker 2011).

---

[1]Tolkien 1966 (1954).

© The Author(s), under exclusive license to Springer Fachmedien Wiesbaden GmbH, part of Springer Nature 2021
A. Colell, *Alternating Current – Social Innovation in Community Energy*, Energiepolitik und Klimaschutz. Energy Policy and Climate Protection, https://doi.org/10.1007/978-3-658-32307-3_12

Locality has been a recurring thematic focus in the literature studying projects implementing energy services 'by' and 'for' locals (Walker and Devine-Wright 2008; Nicholls 2007).[2] Community and locality are discussed, most frequently, with respect to their support or opposition to community-led projects inherently attached to the locality (Warren and McFadyen 2010; Devine-Wright 2009; Vorkinn and Riese 2001). Analysis of the meaning of locality to community energy projects has highlighted how place matters for its physical connotations as much as for the diverse interpretive frames associated with place (Devine-Wright 2009, p. 427), as local values and practices shape energy project outcomes (Pohlmann 2018; Süsser 2016). Evidence of this study suggests that references to locality feature most frequently in community oriented argumentative positions, referring to interpretive frames of locality as a physical and a social space. Locality is framed as a motivator for action, as exemplified by Søren Hermansen rephrasing the infamous proverb of connecting concepts and actions: "think locally, act locally" (Søren Hermansen 12/20/2013). Others highlight the importance of independent localities, an EWS member saying about the cooperative refraining from involvement in other community projects: "if small cooperatives manage to get things done by themselves, well, so much the better" (I2: 179). Local engagement is framed by community energy projects as an expression of shared socio-cultural values, and a shared commitment to fostering community development and developing alternative perspectives for community futures (van Veelen and Haggett 2017, p. 546). This speaks from slogans such as 'Yes to Schönau' as much as it does from the statement of one project member on Samsø that their renewable energy installation was an expression of their commitment to the island as a person not born there: "I am not just some clever guy from Copenhagen" (W1). This underscores arguments made in the literature on the relevance of locality to entrepreneurship and community development. It confirms that the concept of locality extends beyond natural resource potential for community energy (Devine-Wright 2009), to include shared values and socio-cultural identities ('we are also Schönauers and Black Foresters') connected to the locality (van Veelen and Haggett 2017, pp. 546–547; Süsser 2016, p. 99; see also Haggett and Aitken 2015; Bomberg and McEwen 2012; Hoffman and High-Pippert 2010).

This is not to suggest that alternative perspectives on community futures developed locally go uncontested. Social implications of locality also refer to the

---

[2]More recently, community energy projects are emerging wherein technological innovation is employed to realize distributed systems of energy system services or ownership of energy systems is an expression of other associational beliefs, connecting civil society actors through project ownership rather than locality (see for example the initiatives Solar Green Point in the Netherlands or Windfang in Germany, Solar Green Point 2019; Windfang eG 2011).

diversity of meanings associated with place, as indicated in the literature in reference to which aspects of locality are prioritized—e.g. environmental protection vs. economic potential–or which visions of the locality dominate in perspectives of the individual—e.g. images of locality held by residents born locally vs. those who moved there (van Veelen and Haggett 2017, p. 548; see also Kohn 2002; Vorkinn and Riese 2001). Indeed, the framing of energy system change in reference to the above outlined CORE frames can intensify contestation as values of community life beyond a specific project are put on the line (Kohn 2002). This is exemplified by the referendum in Schönau splitting the community in half, the choice of the local restaurant imbued with expressions of fundamental beliefs about what should or should not be the future of the village (see above at 4.2).

Beyond considerations of locality in the configuration of infrastructure governance, frames referencing locality also suggest changes in the role of the individual in energy system services. Aspects of consumerism and citizenship are merged, values of one expressed through actions of the other. This connects to academic debates on the "citizen/consumer paradox" (Aerts 2013, p. 172) wherein values expressed by the citizen are seemingly not reflected in economic actions of the consumer, as well as concepts of ecological or energy citizenship (Devine-Wright 2007; Dobson 2000), linking psychological engagement in infrastructure services and aspects of governance. Empirical evidence across projects supports findings of the literature that community energy engagement may promote responsibility (Frantzeskaki et al. 2013), connecting energy practices and civic duties. The considerably reduced average consumption of EWS customers, independent of their membership in the cooperative, vis-à-vis conventional household customers supports the notion that association with the project may indeed bridge disparities in attitude and behavior (Kalkbrenner and Roosen 2016). A sense of association for the individual, as the example of EWS customers indicates, does not seem to require a connection through locality or even membership. With respect to civic participation, material and argumentative participation appear linked in energy system change. Projects of the study implemented material changes to the energy system at infrastructural and individual levels. At the same time, projects provided alternative narratives for energy system design and designed energy system services accordingly. These opportunities for simultaneous public and material participation created interventions to established power structures of the energy system. Material participation carried public participation independent of membership.

Community energy projects' references to locality also connect to studies on the potential of community energy engagement to revitalize civic engagement in communities more generally (Hoffman and High-Pippert 2010, 2005). Community energy projects could refute assessments of "declining civic engagement" (Hoffman and High-Pippert 2010, p. 7572) by presenting examples of precisely such engagement, as well as mobilizing diverse members of local communities based on a narrative of locality and community (ibid.). Participatory practices and strong community welfare orientations are highlighted to effectively mobilize community support and engagement (ibid.). Locality as a driver of community energy engagement also connects to studies of trust, locality and leadership in civil society initiatives. Rogers and colleagues in their analysis of community energy projects emphasize the role of project initiators for sustained participation of members (Rogers et al. 2008, p. 4224; see also David and Schönborn 2016, p. 48). Various studies indicate that for local entrepreneurs or project leaders to gain larger support for project ideas, their embedding in local values and social networks is key (Süsser 2016, p. 100; Walker and McCarthy 2010; Walker et al. 2010a).

Resource mobilization over the course of project development indicates the importance of distributed leadership within projects. While across projects trust is articulated as an important resource to sustain engagement, this is not attributed to leadership but to procedural characteristics of the project. These procedural characteristics refer to group structures ("the group catches that and lifts them back up", I2: 190), as well as the individual commitment of group members expressed in their participation in group events ("that I took the time to be present", I9: 320) or their visible commitment to overarching targets ("solar heating was my way of saying, I mean it, I am not just some clever guy from Copenhagen" (W2)). This supports the argument for participatory structures in the literature, but moderates the relevance of localized leadership. Rather than suggesting that leadership does not matter, empirical analysis indicates that leadership must be assumed throughout project structures to sustain community energy structures: Leadership, in this sense, is not less but indeed even more relevant. This is underscored by the diversity of resources mobilized for project establishment and maintenance. Empirics support the importance of a locally embedded, trusted individual leader only in one case: Project emergence on the island of Samsø. However, this particular project had not emerged from the community. And as the project arena was created for the project but as yet void of community, the importance of someone 'that people know the parents of' featured differently, than in projects where community energy aspirations had been developed locally.

These results confirm analyses pointing to trust as an underlying, structural resource of community cooperation, reproduced in procedural aspects of project work (Litfin 2014, pp. 146–147; Walker et al. 2010b). Projects, however, also point to trust as an aspect of organizational form. Development Trusts and charitable status, in the Scottish projects, importantly garnered trust through the collectivization of benefits and the institutionalized insurance of being there 'for the right reasons'. Similarly, German projects point to the cooperative form building trust through representation and 'democratic' structures. This underscores the importance of established associative structures, not only as actual associations that can be mobilized for project goals, but as established legal instruments.

The discussion of community and locality therefore also refers to the conceptual question of how resource types interact with project conditions. In the early resource mobilization literature, various resources referring to material and non-material resource types are introduced without discrimination among them (McCarthy and Zald 2017 (1987), p. 22). More recent analyses of resource mobilization in community energy contexts have refined the various types of resources relevant to project development (Schreuer 2015), and highlighted the importance of symbolic resources specifically (Goedkoop and Devine-Wright 2016; Bomberg and McEwen 2012). However, resource types are not related to one another, or to the dynamic (re)distribution of power within energy system change. Patterns of resource mobilization have been shown to indicate how different types of resources shape project development over time and in response to challenges, and how these correspond to frame resonance and respective distribution of power in arenas of negotiation relevant to project formation. Interrelations between resource types and their ties to organizational innovation biographies therefore also connect to a more differentiated understanding of community and locality in energy system change.

Empirical analysis has indicated how community energy projects may form "communities of relevance" (Batel 2017, p. 357), affected by energy system change while relatively powerless at the infrastructural levels of decision-making governing respective system (changes). Empirical evidence also indicates that this may refer to innovations of the energy system, introducing new technologies or practices of energy production or consumption, as much as to exnovations, wherein installations are reversed (Heyen 2017). As discussed, relative power positions within different arenas of negotiations may moderate the potential for communities of relevance to overcome a position of simultaneous affectedness and unaffectedness, as suggested by Batel (ibid.). In addition, projects have been shown to form communities of intention, superseding locality in their establishment of community. The most prominent example in the case lineup is EWS, wherein members bound by a shared goal and belief in procedural qualities are enacting energy system change in material and decision-making structures without sharing geographical proximity. The cooperative as an organizational focus point

has replaced the geographical rallying point typical of community energy engagement. This is not shared across projects. Fintry, for example, has an exclusively
local focus in both membership and project implementation, based on its organizational structure as a Development Trust. Both on Mull and on Samsø, opening
membership to non-locals was initially an expression of support for the island
despite geographic distance. On Samsø, project members have since attempted
to establish membership structures beyond the local. It has proven more difficult
than in the case of Schönau, however, as opportunities of remote material participation in project targets are lacking. The intentional linkages forged between roles
of the customer and the citizen by EWS may point to a key difference between
projects in this respect. While material and public participation are collectivized
in similar ways within the German cooperative, collective mechanisms of public
participation in the Danish case are combined with individualized opportunities
for material participation. Community energy projects may combine material and
public participation. But whether these align in mutual expressions of one another
depends on organizational and procedural aspects.

The discussion of CORE frames and their conceptual implications further
underscores the limited utility of the concept of "communities of the affected"
(Batel 2017, p. 357). Project members did not define their participation in measures energy system change based on immediate adverse effects of the current
energy system. EWS founders were largely unaffected by the nuclear devastation
of Chernobyl which instigated their activities. It was not an immediate material
effect of energy system configuration on the respective communities that spurred
project development.

Analysis also confirms that discursive positions are more diverse across
community energy projects than suggested by niche management or challenger-
incumbent paradigms. This not only affects mobilization of resources for project
realization locally, but also shapes projects' interactions with the actors, policies or regulation, and institutions of their respective environments. Arenas of
negotiation show that project narratives can importantly shape power relations by
determining how a project interacts, and with whom. Community energy projects
will not necessarily act primarily within energy-related arenas, and indeed connections to other arenas can improve their relative positions of power vis-à-vis other
actors in the energy sector. Mapping projects according to arenas of negotiation
therefore does not give a new name to the niche concept, setting projects within
a project arena (=niche) and then conceptualizing its interactions with landscape
and regime actors accordingly, but instead offers a view of power relations within
the research situation that transcends mechanisms of niche/regime interactions.
In her review of situational analysis as a counterpoint to multi-level analyses

of grassroots innovations, Angela Pohlmann (forthcoming) suggests that Clarke's method enables an understanding of the more diverse interactions of niche, landscape and regime. This study confirms and extends these arguments, by showing how relative positions of powers within the various arenas of negotiation that projects engage in, as well as the transfer of resources between said arenas affects the shape of energy system change achieved within community energy projects. This corresponds to Ziegler's concept of 'ecological niches' for community organization wherein innovations form contingent to their surroundings, and are oriented towards maintaining or adjusting their position, rather than upscaling or extending it.

12.2 Stories that Stick: Changing how People Think and Feel about Energy Systems

The discussion of community definitions underpinning the study of energy engagement has related empirical findings to academic debates of participation and citizenship. Community definitions refer to dimensions of locality, affectedness, and intention based on shared stories of energy system change that resonate with members' everyday lives. The literature refers to the establishment of shared frames that remain viable over time as "sticky stories" (van der Stoep 2014, p. 182), based on the concept of 'sticky messages' (Heath and Heath 2007). Frame resonance increases the ability to mobilize resources on behalf of a story, building attention and coalitions in its favor (van Dijk 2011, p. 138). Sticky stories increase their agenda setting capacity by providing opportunities for "listeners or readers [to] stick their own experiences and frames to the story" (van der Stoep 2014, p. 182; Baker 2010). Frame resonance therefore refers not only to the credibility and salience of the frame in addition to the credibility of those advocating in its favor, but also to its ability to incorporate others' stories and to provide an overarching narrative.

The literature offers various interpretations of what creates stickiness, focusing on the inherent qualities of a story (Heath and Heath 2007), a story's ability to interconnect with existing stories (van Dijk 2011; Baker 2010) and the ways by which stories are embedded in processes of joint storytelling (van der Stoep 2014, pp. 185–186; Baker 2010). In the context of community development, authors highlight the importance of informal and formal conversations for building frame resonance (van der Stoep 2014, p. 186; Baker 2010; Huitema and Meijerink 2010). Events or external opportunities are also referenced as potentially beneficial to enhancing the persuasiveness of a story. This is particularly true for authors emphasizing the importance of policy entrepreneurs or storytellers (Levin 2005, pp. 84–85; Chong 2000, 8–9, 116–117; Huitema and Meijerink 2010). Evidence of this study especially emphasizes frame resonance, connections to existing stories, and processes of storytelling.

Links between energy system change and established narratives in the community could either be forged by energy stories being told as stories of community welfare and development (for example in Schönau or on Mull), or stories of community welfare being opened and re-interpreted to encompass energy stories (for example on Samsø). In both configurations, frame resonance depended on projects' references to CORE frames. The struggle of olegeno to mobilize resources in its favor could be interpreted in reference to internal aspects of the story. The literature points to the importance for a sticky story that it be simple, have an element of surprise and be persuasive (Heath and Heath 2007). It should also articulate "clear and stable ambitions" (van der Stoep 2014, p. 182). Not all communities achieved this. olegeno struggled to build a clear storyline about why the cooperative application was necessary to realize CORE frames in energy system change, as opposed to relying on municipal authorities and the energy incumbent. Resource mobilization relied, instead, on knowledge rather than shared symbols of community, ownership, or responsibility—both as a tool in negotiations of the local energy arena, and as a resource to mobilize support within the project arena. Case evidence suggests that stories of energy system change that failed to credibly build a connection to CORE frames did not stick.

Informal and formal conversations are emphasized by various authors explaining why some stories stick more than others. Both van der Stoep (2014) and Baker (2010) point to the importance of conversations as a story-building tool, wherein frames are told and retold and empathetic listening enables integration of established frames into the overarching narrative (see exemplary van der Stoep 2014, p. 185). Van der Stoep highlights the role of informal conversations both within community contexts and between policy makers in building stories supportive of community development and for aligning community project stories with policy agenda setting (2014, pp. 185–186).

Across projects, the importance of conversation and negotiation is emphasized during project emergence. The story of Samsø islanders talking themselves into consensus may be the best example. The founding ad of EWS which showed parents seeking comrades-in-arms who wanted to resist nuclear technologies despite not knowing how to go about this, similarly points to the importance of coming together to develop ideas and strategies to overcome a perceived threat. Building on the storytelling literature, this study shows the continued relevance of these negotiations as projects seek to maintain their positions in community development. Fintry project members point to financial resources continuing to mobilize negotiations for new activities, but are quick to qualify this by pointing to the importance of project actions' continued relevance to community members. Organizational institutionalization of this relevance, for example through charitable status, can hedge against a loss of community trust. Still, project organizers also highlight the importance of continued processes of negotiation. An example of such negotiations during project maintenance is the reduction of members' dividends in Schönau, which was agreed upon only after intense discussions. Meaningful conversations negotiating frame resonance and creating a sticky story require careful mobilization of organizational, symbolic and knowledge resources. The image of talking over coffee to determine energy system change became so dominant in stories of Samsø's transition process that Energy Academy members have since grown weary of it. Some even fear, it might create an overly simplified image of the negotiations required to attain project goals (Papazu 2018).

olegeno's story failed to achieve stickiness within the project arena. The project did, however, manage to foster negotiations on what this story should be and this has ensured the continued existence of the project to date. The difficulty of maintaining such negotiations is underscored by the assessment of the diversity and scope of resource mobilization required for project establishment and maintenance. It also speaks from the literature cautioning against a streamlining of community project development, neglecting how winded the path to project success may indeed be (Papazu 2018; Pohlmann 2018).

Evidence from the cases of this study does not support the notion that there was a strong role for rational policy or norm entrepreneurs as is suggested in the literature (Levin 2005; Chong 2000). Such approaches highlight the importance of individuals who can proactively forge links between project frames and so-called master frames (Snow and Benford 1992, p. 134), collective identities and belief systems, because of their position within the movement. Such entrepreneurs create change by connecting institutions, events and narratives in meaningful ways (ibid.). While individual leadership was referenced within projects of this case study, elevated positions of norm entrepreneurs were not created. Proactive attempts to connect frames of the community energy project to existing narratives were important. But processes of negotiation for frame resonance, and the importance of distributed leadership for resource mobilization and effective storytelling throughout the project arena were emphasized. The argument for what more generally has been called "connectors" (van der Stoep 2014, p. 191) between project specific frames and master frames relates to the general argument underlying agenda-setting research within the multiple streams (Kingdon 1984), or punctuated equilibrium models (Baumgartner and Jones 1993). While institutions or events may make something possible, it is people that make things happen. The argument in favor of policy entrepreneurs often relies on case evidence wherein social movements attempted policy change, or at least required policy change in reference to aspired actions (van der Stoep 2014; Levin 2005; Chong 2000). Analysis of community energy projects in this study confirms results of the community energy literature which emphasize connectors that forge links between institutions, events, existing narratives and project-related frames, but also highlights that there were many different kinds of connectors and they held diverse roles within the project (van der Stoep 2014, pp. 191–193).

The ability to forge these links refers to various connector qualities, including the power to credibly tell project stories or identify other credible "frame articulators" (Benford and Snow 2000, p. 620), to empathetically listen to stories within negotiations for frame resonance and align diverging perspectives, and to identify institutions or events potentially creating opportunities for aligning actors and frames (van der Stoep 2014, pp. 191–193). Connectors, hereby, matter not only for their ability to credibly tell a story, but also for their strong interpersonal networks, and their ability to identify additional storytellers who themselves have strong interpersonal networks (ibid.). Samsø project development, arguably presents a case in point, as community members were explicitly invited into project negotiations with a view to their personal connections to existing organizations and associations on the island. Van der Stoep in her assessment of conditions for stickiness points to connectors and conversations as key enabling conditions, complemented by focusing events (2014, p. 208). The integration of external events, made meaningful to storification through connectors' conversations, formed the steady drip necessary to gradually fill the cup and ultimately create a point of spillover and change (van der Stoep 2014, p. 206). Patterns of resource mobilization suggest that the integration of such events into processes of building frame resonance occurred through projects returning to symbolic resources mobilized upon initiation, hereby creating a self-referential cycle reproducing and reinforcing previously established shared frames. It is therefore indeed people that make something happen, even more so when in conversation.

12.3 Biographies of Social Innovation

Empirical results suggest that social innovations seem to develop in distinct stages. Material and non-material resources are mobilized throughout development stages, although non-material resources dominate upon project emergence, while especially financial resources emerge as more important factors of project work during establishment and maintenance. Social innovations seem to commence as value-oriented interventions to networks of actors, institutions and frames; but following up on these interventions requires additional resources. Key dimensions of initial value orientation, however, remain relevant throughout the projects' biographies, determining deployment of non-material resources. This suggests a consolidated assessment of innovation biographies across social and technological developments.

In their account of innovation biographies of renewable energy technologies in Germany, Bruns and colleagues also refer to different development stages characterized by distinct constellations of technical elements, actors, signs and symbols, and natural elements, as well as the respective interrelations of these elements (Bruns et al. 2011, p. 40; Bruns 2011a, 2011b). Analyzing the development of various technologies of renewable energy generation in Germany, the authors conclude that all technologies undergo pioneering or pilot stages, wherein the technology is launched and established; periods of consolidation; and dynamic and instable phases of crisis which serve as transition phases to achieve new stages of consolidation (Bruns et al. 2011, p. 40). While these stages are observable across technologies, the authors argue that individual innovation biographies vary considerably. This suggests strong similarities between the innovation biographies of renewable energy technologies and those of community energy organizations.

Two differences stand out. First, crises in technology innovations are not the equivalent of challenges to community energy projects. Challenges are not transition periods indicating new consolidation, but can also result in a return to activities of emergence or other stages of project work. Also, while challenges may create moments of instability initially, mobilization patterns show projects returning to "tried and tested" resources, cutting down on all other, potentially still unstable activities. This further underscores the effect of challenges reinforcing and ultimately stabilizing frame resonance described in reference to storification. Secondly, innovation biographies of community energy projects indicate a surge of resource mobilization of all types for emergence that apparently does not occur in technologies' biographies.

Does this suggest a distinct innovation biography of social innovations vis-à-vis technological innovations? No. Rather than modelling biographies of social innovations in general, this study proposes three arguments to be made on the quality of sectoral change based on social innovations. The first two were explored in sections one and two of this chapter. For one, biographies of community energy projects as social organizations provide a more nuanced understanding of sectoral change, and the roles of communities and citizens within processes of change.

Communities of intention create meaningful alliances for sectoral change independent of locality, and extend the power base of bottom up actions by merging material and public participation. Secondly, frames and frame resonance are central resources of social innovations that are produced and reinforced in on-going processes of negotiation within arenas relevant to project actions. The ability to tell these stories and empathize with others in order to forge interconnections between frames is a leadership quality enacted by diverse members of the negotiating arena. Social innovations therefore unfold in interactions, wherein leadership is distributed and must be assumed by many.

This leads to the third argument on social innovations and sector change, and the main theoretical contribution of this study. Social innovations speak of changes in power relations and capabilities. Frames and frame resonance focus on the ability of actors to change how people think and feel about something. Community energy projects, by connecting energy system change to CORE frames of community welfare, changed how people thought and felt about energy systems, mobilizing resources and altering power relations in their favor in the process. Changed power relations did not necessarily affect energy sector design immediately, but could also be realized in alternative arenas of negotiation reconfigured to connect to energy system design. Argumentative positions and arenas of negotiation (frames and frame resonance) therefore offer a conceptually more nuanced understanding of how social innovations intervene in existing relations of actor networks, institutions and interpretive frames.

Do certain argumentative positions matter more than others in this respect? Analyses of community projects conclude that projects must forge links to "wider interests and commitments" to gain and sustain influence in community decision-making (Smith and Stirling 2017, p. 76). Indeed, CORE frames of community energy projects did not originally refer to energy-specific aspects at all. This connects the discussion of social innovations for energy system change to academic debates on the conditions for changes in power relations and capabilities initiated from the bottom up more generally. Van Oers and colleagues in an assessment of community gardening organizations (2018), attribute the ability of projects to mobilize resources successfully to the perceived legitimacy of their claims, measured as the alignment of their targets with socially established values (ibid. p. 56). While community energy projects at first glance rely on similar mechanisms, van Oers and colleagues conclude in reference to community gardening that initiatives rely predominantly on the creation of social capital to sustain their activities (van Oers et al. 2018, p. 65). "Marked by reciprocity, trust and cooperation" (ibid.), cooperation sustains high levels of legitimacy within the

group, but risks exclusion of others beyond the circle of those immediately engaged. Alignment of project frames and shared narratives relating to CORE frames
in community energy projects could also be interpreted in reference to the creation
of legitimacy as a linchpin resource to subsequently build social capital.

However, analysis of frame resonance in arenas of negotiation has underscored
the relevance of altering power relations in the process to obtain and sustain access
to the energy system and subsequently provide alternative services. The combination of material and public participation realized within community energy
settings sustained interrelations between energy system performance and CORE
values in project performance. This combination of material and public participation in combination with changed power relations within social innovations
has been interpreted as an emancipatory experimentalism (Ziegler 2017a, p. 96).
Drawing on the analysis of community water projects in Bavaria, Ziegler argues
that the experimental realization of alternative systems of water governance constituted a case of emancipation from established systems of water governance
on community levels (ibid.). Citizens were empowered to claim delegation of
localized services to citizen-led structures based on critical assessment of civic
capacities (ibid.; drawing on Fricker 2017).

These considerations suggest the following conclusion as one answer to the
questions of who gets what, when and how, in energy system change. The ability
of community energy projects to provide distinct linguistic images of community-
oriented energy system change has created cognitive models wherein new actors
and services were established in the energy sector. Power to create frame resonance for energy services was redistributed, and new alliances were forged. By
attempting to tell stories that would stick, projects invoked feedback loops of
symbolic and organizational resources that reinforced linguistic images of energy
system changes as much as the distributed structures of leadership required for
reproducing them. Social innovations in community energy can therefore shift
both the stories told of energy system change, and the power to own and tell
these stories.

12.4   Fostering Social Innovation: What Role for Policy and Regulation?

The discussion of who gets what, when, and how in community energy contexts concludes with an assessment of the potential to foster social innovations.
In referring to governance of social innovations, this section will reflect on conditions for social innovation both within the regulatory environments of community

projects, and within their community settings. This is not a prescription for successful social innovations. Drawing on the analysis of resources mobilized by projects and their origins within projects and their socio-cultural and regulatory environments, this assessment of stepping and stumbling stones to community energy projects in three European countries rather serves as a contribution to an on-going debate on the potential to foster innovations in socio-technical systems.

Bruns and colleagues argue in their analysis of innovation biographies of renewable energy technologies that innovation processes are determined by a complex interplay of technological, economic and societal factors (Bruns et al. 2011, pp. 40–41). Regulatory factors, while central in their role in this development, could not be singled out with respect to their effects on innovation biographies. Yet, while regulation could not initiate innovation biographies, the authors conclude it could importantly shape their trajectories (ibid.). This is confirmed by the analysis of the role of regulation for innovation biographies of community energy projects within this study. Patterns of resource mobilization have indicated that material, and especially financial resources, are less relevant to project initiation than assumed by analyses of community energy sector development by country. The vibrant community energy sectors of Germany and Denmark, for example, are frequently assigned to the availability of financial support for community energy actors (Seyfang et al. 2013; Brickmann et al. 2012; Boomsma et al. 2012; Lewis and Wiser 2007; Hinshelwood 2001). Yet, with the exception of one project within the case selection, projects were not initiated upon a policy instrument providing financial security. Instead, it was the ability to tell a story that would stick, finding storytellers committed to key frames of the project, and combining these with stories and experiences of their own, that built a group of supporters large and diverse enough to provide necessary knowledge and social networks to sustain energy system services. But while stories of community may correspond to place and culture, analysis also showed that CORE frames were shared across all stories. The literature also points to the importance of storytelling abilities. This resonates with the analysis of resource mobilization, which points to organizational and knowledge resources as key factors structural resources could provide during emergence. Social innovations arise out of community organizing for combinations of organizational, financial and technological solutions that would fit local values of energy system change. Policy support can contribute importantly to this process by not prescribing technological or organizational solutions, but supporting the organizational structures developing these solutions locally. One example is the creation of Community Energy Scotland as a source of knowledge and organizational resources to community energy projects,

or the Scottish Land Reform Act creating a narrative of community development potential even in cases where community ownership of land was not central
to project realization (van Veelen and Haggett 2017). The Samsø Energy Academy is an example where financial support for a moderator importantly fostered
the development of local structures. The call for public administration to provide knowledge resources and consultancy on the establishment of networks is
echoed by assessments of community-led innovations in other sectors such as
water (Ziegler 2017a, pp. 116–117).

Technology innovation studies point to the importance of considering development stages in designing regulatory support for the development of innovations.
While this could refer to the identification and support of processes of innovation
in pioneering and pilot stages, the literature also indicates that support during
later stages increasingly branches out to system integration (Bruns et al. 2011,
pp. 493–497). Similarly, development of community energy projects indicates that
the diversity and extent of resource mobilization increases over the early stages of
project development, and remains consistently high to maintain and extend project activities over time. Potential regulatory measures fostering social innovation
in this respect do not refer to the provision of extensive and diverse structural
resources. Rather, regulatory action could address stumbling blocks to community mobilization of resources. This refers to the acknowledgement of uneven
power relations within the sector. Rather than suggesting niche-regime mechanisms wherein community energy actors are shielded from system incumbents,
regulatory recommendations point to fostering system integration by acknowledging and supporting the transfer of power resources between arenas of negotiation.
This could refer, for example, to the incorporation of frames consistently phrased
by community actors in energy system regulation, such as in the Danish requirement of community involvement based on locality. It could also translate to the
equal consideration of regulatory requirements in energy industry law, referring to
the technological, as well as organizational and financial requirements for energy
system design.

A key challenge with respect to system integration of social innovations when
considering the mechanisms of successful framing of energy system services lies
in the socio-cultural differences between energy sectors and community actors
observed in case studies. The energy sector has repeatedly been characterized
by its strong path dependencies and institutional stickiness. By contrast, development of social innovations is shaped by the ability to build stories that stick
and provide meaningful connections to arenas of negotiation. The development of
frame resonance, in turn, is shaped importantly by informal instruments of communication and exchange. This suggests two approaches of policy and regulation

for community energy. The example of ACCESS has shown the potential for a public moderator, translating between energy sector incumbents and new market entrants and moderating differences in organizational and communicative culture. This further supports introduction of a platform service provider, offering knowledge and organizational resources, as a policy instrument for fostering social innovations. It also points to the importance of creating new forums of exchange between actors of the energy system, including public actors and regulators, as these are embedded in the institutional stickiness of the energy sector. This also holds implications for regulatory development, as the configuration of regulatory requirements can run the risk of endangering social innovations 'by accident'.[3]

Comparing innovation biographies can indicate opportunities and limitations for policy planning and regulatory control. The importance of frame resonance and non-material resources for community energy projects underscore the limited potential for policy planning for social innovation. A biographies perspective instead points to the regulatory moderation of related processes, and the fostering of social innovations while acknowledging limited influence on the direction of their development. This is supported by the literature on technological innovations (Bruns et al. 2011, p. 493). Strategies for policy instruments must instead acknowledge the diversity of actors involved in respective innovation biographies (ibid., p. 475), and informal modes of interaction shaping their stories (van der Stoep 2014, p. 186).

The comparison of technological and organizational innovation biographies, lastly, poses the question of whether organizational innovation biographies can account for variation in the complex socio-technical dynamics embedding technology development. Bruns and colleagues in their analysis of innovation biographies of diverse renewable energy technologies point to the creative environment of civil society initiatives as one of the key drivers of innovation (Bruns et al. 2011, p. 474). The authors highlight that within the German deployment of renewable energy technologies, technological innovations developed in response to visions and narratives for energy system services shared by community organizations implementing these technologies (ibid.). This corresponds to the analysis of resource mobilization, wherein natural and technological resources were mobilized in reference to established interpretive frames, not vice versa.

---

[3]For example, EU induced reform of capital investment laws in Germany (Kapitalanlagegesetzbuch) almost created financial regulation that would have exceeded organizational capacities of cooperatives to the extent that many would have had to be resolved, although small scale cooperatives where not the target of regulation on financial security (Colell and Neumann-Cosel 2016).

Conclusion

13

The overarching research question of this study was how can social innovations in community energy projects explain energy system change in the electricity sector? Energy system change towards more sustainable infrastructures involves both a shift in the resource base and related ways of generating and consuming power. But it also involves new organizational paradigms for public utilities referring to their ownership and operations. Analysis has shown that the concept of social innovations offers insights into the changes to underlying values of the energy system. These interpretive frames guide the experience and performance of energy services, as well as corresponding structures governing decision-making and service provision in the energy system. By reframing energy infrastructures corresponding to shared values of community, ownership, responsibility and empowerment, this study has shown that community energy projects give rise to distinct forms of energy system change. Focusing on innovation biographies of an organizational form offers insights into how technological innovations feature in socio-cultural developments, rather than choosing to understand socio-cultural features as the background to technological development.

Social innovations refer to intentional changes in power relations and the improvement of individual and community capabilities in the process (Nicholls and Ziegler 2017(2015)). Understanding community energy projects in these terms refers the study of community energy actors to concepts of power in the political sciences, such as coercion, agenda setting or framing. In a dynamic and relational understanding of power, this study used resource mobilization approaches to understand how community energy actors gained access to physical power grids and infrastructures of decision-making by reframing 'energy'. This enabled on-par analysis of material and non-material resource types, focusing attention on the relative relevance of financial, natural, and technological resources (material), and organizational, knowledge, and symbolic resources (non-material). Situational analysis maps were used to visualize and understand relationships between actors, institutions and arguments within each research situation, as well as arenas of negotiation and argumentative positions assumed or neglected in the context of each community energy project.

Over the course of project development, patterns of resource mobilization emerged in reference to the distinct innovation biographies of each case. Analysis showed that community energy projects were similar in the resource types they required during different stages of project development, but unique with respect to the configuration of resources within these types. What could be accessed and mobilized as a resource depended on the individual innovation biographies of each case. Comparison of community energy projects served the analysis of conditions for social innovations. The importance of non-material resources stood out, dominating resource mobilization during project emergence as well as in response to challenges. Successful establishment of the project, as well as successful assertion, was characterized by feedback-loops of symbolic and organizational resources. The stories projects told, frames and narratives, importantly determine how local energy system alternatives were set within larger trends of energy system change, or community development. This also determined the arenas of negotiation that projects engaged in, including the power relations they entailed. Based on this narrative, time and labor of project members and supporters could be mobilized, which in turn reinforced shared symbols of the narrative. Storytelling moreover enabled community energy actors to transfer resources between arenas, connecting previously unrelated argumentative positions to energy system change. Social innovation is therefore a useful conceptual approach to understand community energy projects with respect to the changes to the way people think and feel about energy system services, corresponding changes in decision-making infrastructures. In maintaining project activities, material resources gained prominence. Notably, material resources assumed a role within the feedback loop of organizational and symbolic resources, providing the backdrop against which the initial momentum of stories and storytellers could be sustained. Situational analysis mapping proved a powerful tool to visualize these interrelations and power shifts, while recognizing actors, institutions, installations and arguments as equals in the research situation.

The main theoretical argument made in this study lies in the combination of frames and framing as concepts of power to the study of social innovations. Social innovations focus on interventions to the relations between social networks, institutions and interpretive frames. Frames and arenas offer new perspectives on how changes in the way people think and feel about an issue change power relations and create such interventions.

The "transformative potential" of community energy organization is widely acknowledged as a corner stone of sustainability transitions (WBGU 2014, 98, 107). This study emphasizes that transformative potential refers not only to changes in how energy is generated or consumed, but also to the conceptual paradigms governing how energy infrastructures are designed and owned, and the services they perform. Community energy projects can draw power from political and cultural narratives previously unrelated to energy system design, and connect energy system design to respective public debates. This creates the potential for interactions between arguments, but also between social spheres previously neglected. Policy and regulation cannot initiate these kinds of social innovations, but it provides key frameworks for their development. This refers predominantly to the provision of organizational resources in policy and regulatory contexts.

This study has emphasized biographies of social innovations in the transformative processes underpinning infrastructure change in reference to sustainable development. It has shown the potential for comparison of project rather than country cases, pointing to shared conditions of social innovations across differences in the regulatory and economic context of community engagement. Questions remain. These refer to dynamics of community engagement, such as the relative importance of narrative and personal leadership. These also include contextual dynamics. By setting energy services in policy arenas previously unencumbered by related debates, community energy projects could mobilize resources in their favor which otherwise would not have been available. The same holds true for policy makers, seeking to set energy policy making in alternative arenas to surpass economic path dependencies, such as the dominance and related lobbying power of market incumbents, or to outpace political competition or regulatory power imbalances within multi-level governance systems.

Social innovations in community energy projects also touched upon the multiple vulnerabilities of community livelihoods. While some challenges could be answered within community projects, others prevail. These questions are augmented in communities where socio-economic vulnerabilities exceed the European context and are potentially exacerbated by vulnerabilities of environmental degradation and climate change. Future research should therefore not only determine how conditions for social innovations shape the trajectories of energy system change and what opportunities and pitfalls this includes for regulatory strategies beyond the European context. It could also provide answers to how social innovations could give rise to energy systems answering to the environmental, economic and social dimensions of sustainability transitions, and reconciling opportunities for material and public participation as communities establish the infrastructures of their daily livelihoods.

As I book my ticket for the ferry to Samsø in the summer of 2019, I have learned much about exceptions. It might not be about finding the parameters wherein exceptions with extraordinary results become the rule. Rather, it is about defining the rules that deliver exceptions.

Bibliography

ACCESS (2015): About ACCESS. Assisting Communities to Connect to Electric Sustainable Sources. Available online at http://www.accessproject.org.uk/, checked on 3/12/2016.

Aerts, Stef (2013): The consumer does not exist: overcoming the citizen/consumer paradox by shifting focus. In : The ethics of consumption. The citizen, the market and the law: EurSafe 2013, Uppsala, Sweden, 11–14 September 2013 / edited by: Helena Röcklinsberg, Pers Sandin. Wageningen, The Netherlands, pp. 172–176.

Agora Energiewende (2017): Energiewende und Dezentralität. Zu den Grundlagen einer politisierten Debatte. Berlin. Available online at https://www.agora-energiewende. de/fileadmin2/Projekte/2016/Dezentralitaet/Agora_Dezentralitaet_WEB.pdf, checked on 3/21/2019.

Agterbosch, Susanne; Vermeulen, Walter; Glasbergen, Pieter (2004): Implementation of wind energy in the Netherlands: the importance of the social–institutional setting. In *Energy Policy* 32 (18), pp. 2049–2066. https://doi.org/10.1016/s0301-4215(03)00180-0.

Aitken, Mhairi (2010): Why we still don't understand the social aspects of wind power: A critique of key assumptions within the literature. In *Energy Policy* 38 (4), pp. 1834–1841. https://doi.org/10.1016/j.enpol.2009.11.060.

Aitken, Mhairi (2010): Wind power and community benefits: Challenges and opportunities. In *Energy Policy* 38 (10), pp. 6066–6075. https://doi.org/10.1016/j.enpol.2010.05.062.

Ajzen, Icek (1991): The theory of planned behavior. In *Organizational Behavior and Human Decision Processes* 50 (2), pp. 179–211. https://doi.org/10.1016/0749-5978(91)90020-t.

Alexander, Jeffrey A.; Comfort, Maureen E.; Weiner, Bryan J.; Bogue, Richard (2001): Leadership in Collaborative Community Health Partnerships. In *Nonprofit Management Leadership* 12 (2), pp. 159–175. https://doi.org/10.1002/nml.12203.

Alisat, Susan; Riemer, Manuel (2015): The environmental action scale: Development and psychometric evaluation. In *Journal of Environmental Psychology* 43, pp. 13–23. https://doi.org/10.1016/j.jenvp.2015.05.006.

Anderson, J.A. (Ed.) (1988): Communication yearbook. 11th ed. Newbury Park, CA: Sage.

Ankjaergaard, Line (2017): Newsletter No. 15: Without Samsö - no EU? Samsö. Available online at https://arkiv.energiakademiet.dk/en/newsletters/nyhedsbrev-nr-15/, updated on 7/17/2017, checked on 3/28/2019.

Araújo, Kathleen (2014): The emerging field of energy transitions: Progress, challenges, and opportunities. In *Energy Research & Social Science* 1, pp. 112–121. https://doi.org/10.1016/j.erss.2014.03.002.

Arbeitsgruppe Bielefelder Soziologen (Ed.) (1973): Alltagswissen, Interaktion und gesellschaftliche Wirklichkeit. 1 volume. Reinbek.

Arthur, W. Brian (1994): Increasing Returns and Path Dependence in the Economy. Ann Arbor, MI: University of Michigan Press.

Avelino, Flor (2011): Power in transition. Empowering discourses on sustainability transitions. PhD, Rotterdam, Netherlands. Available online at https://repub.eur.nl/pub/30663/PhD%20Thesis%20Flor%20Avelino.pdf, checked on 4/13/2015.

Avelino, Flor (2015): Transitions towards New Economies? A Transformative Social Innovation Perspective (TRANSIT Working Paper, 4). Available online at https://core.ac.uk/download/pdf/46523320.pdf, checked on 4/26/2019.

Avelino, Flor; Wittmayer, Julia M. (2016): Shifting Power Relations in Sustainability Transitions: A Multi-actor Perspective. In *Journal of Environmental Policy & Planning* 18 (5), pp. 628–649. https://doi.org/10.1080/1523908x.2015.1112259.

Bachrach, Peter; Baratz, Morton S. (1962): Two Faces of Power. In *Am Polit Sci Rev* 56 (04), pp. 947–952. https://doi.org/10.2307/1952796.

Land Baden-Württemberg (11/19/1953): Verfassung des Landes Baden-Württemberg, revised 2015. In *Gesetzblatt für Baden-Württemberg* 29, p. 173. Available online at https://www.lpb-bw.de/bwverf/bwverf.htm, checked on 3/21/2019.

Baker, Ann C. (2010): Catalytic Conversations. London: Routledge.

Bakir, Vian (2006): Policy Agenda Setting and Risk Communication. In *Harvard International Journal of Press/Politics* 11 (3), pp. 67–88. https://doi.org/10.1177/1081180x06289213.

Bamberg, Sebastian (2003): How does environmental concern influence specific environmentally related behaviors? A new answer to an old question. In *Journal of Environmental Psychology* 23 (1), pp. 21–32. https://doi.org/10.1016/s0272-4944(02)00078-6.

Barry, John; Ellis, Geraint (2010): Beyond Consensus? Agonism, Republicanism and a Low Carbon Future. In Patrick Devine-Wright (Ed.): Renewable Energy and the Public: From NIMBY to Participation. London: Earthscan, pp. 29–42.

Barry, Martin; Chapman, Ralph (2009): Distributed small-scale wind in New Zealand: Advantages, barriers and policy support instruments. In *Energy Policy* 37 (9), pp. 3358–3369. https://doi.org/10.1016/j.enpol.2009.01.006.

Batel, Susana (2017): A critical discussion of research on the social acceptance of renewable energy generation and associated infrastructures and an agenda for the future. In *Journal of Environmental Policy & Planning* 20 (3), pp. 356–369. https://doi.org/10.1080/1523908x.2017.1417120.

Batel, Susana; Devine-Wright, Patrick (2017): Energy Colonialism and the Role of the Global in Local Responses to New Energy Infrastructures in the UK. A Critical and Exploratory Empirical Analysis. In *Antipode* 49 (1), pp. 3–22. https://doi.org/10.1111/anti.12261.

Batel, Susana; Devine-Wright, Patrick; Tangeland, Torvald (2013): Social acceptance of low carbon energy and associated infrastructures: A critical discussion. In *Energy Policy* 58, pp. 1–5. https://doi.org/10.1016/j.enpol.2013.03.018.

Batley, S.L.; Colbourne, D.; Fleming, P.D; Urwin, P. (2001): Citizen versus consumer: challenges in the UK green power market. In *Energy Policy* 29 (6), pp. 479–487. https://doi.org/10.1016/s0301-4215(00)00142-7.

Baumgartner, Frank; Jones, Bryan (1993): Agendas and Instability in American Politics. Chicago, Ill.: University of Chicago Press.

Baumgartner, Frank R. (1989): Independent and Politicized Policy Communities: Education and Nuclear Energy in France. In *Governance: An International Journal of Policy and Administration* 2 (1), pp. 42–66.

Baur, Nina; Blasius, Jörg (Eds.) (2014): Handbuch Methoden der empirischen Sozialforschung. Wiesbaden. Available online at http://dx.doi.org/10.1007/978-3-531-189 39-0.

Bauwens, Thomas (2016): Explaining the diversity of motivations behind community renewable energy. In *Energy Policy* 93, pp. 278–290. https://doi.org/10.1016/j.enpol.2016. 03.017.

Bauwens, Thomas; Gotchev, Boris; Holstenkamp, Lars (2016): What drives the development of community energy in Europe? The case of wind power cooperatives. In *Energy Research & Social Science* 13, pp. 136–147. https://doi.org/10.1016/j.erss.2015.12.016.

BDEW Bund der Deutschen Energie- und Wasserwirtschaft (2019): Anzahl der Unternehmen am Energiemarkt in Deutschland nach Bereichen im Jahr 2018. Available online at https://de.statista.com/statistik/daten/studie/173884/umfrage/zahl-der-untern ehmen-in-den-einzelnen-marktbereichen-des-energiemarktes/, checked on 3/27/2019.

BDEW Bund der Deutschen Energie- und Wasserwirtschaft (2012): Energiemarkt Deutschland. Zahlen und Fakten zur Gas-, Strom- und Fernwärmeversorgung. Frankfurt am Main (ISBN 978-3-8022-1103-4). Available online at http://docs.dpaq.de/2436-energie-markt_2012d_web.pdf, checked on 3/27/2019.

Beck, Gerald; Kropp, Cordula (Eds.) (2012): Gesellschaft innovativ. Wiesbaden: Springer.

Becker, Peter; Templin, Wolf (2013): Missbräuchliches Verhalten von Netzbetreibern bei Konzessionierungsverfahren udn Netzübernahmen nach §§ 30, 32 EnWG. In *Zeitschrift für Neues Energierecht ZNER* 1, pp. 10–18.

Becker, Sören; Kunze, Conrad; Vancea, Mihaela (2017): Community energy and social entrepreneurship. Addressing purpose, organisation and embeddedness of renewable energy projects. In *Journal of Cleaner Production* 147, pp. 25–36. https://doi.org/10.1016/j.jcl epro.2017.01.048.

Becker, Sören; Gailing, Ludger; Naumann, Matthias (2013): Die Akteure der neuen Energielandschaften - Das Beispiel Brandenburg. In Ludger Gailing, Markus Leibenath (Eds.): Neue Energielandschaften - Neue Perspektiven der Landschaftsforschung. Wiesbaden: Springer VS. pp. 19–31.

Beckert, Jens (2010): How Do Fields Change? The Interrelations of Institutions, Networks, and Cognition in the Dynamics of Markets. In *Organization Studies* 31 (5), pp. 605–627. https://doi.org/10.1177/0170840610372184.

Beisheim, Marianne; Weizsäcker, Ernst Ulrich von; Young, Oran R.; Finger, Matthias (2005): Limits to privatization. How to avoid too much of a good thing. Sterling, VA. Available online at http://www.netLibrary.com/urlapi.asp?action=summary&v=1& bookid=126162.

Benford, Robert D.; Snow, David A. (2000): Framing Processes and Social Movements: An Overview and Assessment. In *Annual Review of Sociology* 26, pp. 611–639.

Bénit-Gbaffou, Claire; Katsaura, Obvious (2014): Community Leadership and the Construction of Political Legitimacy: Unpacking Bourdieu's 'Political Capital' in Post-Apartheid

Johannesburg. In *Int J Urban Reg Res* 38 (5), pp. 1807–1832. https://doi.org/10.1111/1468-2427.12166.

Berlo, Kurt; Wagner, Oliver (2013): Stadtwerke-Neugründungen und Rekommunalisierungen. Energieversorgung in kommunaler Verantwortung. Sondierungsstudie. Wuppertal Institut für Klima, Umwelt, Energie.

Betsill, Michele M.; Bulkeley, Harriet (2006): Cities and the Multilevel Governance of Global Climate Change. In *Global Governance* 12 (2), pp. 141–159.

Bettzieche, Jochen (2009a): Drei gute Jahre. In *neue energie* 10, 2009, pp. 82–85.

Bettzieche, Jochen (2009b): In guter Gesellschaft. In *neue energie* 12, 2009, pp. 97–99.

Bickerstaff, Karen; Agyeman, Julian (2009): Assembling Justice Spaces: The Scalar Politics of Environmental Justice in North-east England. In *Antipode* 41 (4), pp. 781–806. https://doi.org/10.1111/j.1467-8330.2009.00697.x.

Biel, Anders; Thøgersen, John (2007): Activation of social norms in social dilemmas: A review of the evidence and reflections on the implications for environmental behaviour. In *Journal of Economic Psychology* 28 (1), pp. 93–112. https://doi.org/10.1016/j.joep.2006.03.003.

Bijman, Jos; Hanisch, Markus (2012): Support for Farmers' Cooperatives; Developing a typology of cooperatives and producer organisations in the EU. Wageningen: Wageningen UR.

Birsl, Ursula (2016): Anthony Giddens: The Constitution of Society. Outline of the Theory of Structuration, Cambridge: Polity Press 1984, 402 S. (dt. Die Konstitution der Gesellschaft. Grundzüge einer Theorie der Strukturierung, Campus: Frankfurt/New York 1988, 460 S.). In: Klassiker der Sozialwissenschaften. 100 Schlüsselwerke im Portrait. 2. Aufl. Wiesbaden: Springer VS, pp. 346–349.

Blanchet, Thomas (2015): Struggle over energy transition in Berlin: How do grassroots initiatives affect local energy policy-making? In *Energy Policy* 78, pp. 246–254. https://doi.org/10.1016/j.enpol.2014.11.001.

Blumer, Herbert (1973): Der methodologische Standort des Symbolischen Interaktionismus. In Arbeitsgruppe Bielefelder Soziologen (Ed.): Alltagswissen, Interaktion und gesellschaftliche Wirklichkeit. Reinbek: Rohwolt, pp. 80–146.

BNetzA Bundes Netzagentur (2019): Monitoringbericht 2018. Bonn. Available online at https://www.bundesnetzagentur.de/SharedDocs/Downloads/DE/Allgemeines/Bundesnetzagentur/Publikationen/Berichte/2018/Monitoringbericht_Energie2018.pdf;jsessionid=0E284CD9AE045839683823C5CC94C2D9?__blob=publicationFile&v=5.

BNetzA Bundes Netzagentur (2017): Monitoringbericht 2017. Bonn. Available online at https://www.bundesnetzagentur.de/SharedDocs/Downloads/DE/Allgemeines/Bundesnetzagentur/Publikationen/Berichte/2017/Monitoringbericht_2017.pdf;jsessionid=0E284CD9AE045839683823C5CC94C2D9?__blob=publicationFile&v=4, checked on 5/23/2019.

BNetzA Bundes Netzagentur (2016): Monitoringbericht 2016. Bonn. Available online at https://www.bundesnetzagentur.de/SharedDocs/Downloads/DE/Sachgebiete/Energie/Unternehmen_Institutionen/DatenaustauschUndMonitoring/Monitoring/Monitoringbericht2016.pdf;jsessionid=0E284CD9AE045839683823C5CC94C2D9?__blob=publicationFile&v=2, checked on 5/23/2019.

BNetzA Bundes Netzagentur (2015 (2016)): Monitoringbericht 2015. Available online at https://www.bundesnetzagentur.de/SharedDocs/Downloads/DE/Allgemeines/Bundes netzagentur/Publikationen/Berichte/2015/Monitoringbericht_2015_BA.pdf;jsessionid= 0E284CD9AE045839683823C5CC94C2D9?__blob=publicationFile&v=4, checked on 5/23/2019.

Bogner, Alexander; Menz, Wolfgang (2009): The Theory-Generating Expert Interview: Epistemological Interest, Forms of Knowledge, Interaction. In A. Bogner, B. Littig, W. Menz (Eds.). Interviewing Experts (Research Methods Series), pp. 43–80.

Bogner, A.; Littig, B.; Menz, W. (Eds.) (2009): Interviewing Experts. London: Palgrave Macmillan (Research Methods Series).

Bolinger, Mark (2001): Community wind power ownership schemes in Europe and their relevance to the United States. Lawrence Berkeley National Laboratory. Environmental Technologies Division. Available online at: https://www.osti.gov/servlets/purl/827946

Bomberg, Elizabeth; McEwen, Nicola (2012): Mobilizing community energy. In *Energy Policy* 51, pp. 435–444. https://doi.org/10.1016/j.enpol.2012.08.045.

Boomsma, Trine Krogh; Meade, Nigel; Fleten, Stein-Erik (2012): Renewable energy investments under different support schemes: A real options approach. In *European Journal of Operational Research* 220 (1), pp. 225–237. https://doi.org/10.1016/j.ejor.2012.01.017.

Boon, Frank Pieter; Dieperink, Carel (2014): Local civil society based renewable energy organisations in the Netherlands: Exploring the factors that stimulate their emergence and development. In *Energy Policy* 69, pp. 297–307. https://doi.org/10.1016/j.enpol.2014. 01.046.

Borup, Mads; Brown, Nik; Konrad, Kornelia; van Lente, Harro (2006): The sociology of expectations in science and technology. In *Technology Analysis & Strategic Management* 18 (3–4), pp. 285–298. https://doi.org/10.1080/09537320600777002.

Boulding, Kenneth Ewart (1990 (1989)): Three faces of power. 2nd print. Newbury Park: Sage.

Breukers, Sylvia; Wolsink, Maarten (2007): Wind energy policies in the Netherlands: Institutional capacity-building for ecological modernisation. In *Environmental Politics* 16 (1), pp. 92–112. https://doi.org/10.1080/09644010601073838.

Breukers, Sylvia; Wolsink, Maarten (2007): Wind power implementation in changing institutional landscapes: An international comparison. In *Energy Policy* 35 (5), pp. 2737–2750. https://doi.org/10.1016/j.enpol.2006.12.004.

Brickmann, Irene; Kropp, Cordula; Türk, Jana (2012): Aufbruch in den Alpen – Lokales Handeln für eine globale Transformation? In Gerald Beck, Cordula Kropp (Eds.): Gesellschaft innovativ, vol. 2003. Wiesbaden: Springer VS, pp. 65–83.

Brieden, Bianca (2013): Ökostrom-Kunden sparen doppelt. In *pricewise*, 9/11/2013. Available online at http://www.pricewise.de/strom/nachrichten/oekostrom-kunden-sparen-doppelt/, checked on 11/24/2018.

Bright, Jenny; Langston, Rowena; Bullman, Rhys; Evans, Richard; Gardner, Sam; Pearce-Higgins, James (2008): Map of bird sensitivities to wind farms in Scotland. A tool to aid planning and conservation. In *Biological Conservation* 141 (9), pp. 2342–2356. https://doi.org/10.1016/j.biocon.2008.06.029.

Broman Toft, Madeleine; Schuitema, Geertje; Thøgersen, John (2014): The importance of framing for consumer acceptance of the Smart Grid: A comparative study of Denmark,

Norway and Switzerland. In *Energy Research & Social Science* 3, pp. 113–123. https://doi.org/10.1016/j.erss.2014.07.010.

Brown, L. David; Timmer, Vanessa (2006): Civil Society Actors as Catalysts for Transnational Social Learning. In *Voluntas* 17 (1), pp. 1–16. https://doi.org/10.1007/s11266-005-9002-0.

Browne, Anthony (2015): A video guide to the access project. (3:02 minutes). Available online at https://vimeo.com/137118601, checked on 11/13/2017.

Brunnengräber, Achim; Di Nucci, Maria Rosaria (Eds.) (2014): Im Hürdenlauf zur Energiewende. Von Transformationen, Reformen und Innovationen ; zum 70. Geburtstag von Achim Mez. Wiesbaden: Springer.

Bruns, Elke (2011): Renewable energies in Germany's electricity market. A biography of the innovation process. In Elke Bruns, Dörte Ohlhorst, Bernd Wenzel, Johann Köppel (Eds.): Renewable Energies in Germany's Electricity Market. A Biography of the Innovation Process. Springer Netherlands.

Bruns, Elke (2011): Wind Power Generation in Germany. a transdisciplinary view on the innovation biography. In *The Journal of Transdisciplinary Environmental Studies* 10 (1), pp. 45–67.

Bruns, Elke; Ohlhorst, Dörte; Wenzel, Bernd; Köppel, Johann (Eds.) (2011): Renewable Energies in Germany's Electricity Market. A Biography of the Innovation Process. Springer Netherlands.

Bruns, Elke; Köppel, Johann; Ohlhorst, Dörte; Schön, Susanne (2008): Die Innovationsbiographie der Windenergie. Absichten und Wirkungen von Steuerungsimpulsen. Berlin: LIT Verlag (Innovationsforschung, 2).

Buchmann, Ilka (11/18/1999): Was kostet das örtliche Stromnetz? BGH fällt Grundsatzurteil zu Gunsten kaufwilliger Kommunen. Freiburg / Kaufering. Available online at https://idw-online.de/de/news15880, checked on 3/27/2019.

Buffetaut, Stéphane; Coulon, Pierre Jean; Joost, Meelis; Vella, Charles (2017): TEN Section Report on the "Smart Islands" Project. Brussels, Belgium. Available online at https://www.eesc.europa.eu/resources/docs/qe-07-16-088-en-n.pdf, checked on 3/28/2019.

Bukoski, Beth E.; Lewis, Tiffanie C.; Carpenter, Bradley W.; Berry, Matthew S.; Sanders, Kimberly N. (2015): The Complexities of Realizing Community: Assistant Principals as Community Leaders in Persistently Low-Achieving Schools. In *Leadership and Policy in Schools* 14 (4), pp. 411–436. https://doi.org/10.1080/15700763.2015.1021053.

Bulkeley, Harriet; Betsill, Michele (2005): Rethinking Sustainable Cities: Multilevel Governance and the 'Urban' Politics of Climate Change. In *Environmental Politics* 14 (1), pp. 42–63. https://doi.org/10.1080/0964401042000310178.

Bunting, Madeleine (2014): Why independence is on the minds of Scottish islanders. In *the guardian*, 7/7/2014. Available online at https://www.theguardian.com/commentisfree/2014/jul/07/independence-scottish-islands-edinburgh-london-devolution, checked on 3/27/2019.

Bunting, Madeleine (2015): Renewable energy: How wind is changing the fortunes of Lewis islanders. In *the guardian*, 2/12/2015. Available online at https://www.theguardian.com/environment/2015/feb/12/renewable-energy-wind-changed-fortunes-lewis-islanders, checked on 3/27/2019.

Burawoy, Michael (1998): The Extended Case Method. In *Sociological Theory* 16 (1), pp. 4–33. https://doi.org/10.1111/0735-2751.00040.

Burawoy, Michael (2009(1991)): The extended case method. Four countries, four decades, four great transformations, and one theoretical tradition / Michael Burawoy. Berkeley, Calif.: University of California Press.

BürgerEnergie Berlin eG (2018): Geschäftsbericht 2017. Berlin. Available online at https://www.buerger-energie-berlin.de/wp-content/uploads/180605_BEB_GB2017.pdf, checked on 3/25/2019.

BürgerEnergie Berlin eG (2019): Der Ökostrom. Sonnencent für Berlin. Berlin. Available online at https://www.buerger-energie-berlin.de/themen/oekostrom/, checked on 3/28/2019.

Bürgerwerke eG (3/11/2015): Bürgerwerke: Energiegenossenschaften vertreiben bundesweit Ökostrom aus Bürgerhand. Available online at https://buergerwerke.de/wp-content/uploads/Pressemitteilung_Buergerwerke_Bundesweiter_%c3%96kostrom_aus_Buergerhand.pdf, checked on 3/19/2019.

Bürgerwerke eG (9/13/2017): Grüner Strom-Label zertifiziert 100% erneuerbaren Bürgerstrom. Available online at https://buergerwerke.de/wp-content/uploads/170913_Gruener-Strom-Label-fuer-Buergerstrom.pdf, checked on 3/19/2019.

Bürgerwerke eG (12/19/2018): Bürgerwerke liefern ab Januar 2019 BürgerÖkogas aus organischen Reststoffen. Available online at https://buergerwerke.de/wp-content/uploads/Buergerwerke-liefern-BuergerOekogas.pdf, checked on 3/19/2019.

Aichele, Christian; Doleski, Oliver (Eds.) (2014): Smart Market. Vom Smart Grid zum intelligenten Energiemarkt. Wiesbaden: Springer Vieweg.

Callon, Michel; Lascoumes, Pierre; Barthe, Yannick (2009): Acting in an uncertain world. An essay on technical democracy; translated by Graham Burchell. Cambridge, Mass., London: MIT University Press (Inside technology).

Campsie, Alison (2016): How much does it cost to live on a Scottish island? In *The Scotsman*, 5/24/2016. Available online at https://www.scotsman.com/news/how-much-does-it-cost-to-live-on-a-scottish-island-1-4136990.

Carrell, Severin (2008a): £500m project offers jobs and income, but will it devastate the environment? Islanders on Lewis split over plan for arc of 181 tubines, each 140m tall, 2/4/2008. Available online at https://www.theguardian.com/environment/2008/feb/04/windpower.renewableenergy, checked on 3/12/2016.

Carrell, Severin (2008b): Scottish government rejects plans for Lewis wind farm. In *the guardian*, 4/21/2008. Available online at https://www.theguardian.com/environment/2008/apr/21/windpower.renewableenergy, checked on 3/12/2016.

Catney, Philip; Dobson, Andrew; Hall, Sarah Marie; Hards, Sarah; MacGregor, Sherilyn; Robinson, Zoe et al. (2013): Community knowledge networks: an action-orientated approach to energy research. In *Local Environment* 18 (4), pp. 506–520. https://doi.org/10.1080/13549839.2012.748729.

Ceglarz, Andrzej; Beneking, Andreas; Ellenbeck, Saskia; Battaglini, Antonella (2017): Understanding the role of trust in power line development projects: Evidence from two case studies in Norway. In *Energy Policy* 110, pp. 570–580. https://doi.org/10.1016/j.enpol.2017.08.051.

Charmaz, Kathy (2014): Constructing grounded theory. 2nd edition. Los Angeles, Calif.: SAGE (Introducing qualitative methods).

Chiappero-Martinetti, Enrica; Houghton Budd, Christopher; Ziegler, Rafael (2017): Social Innovation and the Capability Approach—Introduction to the Special Issue. In *Journal of*

Human Development and Capabilities 18 (2), pp. 141–147. https://doi.org/10.1080/194 52829.2017.1316002.

Chiappero-Martinetti, Enrica; Jacobi, Nadia von (2015): How can Sen's Capabilities Approach Contribute to Understanding the Role for Social Innovations for the Marginalized? (CRESSI Working Papers, 3). Available online at http://eureka.sbs.ox.ac.uk/7083/ 1/Chapter%203.pdf, checked on 4/30/2019.

Child, Michael; Breyer, Christian (2016): Vision and initial feasibility analysis of a recarbonised Finnish energy system for 2050. In *Renewable and Sustainable Energy Reviews* 66, pp. 517–536. https://doi.org/10.1016/j.rser.2016.07.001.

Child, Michael; Breyer, Christian (2017): Transition and transformation: A review of the concept of change in the progress towards future sustainable energy systems. In *Energy Policy* 107, pp. 11–26. https://doi.org/10.1016/j.enpol.2017.04.022.

Child, Michael; Haukkala, Teresa; Breyer, Christian (2017): The Role of Solar Photovoltaics and Energy Storage Solutions in a 100% Renewable Energy System for Finland in 2050. In *Sustainability* 9 (8), p. 1358. https://doi.org/10.3390/su9081358.

Child, Michael; Nordling, Alexander; Breyer, Christian (2017): Scenarios for a sustainable energy system in the Åland Islands in 2030. In *Energy Conversion and Management* 137, pp. 49–60. https://doi.org/10.1016/j.enconman.2017.01.039.

Chilvers, Jason; Longhurst, Noel (2016): Participation in Transition(s). Reconceiving Public Engagements in Energy Transitions as Co-Produced, Emergent and Diverse. In *Journal of Environmental Policy & Planning* 18 (5), pp. 585–607. https://doi.org/10.1080/1523908x. 2015.1110483.

Chong, Dennis (2000): Rational lives. Norms and values in politics and society / Dennis Chong. Chicago, Ill., London: University of Chicago Press (American politics and political economy).

Christensen, Toke Haunstrup; Friis, Freja (2017): Case study report Denmark. MATCH Project Analysis of smart grid solutions. Aalborg, Denmark. Available online at https://www.match-project.eu/digitalAssets/344/344918_d2.2_danish-case-study-rep ort_match.pdf, checked on 3/28/2019.

Christmann, Gabriela B. (2009): Expert Interviews on the Telephone: A Difficult Undertaking. In A. Bogner, B. Littig, W. Menz (Eds.): Interviewing Experts (Research Methods Series), pp. 157–183.

Clarke, Adele E.; Keller, Reiner (2012): Situationsanalyse. Grounded Theory anch dem Postmodern Turn. Wiesbaden: Springer.

Clarke, Adele E. (2005): Situational analysis. Grounded theory after the postmodern turn. Thousand Oaks, Calif., London: Sage. Available online at http://www.loc.gov/catdir/enh ancements/fy0657/2004022902-d.html.

Clarke, Adele E. (2003): Situational Analyses. Grounded Theory Mapping After the Postmodern Turn. In *Symbolic Interaction* 26 (4), pp. 553–576. https://doi.org/10.1525/si.2003. 26.4.553.

Clarke, Adele E. (1991): Social worlds/arenas theory as organizational theory. In: Maines, David (Ed.): Social Organization and Social Process. Essays in Honor of Anselm Strauss. New York: Aldine De Gruyter, pp. 119–158.

Cohen, Maurie J.; Brown, Halina Szejnwald; Vergragt, Philip (Eds.) (2013): Innovations in sustainable consumption. New economics, socio-technical transitions and social practices. Cheltenham: Edward Elgar Publishing (Advances in ecological economics).

Colell, Arwen; Neumann-Cosel, Luise (2016): Berlin. Cooperative Power and the Transformation of Citizens' Roles in Energy Decision-Making. In Luque-Ayala, Andres; Silver, Jonathan (Ed.): Energy, power and protest on the urban grid. Geographies of the electric city. London, pp. 135–152.

Colell, Arwen; Pohlmann, Angela (2019): Community energy projects redefining energy distribution systems: Examples from Berlin and Hamburg. In Fanny Lopez, Margot Pellegrino, Olivier Coutard (Eds.): Local energy autonomy: spaces, scales, politics. Autonomie énergétique locale: espaces, échelles et politique. London: ISTE.

Combe, Malcolm M. (2014): Land Reform Revisited: The Land of Scotland and the Common Good. In *ELR* 18 (3), pp. 410–413.

Combe, Malcolm M. (2016): The Land Reform (Scotland) Act 2016: another answer to the Scottish land question. In *Juridical Review*, pp. 291–313.

Community Energy Scotland (2019): About us. Available online at http://www.community energyscotland.org.uk/about-us.asp, checked on 3/29/2019.

Cornelissen, Joep P.; Werner, Mirjam d. (2014): Putting Framing in Perspective: A Review of Framing and Frame Analysis across the Management and Organizational Literature. In *ANNALS* 8 (1), pp. 181–235. https://doi.org/10.5465/19416520.2014.875669.

Costanza, Robert (Ed.) (1992 (1991)): Ecological Economics. The Science and Management of Sustainability. 2nd edition. New York: Columbia University Press. Available online at https://ebookcentral.proquest.com/lib/gbv/detail.action?docID=4550050.

Cotton, Matthew David; Devine-Wright, Patrick (2010): NIMBYism and Community Consultation in Electricity Transmission Network Planning. In Patrick Devine-Wright (Ed.): Renewable Energy and the Public: From NIMBY to Participation. London: Earthscan, pp. 115–128.

Cowell, Richard; Bristow, Gill; Munday, Max (2011): Acceptance, acceptability and environmental justice: the role of community benefits in wind energy development. In *Journal of Environmental Planning and Management* 54 (4), pp. 539–557. https://doi.org/10.1080/09640568.2010.521047.

Cowell, Richard; Bristow, Gillian; Munday, Max (2012): Wind energy and justice for disadvantaged communities. York (Viewpoint). Available online at: https://www.hoylakevi sion.org.uk/wp-content/uploads/2012/11/wind-farms-communities-summary.pdf

Cowell, Richard; Ellis, Geraint; Sherry-Brennan, Fionnguala; Strachan, Peter A.; Toke, David (2017): Rescaling the Governance of Renewable Energy: Lessons from the UK Devolution Experience. In *Journal of Environmental Policy & Planning* 19 (5), pp. 480–502. https://doi.org/10.1080/1523908x.2015.1008437.

Cress, Daniel M.; Snow, David A. (1996): Mobilization at the Margins: Resources, Benefactors, and the Viability of Homeless Social Movement Organizations. In *American Sociological Review* 61 (6), p. 1089. https://doi.org/10.2307/2096310.

Cumbers, Andrew (2012): Reclaiming public ownership. Making space for economic democracy. London: Zed Books.

Curtin, Joseph; McInerney, Celine; Johannsdottir, Lara (2018): How can financial incentives promote local ownership of onshore wind and solar projects? Case study evidence from Germany, Denmark, the UK and Ontario. In *Local Economy* 33 (1), pp. 40–62. https://doi.org/10.1177/0269094217751868.

Cynthia Rosenzweig, William D. Solecki, Patricia Romero-Lankao, Shagun Mehrotra, Shobhakar Dhakal, Somayya Ali Ibrahim (2018): Climate Change and Cities: Second

Assessment Report of the Urban Climate Change Research Network. Cambridge, UK: Cambridge University Press.

Dahl, Robert A. (1957): The concept of power. In *Syst. Res.* 2 (3), pp. 201–215. https://doi. org/10.1002/bs.3830020303.

Dahl, Robert A. (1961): Who governs? Democracy and Power in an American City. New Haven: Yale University Press.

Danielsen, Oluf (1995): Large-scale wind power in Denmark. In *Land Use Policy* 12, pp. 60–62.

Danish Energy Agency (n.d.): District heating - Danish experiences. Available online at https:// ens.dk/sites/ens.dk/files/contents/material/file/dh_danish_experiences.pdf, checked on 3/27/2019.

Danish Government (12/27/2008): Promotion of Renewable Energy Act, no. 1392. Available online at https://ens.dk/sites/ens.dk/files/Vindenergi/promotion_of_renewable_ene rgy_act_-_extract.pdf.

Danish Government (February 2011): Energy Strategy 2050 - from coal, oil and gas to green energy. Available online at http://www.danishwaterforum.dk/activities/Climate% 20change/Dansk_Energistrategi_2050_febr.2011.pdf.

Danish Government (3/22/2012): DK Energy Agreement. Available online at https://www. energie-experten.org/uploads/media/DK_Energy_Agreement_March_22_2012.pdf.

Danske Fortidsminder (n.d.): Kanhave vikingetids kanal over Samsø. (Kanhave VIking time canal across Samsø). Available online at http://www.fortidsmindeguide.dk/Kanhave-Kanal.vt005.0.html, checked on 3/14/2016.

David, Martin; Schönborn, Sophia (2016): Die Energiewende als Bottom-up-Innovation. Wie Pionierprojekte das Energiesystem verändern. München (Transformationen, Band 4). Available online at http://api.vlb.de/api/v1/asset/mmo/file/6d6ef477-074b-4101-970f-5b708cd929b1.

Debor, Sarah (2018): Multiplying mighty Davids? The Influence of energy cooperatives on Germany's energy transition. Cham: Springer International Publishing (Contributions to Economics, 1431–1933).

Debor, Sarah (2014): The socio-economic power of renewable energy production cooperatives in Germany: Results of an empirical assessment (Wuppertal Papers, 187).

Delhey, Jan; Newton, Kenneth (2003): Who Trusts? The Origins of Trust in even Societies. In *European Societies* 5 (2), pp. 93–137.

Demski, Christina; Butler, Catherine; Parkhill, Karen A.; Spence, Alexa; Pidgeon, Nick F. (2015): Public values for energy system change. In *Global Environmental Change* 34, pp. 59–69. https://doi.org/10.1016/j.gloenvcha.2015.06.014.

Denzin, Norman K. (Ed.) (1970): Sociological methods: A sourcebook. Chicago, Ill.: University of Chicago Press.

Denzin, Norman K.; Lincoln, Yvonna S. (2003): Collecting and interpreting qualitative materials. 2nd ed. Thousand Oaks, Calif., London: SAGE.

Denzin, Norman K.; Lincoln, Yvonna S. (Eds.) (2011(1994)): The Sage handbook of qualitative research. 4th ed. Thousand Oaks, Calif.: SAGE.

DUKES Department for Business, Energy & Industrial Strategy (2016): DUKES Digest of United Kingdom Energy Statistics. Available online at https://assets.publishing.ser vice.gov.uk/government/uploads/system/uploads/attachment_data/file/577712/DUKES_ 2016_FINAL.pdf.

Department for Business, Energy & Industrial Strategy (2012 (updated 2018)): Digest of UK Energy Statistics DUKES. Chapter 5 Electricity, p. 111–153. Available online at https://assets.publishing.service.gov.uk/government/uploads/system/uploads/att achment_data/file/736152/Ch5.pdf, checked on 1/21/2019.

Department for Business, Energy and Industrial Strategy (2003): Energy White Paper: Our Energy Future—creating a low carbon economy. Available online at https://webarchive.nationalarchives.gov.uk/+tf_/http://www.berr.gov.uk/energy/whitep aper/2003/page21223.html, checked on 6/22/2019.

DECC Department of Energy and Climate Change (2015) Performance and Impact of the Feed-in Tariff Scheme: Review of Evidence. Final Report. Available online at https://assets.publishing.service.gov.uk/government/uploads/system/uploads/att achment_data/file/456181/FIT_Evidence_Review.pdf, checked on 6/22/2019.

DECC Department of Energy and Climate Change (2014) Community Energy Strategy. 27 January 2014. Available online at https://www.gov.uk/government/publications/commun ity-energy-strategy, checked on 6/22/2019.

DECC Department of Energy and Climate Change (2011): Planning our electric future: a white paper for secure, affordable and low-carbon energy. 12 July 2011. Available online at https://www.gov.uk/government/publications/planning-our-electric-future-a-white-paper-for-secure-affordable-and-low-carbon-energy, checked on 6/22/2019.

DECC Department of Energy and Climate Change (2010): Annual Energy Statement. DECC Departmental Memorandum. Available onlien at https://assets.publishing.service.gov.uk/government/uploads/system/uploads/attachment_data/file/47879/237-annual-energy-sta tement-2010.pdf, checked on 6/22/2019.

DECC Department of Energy and Climate Change (2009a): Carbon Valuation in UK Policy Appraisal: A Revised Approach. July 2009. Available online at https://assets.publishing.service.gov.uk/government/uploads/system/uploads/attachment_data/file/245334/1_2 0090715105804_e_carbonvaluationinukpolicyappraisal.pdf, checked 6/22/2019.

DECC Department of Energy and Climate Change (2009b): UK Low Carbon Transition Plan. National Strategy for Climate and Energy. 15 July 2009. Available online at https://www.gov.uk/government/publications/the-uk-low-carbon-transition-plan-national-strategy-for-climate-and-energy, checked on 6/22/2019.

Deutscher Gründerpreis (2007): Ursula und Dr. Maichael (sic!) Sladek, EWS Vertriebs GmbH. Available online at https://www.deutscher-gruenderpreis.de/preistraeger/2007/urs ula-und-dr-michael-sladek-ews-vertriebs-gmbh/, checked on 4/1/2019.

Devine-Wright, Patrick (2011a): Place attachment and public acceptance of renewable energy: A tidal energy case study. In Journal of Environmental Psychology 31 (4), pp. 336–343. https://doi.org/10.1016/j.jenvp.2011.07.001.

Devine-Wright, Patrick (2009): Rethinking NIMBYism: The role of place attachment and place identity in explaining place-protective action. In J. Community. Appl. Soc. Psychol. 19 (6), pp. 426–441. https://doi.org/10.1002/casp.1004.

Devine-Wright, Patrick (2009): Energy Citizenship: Psychological Aspects of Evolution in Sustainable Energy Technologies. In Joseph Murphy (Ed.): Framing the Present, Shaping the Future: Contemporary Governance of Sustainable Technologies. London: Earthscan, pp. 63–86.

Devine-Wright, Patrick (2001): Role of social capital in advancing regional sustainable development. In Impact Assessment and Project Appraisal 19 (2), pp. 161–167.

Devine-Wright, P.; Batel, S. (2017): My neighbourhood, my country or my planet? The influence of multiple place attachments and climate change concern on social acceptance of energy infrastructure. In *Global Environmental Change* 47, pp. 110–120. https://doi. org/10.1016/j.gloenvcha.2017.08.003.

Devine-Wright, Patrick; Batel, Susana; Aas, Oystein; Sovacool, Benjamin; Labelle, Michael Carnegie; Ruud, Audun (2017): A conceptual framework for understanding the social acceptance of energy infrastructure: Insights from energy storage. In *Energy Policy* 107, pp. 27–31. https://doi.org/10.1016/j.enpol.2017.04.020.

Dierkes, Meinolf; Weiler, Hans N.; Antal, Ariane Berthoin (Eds.) (1987): Comparative Policy Research: Learning from Experience. Aldershot: Gower.

Ding, Zhujun; Au, Kevin; Chiang, Flora (2015): Social trust and angel investors' decisions: A multilevel analysis across nations. In *Journal of Business Venturing* 30 (2), pp. 307–321. https://doi.org/10.1016/j.jbusvent.2014.08.003.

Dobson, Andrew (2006): Ecological citizenship: a Defence. In *Environmental Politics* 15 (3), pp. 447–451. https://doi.org/10.1080/09644010600627766.

Dobson, Andrew (2000): Ecological citizenship: a disruptive influence. In: Politics at the Edge. The PSA Yearbook 1999. London: Palgrave Macmillan (Political Studies Association Yearbook Series), pp. 40–62.

Dóci, Gabriella; Vasileiadou, Eleftheria; Petersen, Arthur C. (2015): Exploring the transition potential of renewable energy communities. In *Futures* 66, pp. 85–95. https://doi.org/10. 1016/j.futures.2015.01.002.

Dooley, Larry M. (2002): Case study research and theory building. In *Advances in Developing Human Resources* 4 (3), pp. 335–354.

Droste-Franke, Bert; Carrier, M.; Kaiser, M.; Schreurs, M.; Weber, C.; Ziesemer, Th. (2015): Improving energy decisions. Towards better scientific policy advice for a safe and secure future energy system. Cham: Springer (Ethics of Science and Technology Assessment, volume 42).

Dryzek, John S. (2001): Resistance is Fertile. In *Global Environmental Politics* 1 (1), pp. 11–17. https://doi.org/10.1162/152638001570723.

DTAS (2019): About DTAS. Development Trust Association Scotland. Available online at https://dtascot.org.uk/about-dtas/about-dtas, checked on 3/29/2019.

Dunn, P. d. (1978): Appropriate Technology. Technology with a Human Face. London: Palgrave Macmillan, s.l. Available online at http://dx.doi.org/10.1007/978-1-349-160 64-8.

Dwyer, Patrick C.; Bono, Joyce E.; Snyder, Mark; Nov, Oded; Berson, Yair (2013): Sources of Volunteer Motivation: Transformational Leadership and Personal Motives Influence Volunteer Outcomes. In *Nonprofit Management Leadership* 24 (2), pp. 181–205. https:// doi.org/10.1002/nml.21084.

Dwyer, Patrick C.; Maki, Alexander; Rothman, Alexander J. (2015): Promoting energy conservation behavior in public settings: The influence of social norms and personal responsibility. In *Journal of Environmental Psychology* 41, pp. 30–34. https://doi.org/ 10.1016/j.jenvp.2014.11.002.

Broszewski, Achim; Maeder, Christoph; Nentwich, Julia (Eds.) (2015): Vom Sinn der Soziologie. Festschrift für Thomas S. Eberle. Wiesbaden: Springer VS (Wissen, Kommunikation und Gesellschaft).

Education Scotland (2016): Mull and Iona Community Trust Review. Edinburgh/Livingston, Scotland. Available online at https://education.gov.scot/Documents/MullandIonaComm unityTrustReview310516.pdf, checked on 1/21/2019.

Edwards, Bob; McCarthy, John d. (2004): Resources and Social Movement Mobilization. In: David A. Snow, Sarah A. Soule, Hanspeter Kriesi (Eds.): The Blackwell Companion to Social Movements. Hoboken, New Jersey: Wiley-Blackwell, pp. 116–152.

Elsen, Susanne; Rausch, Günter; Biesecker, Adelheid (Eds.) (2011): Solidarische Ökonomie und die Gestaltung des Gemeinwesens. Perspektiven und Ansätze der ökosozialen Transformation von unten. 1. Aufl. Neu-Ulm: AG-SPAK-Bücher (Münchener Hochschulschriften für angewandte Sozialwissenschaften, M 244: Reihe Gemeinwesenarbeit).

Energiministeriet (1981): Energiplan 81 (Energy Plan 81). Kopenhagen.

Energistyrelsen (2019): Oversigtstabel over vindkraftanlaeg (Overview of windpower installations). Available online at https://ens.dk/sites/ens.dk/files/Vindenergi/oversigtstabeller_uk-dk.xls, updated on 6/3/2019, checked on 6/8/2019.

Energy Academy (2019): Sören Hermansen, Direktör. Available online at https://ene rgiakademiet.dk/omkring/energiakademiet/medarbejdere/soeren-hermansen/, checked on 4/1/2019.

Energy Academy (2016): RIGHT HERE Pioneer Guide. Samsö. Available online at http://www.pioneerguide.com/, checked on 3/28/2019.

Energy4All (2014): Our History. Available online at https://energy4all.co.uk/about-us/our-history/, updated on 2014, checked on 3/20/2019.

Enzensberger, N.; Fichtner, W.; Rentz, O. (2003): Evolution of local citizen participation schemes in the German wind market. In IJGEI 20 (2), p. 191. https://doi.org/10.1504/ijgei.2003.005303.

European Commission (2019): Social Innovation. Available online at http://ec.europa.eu/gro wth/industry/innovation/policy/social_en, checked on 4/30/2019.

European Commission (2010) Europe 2020: A strategy for smart, sustainable and inclusive growth. COM(2010) 2020 final. Available online at http://eur-lex.europa.eu/LexUriServ/LexUriServ.do?uri=COM:2010:2020:FIN:EN:PDF, checked on 4/30/2019.

European Commission (12/19/1996): Directive 96/92/EC concerning common rules for the internal market in electricity. 96/92/EC. In Official Journal L 027, pp. 20–29. Available online at https://eur-lex.europa.eu/LexUriServ/LexUriServ.do?uri=CELEX:31996L0092: EN:HTML, checked on 4/30/2019.

Eurostat Press Office (3/10/2015): Share of renewables in energy consumption up to 15% in the EU in 2013. 43/2015, eurostat-pressoffice@ec.europa.eu. Available online at http://ec.europa.eu/eurostat/documents/2995521/6734513/8-10032015-AP-EN. pdf/3a8c018d-3d9f-4f1d-95ad-832ed3a20a6b, checked on 2/12/2016.

EWS Elektrizitätswerke Schönau (2019): Schönauer Stromrebellen. Available online at https://www.ews-schoenau.de/energiewende-magazin/themenhefte/thema-schoenauer-stromrebellen/, checked on 3/28/2019.

EWS Elektrizitätswerke Schönau (2/27/2019): 200.000 Kunden bei den EWS. Immer mehr Menschen entscheiden sich für echten Oekostrom. Schönau. Available online at https://www.ews-schoenau.de/ews/presse/pressemeldungen/200.000-kunden-bei-den-ews/#parent-id=398032e9-25fb-11e5-9fb8-00155d0ae81a, checked on 3/21/2019.

EWS Elektrizitätswerke Schönau (2018a): Geschäftsbericht 2017. Schönau im Schwarzwald. Available online at https://www.ews-schoenau.de/export/sites/ews/ews/genossenschaft/. files/geschaeftsbericht-2017-ews-eg.pdf, checked on 3/21/2019.

EWS Elektrizitätswerke Schönau (2018b): Geschichte der EWS. Available online at https:// www.ews-schoenau.de/ews/geschichte/, checked on 3/21/2019.

EWS Elektrizitätswerke Schönau (2018c): Der Ökostrom. Printed ad for renewably sourced electricity supply services of EWS. Schönau, Friedrichstraße 53/55, 79677 Schönau im Schwarzwald.

EWS Elektrizitätswerke Schönau (8/28/2018): EWS starten neues Stromspar Projekt "Doppelte Dividende". Förderzusage der Deutschen Bundesstiftung Umwelt. Schönau im Schwarzwald. Available online at https://www.ews-schoenau.de/ews/presse/pressemel dungen/ews-starten-neues-stromsparprojekt-doppelte-dividende/#parent-id=398032e9-25fb-11e5-9fb8-00155d0ae81a, checked on 3/25/2019.

EWS Elektrizitätswerke Schönau (2017): Geschäftsbericht 2016. Schönau im Schwarzwald. Available online at https://www.ews-schoenau.de/export/sites/ews/ews/genossenschaft/. files/geschaeftsbericht-2016-ews-eg.pdf, checked on 3/21/2019.

EWS Elektrizitätswerke Schönau (11/16/2017): EWS führen Mitgliedertarif ein. Mitglieder unserer Genossenschaft erhalten ab 2018 einen vergünstigten Tarif. Schönau im Schwarzwald. Available online at https://www.ews-schoenau.de/ews/presse/pressemeldungen/ ews-fuehren-mitgliedertarif-ein/, checked on 3/21/2019.

EWS Elektrizitätswerke Schönau (11/21/2017): Stromrebellen von ihrer Heimatstadt ausgezeichnet. Bürgermedaille der Stadt Schönau an Ehepaar Sladek. Available online at https://www.ews-schoenau.de/ews/presse/pressemeldungen/stromrebellen-von-ihrer-heimatstadt-ausgezeichnet/, checked on 4/1/2019.

EWS Elektrizitätswerke Schönau (2016): Geschäftsbericht 2015. Schönau im Schwarzwald. Available online at https://www.ews-schoenau.de/export/sites/ews/ews/genossenschaft/. files/geschaeftsbericht-2015-netzkauf-ews-eg.pdf, checked on 3/25/2019.

EWS Elektrizitätswerke Schönau (2015): Geschäftsbericht 2014. Schönau im Schwarzwald. Available online at https://www.ews-schoenau.de/export/sites/ews/ews/genossenschaft/. files/geschaeftsbericht-2014-netzkauf-ews-eg.pdf, checked on 3/25/2019.

EWS Elektrizitätswerke Schönau (6/23/2015): 171.545 Menschen und 30 Verbände gegen AKW Hinkley Point. Unterstützung für die Klage Österreichs. Schönau im Schwarzwald. Available online at https://www.ews-schoenau.de/ews/presse/pressemeldungen/171.545-menschen-und-30-verbaende-gegen-akw-hinkley-point/, checked on 3/21/2019.

EWS Elektrizitätswerke Schönau (6/29/2015): Ursula und Dr. Michael Sladek aus dem Vorstand verabschiedet. Schönau im Schwarzwald. Available online at https://www.ews-schoenau.de/ews/presse/pressemeldungen/ursula-und-dr.-michael-sladek-aus-dem-vor stand-verabschiedet/, checked on 4/1/2019.

EWS Elektrizitätswerke Schönau (11/27/2015): 180.000 Beschwerden gegen AKW-Neubau Hinkley Point C. Schönau im Schwarzwald. Available online at https://www.ews-sch oenau.de/ews/presse/pressemeldungen/180.000-beschwerden-gegen-akw-neubau-hin kley-pointc/, checked on 5/14/2017.

EWS Elektrizitätswerke Schönau (11/15/2014): EWS fördern Photovoltaikanlage bei Fukushima. Japanischer Stromrebell und EWS planen Anlage in der verstrahlten Zone. Schönau im Schwarzwald. Available online at https://www.ews-schoenau.de/ews/presse/pressemel dungen/ews-foerdern-photovoltaikanlage-bei-fukushima/, checked on 3/25/2019.

EWS Elektrizitätswerke Schönau (8/9/2013): Ursula Sladek erhält den Deutschen Umweltpreis 2013. Schönau im Schwarzwald. Available online at https://www.ews-schoenau.de/ews/presse/pressemeldungen/ursula-sladek-erhaelt-den-deutschen-umweltpreis-2013/, checked on 3/1/2019.

Faltin, Thomas; Schulz-Braunschmidt, Wolfgang (2013): Bei der Energieerzeugung spielt die Musik. In *Stuttgarter Zeitung*, 1/31/2013. Available online at https://www.stuttgarter-zei tung.de/inhalt.stadtwerke-stuttgart-die-energiewende-muss-dezentral-stattfinden-page2. 36915bab-e1bc-4c74-bfbf-3eef1e8a016d.html, checked on 3/29/2019.

Farla, Jacco; Markard, Jochen; Raven, Rob; Coenen, Lars (2012): Sustainability transitions in the making. A closer look at actors, strategies and resources. In *Technological Forecasting and Social Change* 79 (6), pp. 991–998. https://doi.org/10.1016/j.techfore.2012.02.001.

FDT Fintry Development Trust (2018a): About. History. Available online at http://fintrydt. org.uk/about/history/, checked on 3/21/2019.

FDT Fintry Development Trust (2018b): AGM Meetings. Available online at http://fintrydt. org.uk/about/agm-minutes/, checked on 3/21/2019.

FDT Fintry Development Trust (2018c): Board Meeting Minutes. Available online at http:// fintrydt.org.uk/about/board-meeting-minutes/, checked on 3/21/2019.

FDT Fintry Development Trust (2018d): How we can help. Available online at http://fintrydt. org.uk/how-we-can-help/, checked on 3/21/2019.

FDT Fintry Development Trust (2018e): Past Projects. Available online at http://fintrydt.org. uk/past-projects/, checked on 3/21/2019.

Federal Republic of Germany (10/13/2016): Gesetz zur Einführung von Ausschreibungen für Strom aus erneuerbaren Energien und zu weiteren Änderungen des Rechts der erneuerbaren Energien. EEG 2017. In *Bundesgesetzblatt* I (49), pp. 2258–2358.

Federal Republic of Germany (7/21/2014): Gesetz zur grundlegenden Reform des Erneuerbare-Energien-Gesetzesund zur Änderung weiterer Bestimmungen des Energiewirtschaftsrechts. EEG 2014. In *Bundesgesetzblatt* I (33), pp. 1066–1132.

Federal Republic of Germany (7/14/2012): Gesetz zur Änderung des Rechtsrahmens für Strom aus solarer Strahlungsenergie und zu weiteren Änderungen im Recht der erneuerbaren Energien. PV Novelle, revised 9/3/2012. In *Bundesgesetzblatt* I, p. 1754.

Federal Republic of Germany (7/28/2011): Gesetz zur Neuregelung des Rechtsrahmens-für die Förderung der Stromerzeugung aus erneuerbaren Energien. EEG 2012. In *Bundesgesetzblatt* I (42), pp. 1634–1678.

Federal Republic of Germany (6/6/2011): Gesetzentwurf der Fraktionen der CDU/CSU und FDP: Entwurf eines Dreizehnten Gesetzes zur Änderung des Atomgesetzes. In *Drucksache 17/6070*. Available online at http://dipbt.bundestag.de/dip21/btd/17/060/1706070. pdf, checked on 3/29/2019.

Federal Republic of Germany (12/13/2010): Elftes Gesetz zur Änderung des Atomgesetzes. In: *Bundesgesetzblatt* I, Nr. 62, S. 1814. Source: https://www.bgbl.de/xaver/bgbl/start.xav?startbk=Bundesanzeiger_BGBl&bk=Bundesanzeiger_BGBl&start=//*%255 B@attr_id=%2527bgbl110s1814.pdf%2527%255D#__bgbl__%2F%2F*%5B%40attr_id%3D%27bgbl110s1814.pdf%27%5D_1561236584748

Federal Republic of Germany (9/28/2010): Energiekonzept für eine umweltschonenede, zuverlässige und bezahlbare Energieversorgung und 10 Punkte Sofortprogramm - Monitoring und Zwischenbericht der Bundesregierung. Drucksache 17/3049. Available online at http://dipbt.bundestag.de/doc/btd/17/030/1703049.pdf, checked on 5/10/2019.

Federal Republic of Germany (10/25/2008): Gesetz zur Neuregelung des Rechts der Erneuerbaren Energien im Strombereich und zur Änderung damit zusammenhängender Vorschriften. EEG 2009. In *Bundesgesetzblatt* I (49), pp. 2047–2100.

Federal Republic of Germany (2006): Genossenschaftsgesetz in der Fassung der Bekanntmachung vom 16. Oktober 2006 (BGBl. I S. 2230), das zuletzt durch Artikel 8 des Gesetzes vom 17. Juli 2017 (BGBl. I S. 2541) geändert worden ist. GenG. Available online at https://www.gesetze-im-internet.de/geng/BJNR000550889.html, checked on 3/19/2019.

Federal Republic of Germany (7/7/2005): Gesetz über die Elektrizitaets- und Gasversorgung (Energiewirtschaftsgesetz - EnWG). EnWG, revised 12/17/2018. Available online at https://www.gesetze-im-internet.de/enwg_2005/EnWG.pdf.

Federal Republic of Germany (7/21/2004): Gesetz zur Neuregelung des Rechts der Erneuerbaren Energien im Strombereich. EEG 2004. In *Bundesgesetzblatt* I (40).

Federal Republic of Germany (3/29/2000): Gesetz für den Vorrang Erneuerbarer Energien(Erneuerbare-Energien-Gesetz – EEG) sowie zur Änderung des Energiewirtschaftsgesetzes und des Mineralölsteuergesetzes. EEG, revised 2004, 2009, 2012, 2014, 2017. In *Bundesgesetzblatt* I (13), pp. 305–309. Available online at https://www.bgbl.de/xaver/bgbl/start.xav#_bgbl_%2F%2F*%5B%40attr_id%3D%27bgbl100s0305.pdf%27%5D_1558601583165, checked on 3/14/2018.

Federal Republic of Germany (12/7/1990): Gesetz über die Einspeisung von Strom aus erneuerbaren Energien in das öffentliche Netz. Stromeinspeisungsgesetz, StrEG. In *Bundesgesetzblatt* I (67), pp. 2633–2634. Available online at https://www.bgbl.de/xaver/bgbl/start.xav?start=//*%5B@attr_id=%27bgbl190s2633b.pdf%27%5D#_bgbl_%2F%2F*%5B%40attr_id%3D%27bgbl190s2633b.pdf%27%5D_1558601762873, checked on 3/14/2018.

Finch, Moray (2015): Construction nearing completion! Blog entry, March 6, 2015. Available online at http://www.garmonyhydro.info/, checked on 6/22/2019.

Finnemore, Martha (1996): Norms, culture, and world politics: insights from sociology's institutionalism. In *Int. Org.* 50 (02), pp. 325–347. https://doi.org/10.1017/s0020818300028587.

Finnemore, Martha; Sikkink, Kathryn (1998): International Norm Dynamics and Political Change. In *International Organization* 52 (4), pp. 887–917. https://doi.org/10.1162/002081898550789.

Firestone, Jeremy; Kirk, Hannah (2019): A strong relative preference for wind turbines in the United States among those who live near them. In *Nature Energy*. https://doi.org/10.1038/s41560-019-0347-9.

Flanagan, Constance (2003): Trust, Identity, and Civic Hope. In *Applied Developmental Science* 7 (3), pp. 165–171. https://doi.org/10.1207/s1532480xads0703_7.

Flemming, Bo (2013): From Best to Next Practice - Samsoe Energy Academy. Video documentation from the Symposium at Samsoe Energy Academy "From Best to Next Practice" September 18th-20th 2013. Denmark, 03:55 minutes.

Fletcher, Martin (2016): The impoverished Scottish community who bought their island back from landowners - and how 'Wendy', 'Fanny' and 'Blowy' reversed their fortunes, 7/29/2016. Available online at https://www.telegraph.co.uk/news/2016/07/29/the-impoverished-scottish-community-who-bought-their-island-back/, checked on 3/27/2019.

Flick, Uwe (2009): Qualitative Methoden in der Evaluationsforschung. In *Zeitschrift für Qualitative Forschung* 10 (1), pp. 9–18.

Flieger, Burghard; Klemisch, Herbert (2008): Eine andere Energiewirtschaft ist möglich - Neue Energiegenossenschaften. In *Widerspruch* 54, pp. 105–110.

Flieger, Burghard (2011): Energiegenossenschaften. Eine klimaverantwortliche, bürgernahe Energiewirtschaft ist möglich. In: Elsen, Susanne; Rausch, Günter (Eds.) Solidarische Ökonomie und die Gestaltung des Gemeinwesens: Perspektiven und Ansätze der ökosozialen Transformation von unten. Neu-Ulm: AG-SPAK-Bücher, pp. 315–338.

Fligstein, Neil; McAdam, Doug (2012): A theory of fields. Oxford, New York: Oxford University Press.

Florin, Paul; Wandersman, Abraham (1990): An introduction to citizen participation, voluntary organizations, and community development: Insights for empowerment through research. In *American Journal of Community Psychology* 18 (1), pp. 41–54. https://doi.org/10.1007/bf00922688.

Foster-Fishman, Pennie G.; Collins, Charles; Pierce, Steven J. (2013): An investigation of the dynamic processes promoting citizen participation. In *American Journal of Community Psychology* 51 (3-4), pp. 492–509. https://doi.org/10.1007/s10464-012-9566-y.

Foxon, Timothy J.; Gross, R.; Chase, A.; Howes, J.; Arnall, A.; Anderson, D. (2005): UK innovation systems for new and renewable energy technologies: drivers, barriers and systems failures. In *Energy Policy* 33 (16), pp. 2123–2137. https://doi.org/10.1016/j.enpol.2004.04.011.

Fram, Sheila (2013): The Constant Comparative Analysis Method outside of Grounded Theory. In *The Qualitative Report* 18, pp. 1–25.

Frankfurter Allgemeine Zeitung (2011): Alte Atomkraftwerke werden abgeschaltet. Drei Monate Moratorium. In *Frankfurter Allgemeine Zeitung*, 3/15/2011. Available online at http://dipbt.bundestag.de/dip21/btd/17/060/1706070.pdf, checked on 3/14/2019.

Frantzeskaki, Niki; Avelino, Flor; Loorbach, Derk (2013): Outliers or Frontrunners? Exploring the (Self-) Governance of Community- Owned Sustainable Energy in Scotland and the Netherlands. In : Renewable energy governance. Complexities and challenges / Evanthie Michalena, Jeremy Maxwell Hills, editors, vol. 23. London (Lecture Notes in Energy), pp. 101–116.

Fricker, Miranda (2017): Epistemic Contribution as a Central Human Capability. In George Hull (Ed.): Equal society - essays on equality in theory and practice. Lanham, Maryland: Lexington Books. pp. 73–89.

Friends of the Earth Scotland (2014): Community Power Scotland. From Remote Island Grids to Urban Solar Co-operatives. Edinburgh. Available online at https://foe.scot/wp-content/uploads/2017/07/CommunityPower-2.pdf, checked on 3/21/2019.

Fuchs, Gerhard (2016): Die Energiewende im Ländervergleich: Deutschland und das Vereinigte Königreich verfolgen eigene Transformationspfade. In *GAIA - Ecological Perspectives for Science and Society* 25 (3), pp. 222–224. https://doi.org/10.14512/gaia.25.3.22.

Fuchs, Gerhard; Hinderer, Nele (2014): Situative governance and energy transitions in a spatial context: case studies from Germany. In *Energ Sustain Soc* 4 (1), p. 331. https://doi.org/10.1186/s13705-014-0016-6.

Funtowicz, Silvio O.; Ravetz, Jerome R. (1992 (1991)): A new scientific methodology for global environmental issues. In Robert Costanza (Ed.): Ecological Economics. The Science and Management of Sustainability. 2nd edition. New York: Columbia University Press, pp. 137–152.

FUSS (2007): Das Schönauer Gefühl. Die Geschichte der Stromrebellen aus dem Schwarz-wald. Förderverein für umweltfreundliche Stromverteilung und Energieerzeugung Schö-nau im Schwarzwald eV, Documentary, 2007. Available online at https://www.ews-sch oenau.de/service/online-formulare/dvd-bestellen/#js-form-layer, checked on 3/27/2019.

Gadenne, David; Sharma, Bishnu; Kerr, Don; Smith, Tim (2011): The influence of consumers' environmental beliefs and attitudes on energy saving behaviours. In *Energy Policy* 39 (12), pp. 7684–7694. https://doi.org/10.1016/j.enpol.2011.09.002.

Gailing, Ludger; Leibenath, Markus (Eds.) (2013): Neue Energielandschaften - Neue Perspektiven der Landschaftsforschung. Wiesbaden: Springer.

Gamson, William A. (1992): Talking politics. Cambridge: Cambridge University Press.

Gamson, William A., David Croteau, William Hoynes, and Theodore Sasson (1992): Media Images and the Social Construction of Reality. In *Annual Review of Sociology* 18, pp. 373–393. Available online at http://www.jstor.org/stable/2083459.

García-Valiñas, María A.; Macintyre, Alison; Torgler, Benno (2012): Volunteering, pro-environmental attitudes and norms. In *The Journal of Socio-Economics* 41 (4), pp. 455–467. https://doi.org/10.1016/j.socec.2011.07.001.

Garud, Raghu; Karnøe, Peter (Eds.) (2001): Path Dependence and Creation. London: Routledge.

Garud, Raghu; Karnøe, Peter (2003): Bricolage versus breakthrough: distributed and embed-ded agency in technology entrepreneurship. In *Research Policy* 32 (2), pp. 277–300. https://doi.org/10.1016/s0048-7333(02)00100-2.

Garud, Raghu; Kumaraswamy, Arun; Karnøe, Peter (2010): Path Dependence or Path Crea-tion? In *Journal of Management Studies* 47 (4), pp. 760–774. https://doi.org/10.1111/j.1467-6486.2009.00914.x.

Gawel, Erik; Korte, Klaas; Tews, Kerstin (2015): Distributional Challenges of Sustainability Policies—The Case of the German Energy Transition. In *Sustainability* 7 (12), pp. 16599–16615. https://doi.org/10.3390/su71215834.

Geels, Frank; Verhees, B. (2011): Cultural legitimacy and framing struggles in innovation journeys: A cultural-performative perspective and a case study of Dutch nuclear energy (1945–1986). In *Technological Forecasting and Social Change* 78 (6), pp. 910–930. https://doi.org/10.1016/j.techfore.2010.12.004.

Geels, Frank W. (2006): Multi-Level Perspective on System Innovation: Relevance for Indus-trial Transformation. In Xander Olsthoorn, Anna J. Wieczorek (Eds.): Understanding industrial transformation. Views from different disciplines / edited by Xander Olsthoorn and Anna J. Wieczorek, vol. 44. Dordrecht, Great Britain (Environment & Policy, v. 44), pp. 163–186.

Geels, Frank W. (2004): From sectoral systems of innovation to socio-technical systems. In *Research Policy* 33 (6-7), pp. 897–920. https://doi.org/10.1016/j.respol.2004.01.015.

Geels, Frank W. (2002): Technological transitions as evolutionary reconfiguration processes: a multi-level perspective and a case-study. In *Research Policy* 31 (8-9), pp. 1257–1274. https://doi.org/10.1016/s0048-7333(02)00062-8.

GEM Green Energy Mull (2014) Garmony Hydro. Share Offer Document, January 2014. Available online at http://www.garmonyhydro.info/wp-content/uploads/2014/01/Green-Energy-Mull-Share-Offer-V1.2.pdf, checked on 6/22/2019.

Giddens, Anthony (1984): The constitution of society. Outline of the theory of structuration. Cambridge: polity.

Gifford, Robert; Nilsson, Andreas (2014): Personal and social factors that influence pro-environmental concern and behaviour: a review. In *International journal of psychology : Journal international de psychologie* 49 (3), pp. 141–157. https://doi.org/10.1002/ijop.12034.

Glachant, Jean-Michel; Finon, Dominique (Eds.) (2003): Competition in European electricity markets. A cross-country comparison / edited by J.-M. Glachant, D. Finon. Cheltenham: Edward Elgar Publishing.

Gladwell, Malcolm (2000): The tipping point: How little things can make a big difference. New York: Little Brown.

Glaser, Barney G. (1992): Basics of grounded theory analysis: Emergence vs. forcing. Mill Valley, Calif.: Sociology Press.

Glaser, Barney G. (1965): The Constant Comparative Method of Qualitative Analysis. In *Social Problems* 12 (4), pp. 436–445. https://doi.org/10.2307/798843.

Glaser, Barney G.; Strauss, Anselm (1970): Theoretical sampling. In Norman K. Denzin (Ed.): Sociological methods: A sourcebook. Chicago, Ill.: University of Chicago Press, pp. 105–114.

Gläser, Jochen; Laudel, Grit (2010): Experteninterviews und qualitative Inhaltsanalyse. 4th edition. Wiesbaden: Springer VS.

GLS Gemeischaftsbank für Leihen und Schenken (2019): Branchen. Erneuerbare Energien. Sonne, Wind und mehr. Das spricht dafür: Energiewerke Schönau. Available online at https://www.gls.de/gemeinnuetzige-kunden/branchen/erneuerbare-energien/, checked on 3/27/2019.

Goedkoop, Fleur; Devine-Wright, Patrick (2016): Partnership or placation? The role of trust and justice in the shared ownership of renewable energy projects. In *Energy Research & Social Science* 17, pp. 135–146. https://doi.org/10.1016/j.erss.2016.04.021.

Goffman, Erving (1974): Frame analysis. An essay on the organization of experience / Erving Goffman. Cambridge, Mass.: Harvard University Press.

Goldthau, Andreas (2014): Rethinking the governance of energy infrastructure: Scale, decentralization and polycentrism. In *Energy Research & Social Science* 1, pp. 134–140. https://doi.org/10.1016/j.erss.2014.02.009.

Gomm, Roger; Hammersley, Martyn; Foster, Peter (Eds.) (2000): Case study method. Key issues, key texts. London, Thousand Oaks, Calif.: SAGE.

Goodwin, Jeff; Jasper, James M. (1999): Caught in a winding, snarling vine: The structural bias of political process theory. In *Sociological Forum* 14 (1), pp. 27–54. https://doi.org/10.1023/a:1021684610881.

Göpel, Maja (2016): The great mindshift. How a new economic paradigm and sustainability transformations go hand in hand. Springer Open (The anthropocene, 2367-4024, volume 2).

Grashof, Katherina (2019): Are auctions likely to deter community wind projects? And would this be problematic? In *Energy Policy* 125, pp. 20–32. https://doi.org/10.1016/j.enpol.2018.10.010.

Greenacre, Philip; Gross, Robert; Heptonstall, Phil (2010): Great Expectations: The cost of offshore wind in UK waters—understanding the past and projecting the future. UK Energy Research Centre Report, September 2010. Available online at https://core.ac.uk/download/pdf/19456320.pdf, checked on 22/6/2019.

Greenberg, Michael R. (2014): Energy policy and research: The underappreciation of trust. In *Energy Research & Social Science* 1, pp. 152–160. https://doi.org/10.1016/j.erss.2014. 02.004.

Grimes, Marcia (2006): Organizing consent. The role of procedural fairness in political trust and compliance. In *European Journal of Political Research* 45 (2), pp. 285–315. https:// doi.org/10.1111/j.1475-6765.2006.00299.x.

Gross, Catherine (2007): Community perspectives of wind energy in Australia. The application of a justice and community fairness framework to increase social acceptance. In *Energy Policy* 35 (5), pp. 2727–2736. https://doi.org/10.1016/j.enpol.2006.12.013.

Gross, Robert; Heptonstall, Phil; Blyth, Will (2007): Investment in Electricity Generation - the role of costs, incentives and risk, UK Energy Research Centre, London. Available online at http://www.ukerc.ac.uk/publications/investment-in-electricity-generation-the-role-of-costs-incentives-and-risks.html, checked on 6/22/2019.

Gründinger, Wolfgang (2017): Drivers of Energy Transition. How Interest Groups Influenced Energy Politics in Germany. Wiesbaden (Energiepolitik und Klimaschutz. Energy Policy and Climate Protection). Available online at http://dx.doi.org/10.1007/978-3-658-17691-4.

Gu, Weidong (Ed.) (2009): World Non-Grid-Connected Wind Power and Energy Conference, 2009. WNWEC 2009; Nanjing, China, 24-26 Sept. 2009. 2009 World Non-Grid-Connected Wind Power and Energy Conference (WNWEC 2009). Nanjing, China, 9/24/2009 - 9/26/2009. Piscataway, NJ: IEEE Press.

Gubbins, Nicholas (2007): Community Energy in Practice. In *Local Economy* 22 (1), pp. 80–84.

Gullberg, Anne Therese; Ohlhorst, Dörte; Schreurs, Miranda (2014): Towards a low carbon energy future—Renewable energy cooperation between Germany and Norway. In *Renewable Energy* 68, pp. 216–222. https://doi.org/10.1016/j.renene.2014.02.001.

Hadjilambrinos, Constantine (2000): Understanding technology choice in electricity industries: a comparative study of France and Denmark. In *Energy Policy* 28 (15), pp. 1111–1126. https://doi.org/10.1016/s0301-4215(00)00067-7.

Hager, Carol J.; Haddad, Mary Alice (2015): NIMBY is beautiful. Cases of local activism and environmental innovation around the world. New York: Berghahn Books.

Haggett, Claire; Aitken, Mhairi (2015): Grassroots Energy Innovations: the Role of Community Ownership and Investment. In *Curr Sustainable Renewable Energy Rep* 2 (3), pp. 98–104. https://doi.org/10.1007/s40518-015-0035-8.

Haggett, Claire; Creamer, Emily; Harnmeijer, Jelte; Parsons, Matthew; Bomberg, Elizabeth (2013): Community Energy in Scotland: the Social Factors for Success. Available online at http://www.deg.wales/wp-content/uploads/2015/09/CommunityEnergyinScotland-the SocialFactorsforSuccess.pdf, checked on 4/25/2019.

Hallberg, Lillemor R-M. (2006): The "core category" of grounded theory: Making constant comparisons. In *International Journal of Qualitative Studies on Health and Well-being* 1, pp. 141–148.

Hannes Stephan (2012): Revisiting the Transatlantic Divergence over GMOs: Toward a Cultural-Political Analysis. In *Global Environmental Politics* 12 (4), pp. 104–124.

Haraway, Donna (1988): Situated Knowledges: The Science Question in Feminism and the Privilege of Partial Perspective. In *Feminist Studies* 14 (3), pp. 575–599. https://doi.org/ 10.2307/3178066.

Hargreaves, Tom; Hielscher, Sabine; Seyfang, Gill; Smith, Adrian (2013): Grassroots innovations in community energy: The role of intermediaries in niche development. In *Global
Environmental Change* 23 (5), pp. 868–880. https://doi.org/10.1016/j.gloenvcha.2013.
02.008.

Harnmeijer, Anna; Harnmeijer, Jelte; McEwen, Nicola; Bhopal, Vijay (2014): A report on
community renewable energy in Scotland. Scene Connect Report. Sustainable Community Network. Available online at http://static1.squarespace.com/static/536b92d8e
4b0750dff7e241c/t/53f2251ce4b04928a223d78f/1408378140506/SCENE_Connect_R
eport_Scotland.pdf, checked on 4/25/2019.

Hauser, Eva; Weber, Andreas; Zipp, Alexander; Leprich, Uwe (2014): Bewertung von Aus-
schreibungsverfahren als Finanzierungsmodell für Anlagen erneuerbarer Energienutzung.
Saarbrücken. Available online at https://www.bee-ev.de/fileadmin/Publikationen/Studien/
IZES20140627IZESBEE_EE-Ausschreibungen.pdf, checked on 5/21/2019.

Hawkins, Virgil (2002): The Other Side of the CNN Factor: the media and conflict. In
Journalism Studies 3 (2), pp. 225–240. https://doi.org/10.1080/14616700220129991.

Hayward, Tim (2006): Ecological citizenship: a rejoinder. In *Environmental Politics* 15 (3),
pp. 452–453. https://doi.org/10.1080/09644010600627782.

Heath, Chip; Heath, Dan (2007): Made to stick. Why some ideas survive and others die. 1st
ed. New York: Random House.

Heiskala, Risto (2016): The evolution of the sources of social power, and some extensions. In :
Global powers. Michael Mann's anatomy of the twentieth century and beyond. Cambridge,
New York: Cambridge University Press, pp. 11–37.

Heiskala, Risto (2014): Evidence and interest in social theory. In *Acta Sociologica* 57 (4),
pp. 279–292. https://doi.org/10.1177/0001699314543568.

Heiskala, Risto (2001): Theorizing power: Weber, Parsons, Foucault and neostructuralism. In
Social Science Information 40 (2), pp. 241–264. https://doi.org/10.1177/053901801040
002003.

Heiskanen, Eva; Hyysalo, Sampsa; Kotro, Tanja; Repo, Petteri (2010a): Constructing innovative users and user-inclusive innovation communities. In *Technology Analysis & Strategic
Management* 22 (4), pp. 495–511. https://doi.org/10.1080/09537321003714568.

Heiskanen, Eva; Johnson, Mikael; Robinson, Simon; Vadovics, Edina; Saastamoinen, Mika
(2010b): Low-carbon communities as a context for individual behavioural change. In
Energy Policy 38 (12), pp. 7586–7595. https://doi.org/10.1016/j.enpol.2009.07.002.

Hergesell, Jannis; Maibaum, Arne; Minnetian, Clelia; Sept, Ariane (Eds.) (2018): Innovationsphänomene. Modi und Effekte der Innovationsgesellschaft. Wiesbaden: Springer.

Héritier, Adrienne (2002): Common Goods. Reinventing European Integration Governance. Lanham, Maryland: Lexington Books (Governance in Europe Series). Available
online at http://gbv.eblib.com/patron/FullRecord.aspx?p=1380498.

Hermansen, Søren (2007): Samsø, a Renewable Energy Island: 10 years of development and
evaluation. 10 year report. Samsø, Vedvarende Energi-Ö: 10 ars Udvikling og Evaluering
(10arsrapport). Samsø. Available online at http://arkiv.energiinstituttet.dk/101/, checked
on 3/27/2019.

Hermansen, Søren (2013): Commonity = Common + Community, 12/20/2013. Available online at http://tedxcopenhagen.dk/talks/commonity-commoncommunity, checked on
3/21/2019.

Herrberg, Anne (2009): Die Schönauer Stromrebellen - "Es geht um viel mehr als nur Strom". In *Deutsche Welle*, 3/9/2009. Available online at https://www.dw.com/de/die-sch% C3%B6nauer-stromrebellen-es-geht-um-viel-mehr-als-nur-strom/a-4083873, checked on 4/1/2019.

Hess, David J. (2013): Industrial fields and countervailing power: The transformation of distributed solar energy in the United States. In *Global Environmental Change* 23 (5), pp. 847–855. https://doi.org/10.1016/j.gloenvcha.2013.01.002.

Heyen, Dirk Arne (2017): Governance of exnovation: phasing out non-sustainable structures. Oeko-Institut Working Paper 2/2017. Freiburg. Available online at https://www.oeko.de/ fileadmin/oekodoc/WP-Exnovation-EN.pdf, checked on 4/12/2019.

Heywood, Andrew (2002): Politics. 2nd ed. London, Basingstoke: Palgrave Macmillan (Palgrave foundations).

HIE (2018): Highlands and Islands Enterprise Annual Report and Accounts 2017-18. Edinburgh (SG/2018/142). Available online at http://news.hie.co.uk/media/1441/hieplusannua lplusaccountsplus2017-18.pdf, checked on 5/19/2019.

Hielscher, Sabine; Seyfang, Gill; Smith, Adrian (2013): Grassroots innovations for sustainable energy: Exploring niche-development processes among community-energy initiatives. In Maurie J. Cohen, Halina Szejnwald Brown, Philip Vergragt (Eds.): Innovations in sustainable consumption. New economics, socio-technical transitions and social practices. Cheltenham: Edward Elgar Publishing (Advances in ecological economics), pp. 133–158.

Hirschl, Bernd (2016): Energiepolitische Grabenkämpfe - reloaded. In *energiezukunft.eu*, 6/13/2016. Available online at https://www.energiezukunft.eu/meinung/meinung-der-woche/energiepolitische-grabenkaempfe-reloaded-gn104122/, checked on 4/1/2019.

Hitzler, Ronald (2014): Wohin des Wegs? In : Qualitative research analysis and discussion - 10 Years Berlin Methodentreffen. Wiesbaden: Springer Fachmedien, pp. 55–72.

Hitzler, Ronald (2015): Sinngemäßes. In Achim Brosziewski, Christoph Maeder, Julia Nentwich (Eds.): Vom Sinn der Soziologie. Festschrift für Thomas S. Eberle. Wiesbaden: Springer (Wissen, Kommunikation und Gesellschaft), pp. 115–135.

Holstenkamp, Lars; Müller, Jakob (2013): Zum Stand von Energiegenossenschaften in Deutschland. Ein statistischer Überblick zum 31.12.2012. Lüneburg (Arbeitspapierreihe Wirtschaft & Recht, 14). Available online at https://www.leuphana.de/fileadmin/user_u pload/PERSONALPAGES/_ijkl/janner_steve/Homepage_Master/wpbl_14.pdf, checked on 4/26/2019.

Holstenkamp, Lars; Ulbrich, Stefanie (2010): Bürgerbeteiligung mittels Fotovoltaikgenossenschaften. Marktüberblick und Analyse der Finanzierungsstruktur. Leuphana Universität Lüneburg (Arbeitspapierreihe Wirtschaft & Recht, Nr. 20). Available online at https://www.leuphana.de/fileadmin/user_upload/Forschungseinrichtungen/ifwr/ files/Arbeitpapiere/WPBL8-101215.pdf, checked on 4/26/2019.

House of Commons, Energy and Climate Change Committee (2010): The proposals for national policy statements on energy. Third Report of Session 2009-2010. London: The Stationery Office Ltd.

Huber, Stefanie; Horbaty, Robert; Ellis, Geraint (2016 (2012)): Social Acceptance of Wind Power Projects: Learning from Trans-National Experience. In: Learning from wind power. Governance, societal and policy perspectives on sustainable energy. Basingstoke (Energy, climate and the environment), pp. 215–234.

Hinshelwood, Emily (2001): Power to the People: community-led wind energy - obstacles and opportunities in a South Wales Valley. In *Community Development Journal* 36 (2), pp. 96–110. https://doi.org/10.1093/cdj/36.2.96.

HM (Her Majesty's) Government (2008) Climate Change Act (C.27). Source: https://www.legislation.gov.uk/ukpga/2008/27/contents

HM (Her Majesty's) Government (1998): Scotland Act No. 3178 (C.79) (S.193). Source: http://www.legislation.gov.uk/uksi/1998/3178/contents/made.

Hobson, K. (2002): Competing Discourses of Sustainable Consumption. Does the 'Rationalisation of Lifestyles' Make Sense? In *Environmental Politics* 11 (2), pp. 95–120. https://doi.org/10.1080/714000601.

Hoffman, Steven M.; High-Pippert, Angela (2010): From private lives to collective action: Recruitment and participation incentives for a community energy program. In *Energy Policy* 38 (12), pp. 7567–7574. https://doi.org/10.1016/j.enpol.2009.06.054.

Hoffman, Steven M.; High-Pippert, Angela (2005): Community Energy: A Social Architecture for an Alternative Energy Future. In *Bulletin of Science, Technology & Society* 25 (5), pp. 387–401. https://doi.org/10.1177/0270467605278880.

Hogan, John; Howlett, Michael (Eds.) (2015): Policy Paradigms in Theory and Practice. Discourses, Ideas and Anomalies in Public Policy Dynamics. London: Palgrave Macmillan (Studies in the Political Economy of Public Policy).

Hollstein, Bettina; Ullrich, Carsten (2003): Einheit trotz Vielfalt? Zum konstitutiven Kern qualitativer Forschung. In *Soziologie* 32 (4), pp. 29–43.

Hollweck, Trista (2016): Robert K. Yin. (2014). Case Study Research Design and Methods (5th ed.). Thousand Oaks, CA: Sage. 282 pages. In *CJPE*. https://doi.org/10.3138/cjpe.30.1.108.

Holstenkamp, Lars; Radtke, Jörg (Eds.) (2018): Handbuch Energiewende und Partizipation. Wiesbaden: Springer VS Handbuch.

Holton, Judith (2010): The coding process and its challenges. In *The Grounded Theory Review* 9 (1), pp. 21–40.

Howden, Bruce (1998): Using action research to enhance the teaching of writing. In *QJER* 14. Available online at http://www.iier.org.au/qjer/qjer14/howden.html, checked on 11/20/2015.

Huitema, Dave; Meijerink, Sander (2010): Realizing water transitions: the role of policy entrepreneurs in water policy change. In *Ecology and Society* 15 (2).

Hurding, Gillian (2018): Engaging Communities in a Smart Energy Future. The experience of Scottish Communities. Conference. Edinburgh, 9/25/2018. Available online at https://www.regen.co.uk/wp-content/uploads/Community-Energy-Innovation-Regen_GHurding.pdf, checked on 4/30/2019.

Hull, George (Ed.) (2017): Equal society - essays on equality in theory and practice. Lanham, Maryland: Lexington Books.

Huybrechts, Benjamin; Mertens, Sybille (2014): The Relevance of the Cooperative Model in the Field of Renewable Energies. In *Annals of Public and Cooperative Economics* 85 (2), pp. 193–212. https://doi.org/10.1111/apce.12038.

Icek, Ajzen (2005): Attitudes, Personality and Behavior. Second Edition. Maidenhead, New York: Open University Press.

IEA (2017): Energy Policies of IEA Countries: Denmark 2017. Paris (Energy Policies of IEA Countries).

IPCC (2018): Global Warming of 1.5°C. An IPCC Special Report on the impacts of global warming of 1.5°C above pre-industrial levels and related global greenhouse gas emission pathways, in the context of strengthening the global response to the threat of climate change, sustainable development, and efforts to eradicate poverty [Masson-Delmotte, V., P. Zhai, H.-O. Pörtner, D. Roberts, J. Skea, P.R. Shukla, A. Pirani, W. Moufouma-Okia, C. Péan, R. Pidcock, S. Connors, J.B.R. Matthews, Y. Chen, X. Zhou, M.I. Gomis, E. Lonnoy, T. Maycock, M. Tignor, and T. Waterfield (eds.)]. In Press. Available online at https://www.ipcc.ch/sr15/, checked on 12/18/2018.

IRENA (2013): 30 Years of Policies for WInd Energy. Lessons from 12 Wind Energy Markets. Available online at https://www.irena.org/-/media/Files/IRENA/Agency/Public ation/2013/GWEC_WindReport_All_web-display.pdf, checked on 4/12/2016.

Jacobi, Nadia von; Nicholls, Alex; Chiappero-Martinetti, Enrica (2017): Theorizing Social Innovation to Address Marginalization. In *Journal of Social Entrepreneurship* 8 (3), pp. 265–270. https://doi.org/10.1080/19420676.2017.1380340.

Jacobsson, Steffan; Lauber, Volkmar (2006): The politics and policy of energy system transformation—explaining the German diffusion of renewable energy technology. In *Energy Policy* 34 (3), pp. 256–276. https://doi.org/10.1016/j.enpol.2004.08.029.

Janzing, Bernward (2008): Störfall mit Charme. Die Schönauer Stromrebellen im Widerstand gegen die Atomkraft; wie eine Elterninitiative, die sich nach Tschernobyl gründet, zu einem bundesweiten Stromversorger wird. Vöhrenbach: doldverlag.

Janzing, Bernward (2016a): "Atomkraft einfach wegsparen". Teil 2 der EWS Geschichte von Bernward Janzing. EWS Energiewende Magazin. Available online at https://www.ews-schoenau.de/energiewende-magazin/zur-ews/geschichte-02-ein-stromsparwettbe werb/, checked on 3/25/2019.

Janzing, Bernward (2016b): Der Kampf ums Stromnetz beginnt. Teil 3 der EWS Geschichte. Schönau. Available online at https://www.ews-schoenau.de/energiewende-magazin/zur-ews/geschichte-03-kampf-ums-netz/, checked on 3/27/2019.

Janzing, Bernward (2017a): Durchbruch nach zweitem Anlauf. Teil 4 der EWS Geschichte. Available online at https://www.ews-schoenau.de/energiewende-magazin/zur-ews/geschi chte-04-durchbruch-nach-zweitem-anlauf/, checked on 3/27/2019.

Janzing, Bernward (2017b): Eine Kampagne ebnet den Weg. Teil 5 der EWS Geschichte. Schönau. Available online at https://www.ews-schoenau.de/energiewende-magazin/zur-ews/geschichte-05-eine-kampagne-ebnet-den-weg/, checked on 3/27/2019.

Jasanoff, Sheila; Kim, Sang Hyun (Eds.) (2015): Dreamscapes of modernity. Sociotechnical imaginaries and the fabrication of power. Chicago, London: University of Chicago Press.

Jasper, James T.; Poulsen, Jane D. (1995): Recruiting Strangers and Friends: Moral Shocks and Social Networks in Animal Rights and Anti-Nuclear Protests. In *Social Problems* 42 (4), pp. 493–512.

Jenkins, Craig J. (2001): Social Movements: Resource Mobilization Theory. In *Annual Review of Sociology*, pp. 14368–14371.

Jenkins, Craig J. (1983): Resource Mobilization Theory and the Study of Social Movements. In *Annual Review of Sociology* 9 (1), pp. 527–553. https://doi.org/10.1146/annurev.so.09. 080183.002523.

Jobert, Arthur; Laborgne, Pia; Mimler, Solveig (2007): Local acceptance of wind energy: Factors of success identified in French and German case studies. In *Energy Policy* 35 (5), pp. 2751–2760. https://doi.org/10.1016/j.enpol.2006.12.005.

John Muir Trust (10/30/2015): Trust welcomes collapse of Rannoch wind farm application. Available online at https://www.johnmuirtrust.org/latest/news/694-trust-welcomes-collapse-of-rannoch-wind-farm-application, checked on 2/23/2016.

John Muir Trust (11/2/2015): Trust continues fight for wild land in far north of Scotland. Available online at https://www.johnmuirtrust.org/latest/news/693-trust-continues-fight-for-wild-land-in-far-north-of-scotland, checked on 2/23/2016.

Jolivet, Eric; Heiskanen, Eva (2010): Blowing against the wind—An exploratory application of actor network theory to the analysis of local controversies and participation processes in wind energy. In Energy Policy 38 (11), pp. 6746–6754. https://doi.org/10.1016/j.enpol.2010.06.044.

Jordan, Steven; Kapoor, Dip (2016): Re-politicizing participatory action research: unmasking neoliberalism and the illusions of participation. In Educational Action Research 24 (1), pp. 134–149. https://doi.org/10.1080/09650792.2015.1105145.

Jørgensen, Ulrik (2012): Mapping and navigating transitions—The multi-level perspective compared with arenas of development. In Research Policy 41 (6), pp. 996–1010. https://doi.org/10.1016/j.respol.2012.03.001.

Jørgensen, Ulrik; Karnøe, Peter (1995): The Danish wind-turbine story: technical solutions to political visions? http://www.ipl.dtu.dk/publikation/200/dk/ pp. 57–81.

Julian, Caroline (2014): Creating Local Energy Economies. Lessons from Germany. Available online at https://www.wcmt.org.uk/sites/default/files/migrated-reports/1155_1.pdf, checked on 5/23/2019.

Kabisch, Sigrun; Kunath, Anna K.; Schweizer-Ries, Petra; Steinführer, Annett (Eds.) (2012): Vulnerability, Risks, and Complexity. Impacts of Global Change on Human Habitats. 1st ed. Ashland: Hohgrefe Publishing (Advances in People-Environment Studies, v.3). Available online at https://ebookcentral.proquest.com/lib/gbv/detail.action?docID=563 1200.

Kahla, Franziska; Holstenkamp, Lars; Müller, Jakob R.; Degenhart, Heinrich (2017): Entwicklung und Stand von Bürgerenergiegesellschaften und Energiegenossenschaften in Deutschland. Leuphana Universität Lüneburg (Working Paper Series in Business and Law, 27). Available online at https://www.buendnis-buergerenergie.de/fileadmin/user_u pload/wpbl27_BEG-Stand_Entwicklungen.pdf, checked on 4/10/2019.

Kakabadse, Nada K.; Kakabadse, Andrew P.; Kalu, Kalu N. (2007): Communicative Action through Collaborative Inquiry: Journey of a Facilitating Co-Inquirer. In Syst Pract Act Res 20 (3), pp. 245–272. https://doi.org/10.1007/s11213-006-9061-1.

Kalkbrenner, Bernhard J.; Roosen, Jutta (2016): Citizens' willingness to participate in local renewable energy projects. The role of community and trust in Germany. In Energy Research & Social Science 13, pp. 60–70. https://doi.org/10.1016/j.erss.2015.12.006.

Kates, Robert W.; Clark, William; Corell, RObert; Hall, Michael; Jaeger, Carlo; Lowe, Ian et al. (2001): Environment and Development: Sustainability Science. In Science 292 (5517), pp. 641–642. https://doi.org/10.1126/science.1059386.

Kemp, René (1994): Technology and the transition to environmental sustainability. In Futures 26 (10), pp. 1023–1046. https://doi.org/10.1016/0016-3287(94)90071-x.

Kemp, René; Schot, Johan; Hoogma, Remco (1998): Regime shifts to sustainability through processes of niche formation. The approach of strategic niche management. In Technology Analysis & Strategic Management 10 (2), pp. 175–198. https://doi.org/10.1080/095373 29808524310.

Kemp, René; Rip, Arie; Schot, Johan (2001): Constructing transition paths through the management of niches. In Raghu Garud, Peter Karnøe (Eds.): Path Dependence and Creation. London: Routledge, pp. 269–299.

Kern, Florian; Kuzemko, Caroline; Mitchell, Catherine (2015): How and Why Do Policy Paradigms Change; and Does It Matter? The Case of UK Energy Policy. In John Hogan, Michael Howlett (Eds.): Policy Paradigms in Theory and Practice. Discourses, Ideas and Anomalies in Public Policy Dynamics, vol. 26. London: Palgrave Macmillan (Studies in the Political Economy of Public Policy), pp. 269–291.

Kessi, Shose; Howarth, Caroline (2015): Social change and social steadiness: Reflective community? In Giuseppina Marsico, Ruggero Andrisano Ruggieri, Sergio Salvatore (Eds.): Reflexivity and psychology, vol. 6. Charlotte, NC: Information Age Publishing (YIS: The yearbook of idiographic science), pp. 343–363.

Khwaja, Asim Ijaz (2009): Can good projects succeed in bad communities? In *Journal of Public Economics* 93 (7-8), pp. 899–916. https://doi.org/10.1016/j.jpubeco.2009.02.010.

Kingdon, John W. (1984): Agendas, alternatives and public policies. Boston: Little Brown.

Klemisch, Herbert (2014): Energiegenossenschaften als regionale Antwort auf den Klimawandel. In Carolin Schröder, Heike Walk: Genossenschaften und Klimaschutz. Akteure für zukunftsfähige, solidarische Städte. Wiesbaden: Springer (Bürgergesellschaft und Demokratie, Band 41), pp. 149–166.

Klemisch, Herbert; Boddenberg, Moritz (2016): Energiegenossenschaften und Nachhaltigkeit. Aktuelle Tendenzen und soziologische Überlegungen. In *Soziologie und Nachhaltigkeit* 6. Available online at https://www.google.com/url?sa=t&rct=j&q=& esrc=s&source=web&cd=9&cad=rja&uact=8&ved=2ahUKEwi-jNGFlcXhAhUPmhQ KHcHgCHQQFjAIegQICRAC&url=https%3A%2F%2Fwww.uni-muenster.de%2FEjou rnals%2Findex.php%2Fsun%2Farticle%2Fview%2F1845%2F1790&usg=AOvVaw2EC HMwA4NhlkfQyaeXOPrG, checked on 4/10/2019.

Klemisch, Herbert; Maron, Helene (2010): Genossenschaftliche Lösungsansätze zur Sicherung der kommunalen Daseinsvorsorge. In *Zeitschrift für das gesamte Genossenschaftswesen* 60 (1), pp. 3–13.

Klimaretter.info (2007): Umweltschützer fordern Verbot von Kohlekraftwerken. In *klimaretter.info*, 9/23/2007. Available online at http://www.klimaretter.info/index.php?option= com_content&view=article&id=207&Itemid=154, checked on 3/19/2019.

Kochems, Johannes; Hauser, Eva; Grashof, Katherina (2015): Auswertung von Ausschreibungen erneuerbarer Energien im Ausland. Fallstudie 3: Ausschreibungen für erneuerbare Energien in Südafrika. Saarbrücken. Available online at https://www.iass-potsdam.de/sites/default/files/files/2015-11-10_izes_ee-ausschreibun gen_iass_suedafrika_endbericht_final.pdf.

Kohn, Tamara (2002): Becoming an Islander through Action in The Scottish Hebrides. In *Journal of the Royal Anthropological Institute* 8 (1), pp. 143–158. https://doi.org/10.1111/ 1467-9655.00103.

Köppel, Johann (2016): Energiewende - Manifestierung des Ausgangspfades? In Werner Rammert, Arnold Windeler, Hubert Knoblauch, Michael Hutter (Eds.): Innovationsgesellschaft heute. Wiesbaden: Springer VS, pp. 301–322.

Korhonen, Jouni; Nuur, Cali; Feldmann, Andreas; Birkie, Seyoum Eshetu (2018): Circular economy as an essentially contested concept. In *Journal of Cleaner Production* 175, pp. 544–552. https://doi.org/10.1016/j.jclepro.2017.12.111.

Kosicki, Gerald M. (1993): Problems and Opportunities in Agenda-Setting Research. In *Journal of Communication* 43 (2), pp. 100–127. https://doi.org/10.1111/j.1460-2466.1993.tb0 1265.x.

Kreiss, J.; Ehrhart, K-M; Hanke, A-K (2017): Auction-theoretic analyses of the first offshore wind energy auction in Germany. In *Journal of Physics: Conference Series* 926, p. 12015. https://doi.org/10.1088/1742-6596/926/1/012015.

Kristensen, Michael (2015): Cirkulaer økonomi er blevet en del af Samsøs udbuds- of ind købspolitik. Cirkulaer Ökonomi. (in Danish). Samsø. Available online at http://arkiv.ene rgiinstituttet.dk/id/eprint/532, checked on 6/13/2016.

Krohn, Søren (2000): The wind turbine market in Denmark. Kopenhagen. Available online at http://ele.aut.ac.ir/~wind/en/articles/wtmindk.html, checked on 3/21/2019.

Krott, Max; Bader, Axel; Schusser, Carsten; Devkota, Rosan; Maryudi, Ahmad; Giessen, Lukas; Aurenhammer, Helene (2014): Actor-centred power: The driving force in decentralised community based forest governance. In *Forest Policy and Economics* 49, pp. 34–42. https://doi.org/10.1016/j.forpol.2013.04.012.

Kubal, Timothy; Becerra, Rene (2014): Social Movements and Collective Memory. In *Sociology Compass* 8 (6), pp. 865–875. https://doi.org/10.1111/soc4.12166.

Kuchta, Thorsten (2015): Noch mal Streit um Konzession. In *Nordwest Zeitung online*. Available online at https://www.nwzonline.de/oldenburg/politik/noch-mal-streit-um-kon zession_a_25,0,1704153769.html, checked on 2/13/2017.

Kuchta, Thorsten (2014): Netz: EWE bekommt Zuschlag. In *Nordwest Zeitung online*. Available online at https://www.nwzonline.de/oldenburg/wirtschaft/netz-ewe-bekommt-zuschlag_a_12,5,2546149542.html, checked on 2/13/2017.

Kuchta, Thorsten (2013a): Energie-Genossen werben knapp 42.000 Euro ein. In *Nordwest Zeitung online*, 1/11/2013. Available online at https://www.nwzonline.de/oldenb urg/wirtschaft/energie-genossen-werben-knapp-42000-euro-ein_a_2,0,632281323.html, checked on 3/20/2019.

Kuchta, Thorsten (2013b): Kriterien für Vergabe der Netze stehen fest. In *Nordwest Zeitung online*, 6/6/2013. Available online at https://www.nwzonline.de/oldenburg/kriterien-fuer-vergabe-der-netze-stehen-fest_a_6,1,3476486078.html, checked on 3/19/2019.

Kuchta, Thorsten (2011a): EWE-Strommix ändert sich ständig. In *Nordwest Zeitung online*, 5/20/2011. Available online at https://www.nwzonline.de/oldenburg/wirtschaft/ewe-str ommix-aendert-sich-staendig_a_1,0,611095031.html, checked on 3/21/2019.

Kuchta, Thorsten (2011b): Strom für 570 Haushalte aus der Hunte. In *Nordwest Zeitung online*, 8/6/2011. Available online at https://www.nwzonline.de/oldenburg/wirtschaft/strom-fuer-570-haushalte-aus-der-hunte_a_1,0,591084619.html, checked on 3/27/2019.

Kuchta, Thorsten (2011c): Energienetz - Initiative will erst EWE-Daten sichten. In *Nordwest Zeitung online*, 10/14/2011. Available online at https://www.nwzonline.de/oldenb urg/wirtschaft/initiative-will-erst-ewe-daten-sichten_a_1,0,568795458.html, checked on 3/21/2019.

Kuchta, Thorsten (2011d): Stromnetz - Bewerber mit Experten in der Hinterhand. In *Nordwest Zeitung online*, 12/16/2011. Available online at https://www.nwzonline.de/oldenburg/ wirtschaft/bewerber-mit-experten-in-der-hinterhand_a_1,0,548604577.html, checked on 3/21/2019.

Kuchta, Thorsten (2010): Oldenburg: Grüne wollen Alternative zu EWE. In *Nordwest Zeitung online*, 10/4/2010. Available online at https://www.nwzonline.de/wirtschaft/weser-ems/oldenburg-gruene-wollen-alternative-zu-ewe_a_1,0,739720095.html, checked on 3/21/2019.

Kungl, Gregor (2014): The Incumbent German Power Companies in a Changing Environmnet. A Comparison of E.ON, RWE, EnBW and Vattenfall from 1998 to 2013. Stuttgart (SOI Discussion Paper, 2014–03). Available online at https://www.sowi.uni-stuttgart.de/dokumente/forschung/soi/soi_2014_3_Kungl_The_ Incumbent_German_Power_Companies.pdf, checked on 3/19/2018.

Kunze, Conrad (2011): Soziographie ländlicher Energieprojekte: Eine vergleichende explorative Untersuchung über ländliche partizipative Initiativen zur Entwicklung regionaler Energie-Infrastrukturen mittels regenerativer Energien am Beispiel von sieben Kommunen in einem neuen Bundesland, Dissertationsschrift. Cottbus - Senftenberg; Brandenburgische Technische Universität Cottbus.

Lamnek, Siegfried (1995): Qualitative Sozialforschung. Methoden und Techniken. Weinheim: Beltz.

Land Reform Review Group (2014): The land of Scotland and the common good: a report. Final report. Available online at https://www.gov.scot/publications/land-reform-review-group-final-report-land-scotland-common-good/.

Lasswell, Harold Dwight (1936): Politics: Who gets what, when, how. New York, London, Whittlesey House: McGraw-Hill Book Co.

Latour, Bruno (1996): On Actor-Network Theory: A few clarifications. In *Soziale Welt* 47 (4), pp. 369–381.

Latour, Bruno Technology is Society made Durable. In John Law (Ed.): A Sociology of Monsters: Essays on Power, Technology and Domination. London: Routledge, pp. 103–132.

Latour, Bruno (1987): Science in Action: How to Follow Scientists and Engineers through Society. Cambridge, MA: Harvard University Press.

Lauber, Volkmar (2005): The Politics of European Union Policy on Support Schmes for Electricity from Renewable Energy Sources. Available online at http://www.windworks.org/cms/uploads/media/Lauber_-_EU_policy_on_support_schemes_for_electr icity_from_renewable_energy_sources_-_E_E_-_18_Oct_05.pdf.

Lauber, Volkmar; Mez, Lutz (2006): Renewable Electricity Policy in Germany, 1974 to 2005. In *Bulletin of Science, Technology & Society* 26 (2), pp. 105–120. https://doi.org/10.1177/ 0270467606287070.

Lauber, Volkmar; Mez, Lutz (2004): Three Decades of Renewable Electricity Policies in Germany. In *Energy & Environment* 15 (4), pp. 599–623. https://doi.org/10.1260/095830 5042259792.

Lauber, Volkmar: Renewable Energy at the level of the European Union. In Danyel Reiche (ed.): Handbook of Rneweable Energies in the European Union. Frankfurt a.M., Berlin, Bern, Bruxelles, New York, Oxford, Wien: Peter Lang, pp. 39–54.

Laursen, Lea Holst (2011): Differentiated decline in the Danish outskirt areas. Spatial restructuring and citizen-based development in the village of Klokkerholm. In *Danish Journal of Geoinformatics and Land Management* 46 (1), pp. 96–113.

Law, J. (Ed.) (1991): A Sociology of Monsters: Essays on Power, Technology and Domination. London: Routledge.

Legard, Robin; Keegan, Jill; Ward, Kit (2003): In-depth interview. In Jane Ritchie, Jane Lewis, Carol McNaughton Nicholls, Rachel Ormston (Eds.): Qualitative research practice. A guide for social science students and researchers. London, Thousand Oaks: SAGE, 138-169b.

Leibenath, Markus (2013): Konstruktivistische, interpretative Landschaftsforschung: Prämissen und Perspektiven. In Markus Leibenath, Stefan Heiland, Heiderose Kilper, Sabine Tzschaschel (Eds.): Wie werden Landschaften gemacht? Sozialwissenschaftliche Perspektiven auf die Konstituierung von Kulturlandschaften. Bielefeld: Transcript.

Leibenath, Markus; Heiland, S.; Kilper, H.; Tzschaschel, S. (Eds.) (2013) Wie werden Landschaften gemacht?. Sozialwissenschaftliche Perspektiven auf die Konstituierung von Kulturlandschaften. Bielefeld: Transcript.

Leibenath, Markus; Otto, Antje (2014): Competing Wind Energy Discourses, Contested Landscapes. In LO, pp. 1–18. https://doi.org/10.3097/lo.201438.

Lenzen, K. (2007): Die innovationsbiographische Rekonstruktion technischer Identitäten am Beispiel der Augmented Reality Technologie (Technology Studies Working Paper, TUTS-WP-7-2007).

Levin, David (2005): Framing Peace Policies: The Competition for Resonant Themes. In Political Communication 22 (1), pp. 83–108. https://doi.org/10.1080/105846005909 08456.

Lewis, Joanna I.; Wiser, Ryan H. (2007): Fostering a renewable energy technology industry: An international comparison of wind industry policy support mechanisms. In Energy Policy 35 (3), pp. 1844–1857. https://doi.org/10.1016/j.enpol.2006.06.005.

Lillevang, Lasse; Nielsen, Aage Johnsen (2002): Samsoe: The renewable energy island of Denmark. In News from DBDH 3, 2002. Available online at https://dbdh.dk/download/ hot-cool-magasin/renewable_energy/samsoe-renewable.pdf, checked on 3/14/2017.

Lincoln, Yvonna S.; Denzin, Norman K. (2003): Turning points in qualitative research. Tying knots in a handkerchief / edited by Yvonna S. Lincoln, Norman K. Denzin. Walnut Creek, Calif., Oxford (Crossroads in qualitative inquiry v. 3).

Lipp, Judith (2007): Lessons for effective renewable electricity policy from Denmark, Germany and the United Kingdom. In Energy Policy 35 (11), pp. 5481–5495. https://doi.org/ 10.1016/j.enpol.2007.05.015.

Litfin, Karen (2014): Ecovillages. Lessons for sustainable community. Cambridge: polity.

Lockwood, Matthew; Kuzemko, Caroline; Mitchell, Catherine; Hoggett, Richard (2017): Historical institutionalism and the politics of sustainable energy transitions: A research agenda. In Environment and Planning C: Politics and Space 35 (2), pp. 312–333. https:// doi.org/10.1177/0263774x16660561.

Lopez, Fanny; Pellegrino, Margot; Coutard, Olivier (Eds.) (2019): Local energy autonomy: spaces, scales, politics. Autonomie énergétique locale: espaces, échelles et politique. London: ISTE.

Lovins, Amory B. (1979 (1977)): Soft Energy Paths: Towards a Durable Peace. New York: Harper & Row.

Lucas, Nigel (1985): Western European energy policies. A comparative study of the influence of institutional structure on technical change. Oxford: Clarendon Press.

Lukes, Steven (2005): Power. A radical view. 2nd ed. Basingstoke: Palgrave Macmillan. Available online at http://www.loc.gov/catdir/bios/hol059/2004057346.html.

Lundberg, Liv (2019): Auctions for all? Reviewing the German wind power auctions in 2017. In Energy Policy 128, pp. 449–458. https://doi.org/10.1016/j.enpol.2019.01.024.

Lundén, Malene Anniki (2003): Vingesus. Whisper of wings. Århus: OVE.

Lundy, Patricia; McGovern, Mark (2006): The ethics of silence. In Action Research 4 (1), pp. 49–64. https://doi.org/10.1177/1476750306060542.

Luque-Ayala, Andres; Silver, Jonathan (Eds.) (2016): Energy, power and protest on the urban grid. Geographies of the electric city. London: Routledge.

Lynch, Marc (1999): State Interests and Public Spheres: The International Politics of Jordan's Identity. New York: Columbia University Press.

Maegaard, Preben (2013): Wind Power for the World. The Rise of Modern Wind Energy. With Anna Krenz, Wolfgang Palz. Boca Raton, FL: Taylor & Francis Group: CRC Press

Maegaard, Preben (2009): Wind energy development and application prospects of non-grid-connected wind power. In Weidong Gu (Ed.): World Non-Grid-Connected Wind Power and Energy Conference, 2009. WNWEC 2009 ; Nanjing, China, 24-26 Sept. 2009. 2009 World Non-Grid-Connected Wind Power and Energy Conference (WNWEC 2009). Nanjing, China, 9/24/2009 - 9/26/2009. Piscataway, NJ, pp. 1–3.

Maines, David (Ed.) (1991): Social Organization and Social Process. Essays in Honor of Anselm Strauss. New York: Aldine De Gruyter.

Maiter, Sarah; Simich, Laura; Jacobson, Nora; Wise, Julie (2008): Reciprocity. In *Action Research* 6 (3), pp. 305–325. https://doi.org/10.1177/1476750307083720.

Manheim, Jarol B. (1987): A Model of Agenda Dynamics. In *Annals of the International Communication Association* 10 (1), pp. 499–516. https://doi.org/10.1080/23808985.1987.11678659.

Mann, Michael (1986): The sources of social power. Volume 1, A History of Power from the Beginning to AD 1760. Cambridge, MA: Cambridge University Press.

Mann, Michael (1993): The sources of social power. Volume 2, The Rise of Classes and Nation States 1760-1914. Cambridge, MA: Cambridge University Press.

Mann, Michael (2012): The sources of social power. Volume 3, Global Empires and Revolution 1890-1945. New York, Cambridge: Cambridge University Press.

Mann, Michael (2013): The Sources of Social Power. Volume 4, Globalizations 1945-2011. Cambridge, MA: Cambridge University Press.

Markantoni, Marianna (2016): Low Carbon Governance: Mobilizing Community Energy through Top-Down Support? In *Env. Pol. Gov.* 26 (3), pp. 155–169. https://doi.org/10.1002/eet.1722.

Marmot, Michael (2007): Achieving health equity: from root causes to fair outcomes. In *The Lancet* 370 (9593), pp. 1153–1163. https://doi.org/10.1016/s0140-6736(07)61385-3.

Marres, Noortje (2015 (2012)): Material Participation. 2nd ed. London: Palgrave Macmillan. Available online at http://gbv.eblib.com/patron/FullRecord.aspx?p=4720609.

Marsico, Giuseppina; Ruggieri, Ruggero Andrisano; Salvatore, Sergio (Eds.) (2015): Reflexivity and psychology. Charlotte, NC: Information Age Publishing (YIS: The yearbook of idiographic science).

Martins, Ana Cravinho; Marques, Rui Cunha; Cruz, Carlos Oliveira (2011): Public–private partnerships for wind power generation: The Portuguese case. In *Energy Policy* 39 (1), pp. 94–104. https://doi.org/10.1016/j.enpol.2010.09.017.

Martiskainen, Mari (2017): The role of community leadership in the development of grassroots innovations. In *Environmental Innovation and Societal Transitions* 22, pp. 78–89. https://doi.org/10.1016/j.eist.2016.05.002.

Maruyama, Yasushi; Nishikido, Makoto; Iida, Tetsunari (2007): The rise of community wind power in Japan: Enhanced acceptance through social innovation. In *Energy Policy* 35 (5), pp. 2761–2769. https://doi.org/10.1016/j.enpol.2006.12.010.

Mathiesen, Brian Vad; Hansen, Kenneth; Ridjan, Iva; Lund, Henrik; Nielsen, Steffen (2015): Samsö Energy Vision 2010. Converting Samsö to 100% Renewable Energy. Available online at https://vbn.aau.dk/ws/portalfiles/portal/220291121/Sams_report_20151012.pdf, checked on 5/8/2019.

Mattes, Jannika; Huber, Andreas; Koehrsen, Jens (2015): Energy transitions in small-scale regions—What we can learn from a regional innovation systems perspective. In *Energy Policy* 78, pp. 255–264. https://doi.org/10.1016/j.enpol.2014.12.011.

Mautz, Rüdiger (2008): Auf dem Weg zur Energiewende: die Entwicklung der Stromproduktion aus erneuerbaren Energien in Deutschland. A study of the Sociological Research Institute Göttingen (SOFI). Göttingen. Available online at https://www.ssoar.info/ssoar/bitstream/handle/document/27291/ssoar-2008-mautz_et_al-auf_dem_weg_zur_energiewende.pdf?sequence=1, checked on 3/21/2019.

Maxwell, Douglas (2016): Human Rights and Land Reform: Unanswered Questions. In *Journal of the Law Society of Scotland* 61 (1), pp. 22–24.

Mayer, Axel (2005): AKW Wyhl: 30 Jahre nach der Bauplatzbesetzung. Ein Rückblick von Axel Mayer. Available online at http://www.bund-rvso.de/akw-bauplatzbesetzung-wyhl. html, updated on 10/27/2015, checked on 3/27/2019.

Mazzucato, Mariana; Semieniuk, Gregor (2018): Financing renewable energy: Who is financing what and why it matters. In *Technological Forecasting and Social Change* 127, pp. 8–22. https://doi.org/10.1016/j.techfore.2017.05.021.

McCarthy, John D.; Zald, Mayer N. (1977): Resource Mobilization and Social Movements: A Partial Theory. In *American Journal of Sociology* 82 (6), pp. 1212–1241.

McCarthy, John D.; Zald, Mayer N. (2017 (1987)): Social Movements in an Organizational Society. Collected Essays. Introduction by William A. Gamson. New York: Columbia University Press.

McCarthy, John D.; Zald, Mayer N. (2001): The Enduring Vitality of the Resource Mobilization Theory of Social Movements. In Jonathan H. Turner (Ed.): Handbook of sociological theory, vol. 56. New York, London: Springer (Handbooks of Sociology and Social Research), pp. 533–565.

McCombs, Maxwell E.; Shaw, Donald L. (1972): The Agenda-Setting Function of Mass Media. In *Public Opinion Quarterly* 36 (2), p. 176. https://doi.org/10.1086/267990.

McDonald, Rachel I.; Crandall, Christian S. (2015): Social norms and social influence. In *Current Opinion in Behavioral Sciences* 3, pp. 147–151. https://doi.org/10.1016/j.cob eha.2015.04.006.

Meadows, Donella (1999): Leverage Points. Places to Intervene in a System. Available online at http://www.donellameadows.org/wp-content/userfiles/Leverage_Points.pdf, checked on 5/25/2019.

Mendonça, Miguel; Lacey, Stephen; Hvelplund, Frede (2009): Stability, participation and transparency in renewable energy policy: Lessons from Denmark and the United States. In *Policy and Society* 27 (4), pp. 379–398. https://doi.org/10.1016/j.polsoc.2009.01.007.

Meuser, Michael; Nagel, Ulrike (2009): Das Experteninterview — konzeptionelle Grundlagen und methodische Anlage. In Susanne Pickel, Gert Pickel, Hans-Joachim Lauth, Detlef Jahn (Eds.): Methoden der vergleichenden Politik- und Sozialwissenschaft. Neue Entwicklungen und Anwendungen, 27/1. 1. Aufl. Wiesbaden: Springer VS (Lehrbuch), pp. 465–479.

Mey, Franziska; Diesendorf, Mark (2018): Who owns an energy transition? Strategic action fields and community wind energy in Denmark. In *Energy Research & Social Science* 35, pp. 108–117. https://doi.org/10.1016/j.erss.2017.10.044.

Mey, Günter; Mruck, Katja (Eds.) (2014): Qualitative research analysis and discussion - 10 Years Berlin Methodentreffen. Wiesbaden: Springer.

Mey, Günter; Mruck, Katja (2014): Qualitative Forschung: Analysen und Diskussionen. In Mey, Günter; Mruck, Katja (Eds.): Qualitative research analysis and discussion - 10 Years Berlin Methodentreffen. Fachmedien [Germany], Wiesbaden: Springer [Germany], pp. 9–32.

Meyer, Julienne (2000): Using qualitative methods in health related action research. In *BMJ : British Medical Journal* 320 (7228), pp. 178–181.

Meyer, Christian; Meier zu Verl, Christian: Ergebnispräsentation in der qualitativen Forschung. In Nina Baur, Jörg Blasius (eds.): Handbuch Methoden der empirischen Sozialforschung. Wiesbaden: Springer VS, pp. 245–257.

Meyer, Uli; Schubert, Cornelius (2007): Integrating path dependency and path creation in a general understanding of path constitution. The role of agency and institutions in the stabilisation of technological innovations. In *Science, Technology and Innovation Studies* 3, pp. 23–44.

Mez, Lutz (2003): New Corporate Strategies in the German Electricity Supply Industry. In Jean-Michel Glachant, Dominique Finon (Eds.): Competition in European electricity markets. A cross-country comparison. Cheltenham: Edward Elgar Publishing, pp. 193–216.

Michalena, Evanthie; Hills, Jeremy Maxwell (Eds.) (2013): Renewable energy governance. Complexities and challenges. London: Springer (Lecture Notes in Energy).

Middlemiss, Lucie; Parrish, Bradley d. (2010): Building capacity for low-carbon communities: The role of grassroots initiatives. In *Energy Policy* 38 (12), pp. 7559–7566. https://doi.org/10.1016/j.enpol.2009.07.003.

Miljø- og Fødevareministeriet (2005): Denmark's Climate Policy Objectives and Achievements. Report on Demonstrable Progress in 2005 under the Kyoto Protocol. Available online at https://unfccc.int/resource/docs/dpr/den1.pdf.

Dansk energipolitik 1976 (Danish Energy Politics) (1976). København.

Minkler, Meredith (2004): Ethical challenges for the "outside" researcher in community-based participatory research. In *Health education & behavior : the official publication of the Society for Public Health Education* 31 (6), pp. 684–697. https://doi.org/10.1177/109 0198104269566.

Mitchell, Catherine (2015): Designing energy policy under uncertainty. In *Nature Climate Change* 5 (6), pp. 517–518. https://doi.org/10.1038/nclimate2662.

Mitchell, Catherine (2010): Breaking Free of the Band of Iron. In : The political economy of sustainable energy. New York (Energy, climate and the environment), pp. 1–20.

Mitchell, Catherine (2010): The Difficulty of Delivering the 'Right' Change Quickly Enough. In: The political economy of sustainable energy. New York (Energy, climate and the environment), pp. 61–95.

Mitchell, Catherine (1996): Future Support of Renewable Energy in the UK—Options and Merits. In *Energy & Environment* 7 (3), pp. 267–284. https://doi.org/10.1177/0958305x9 600700303.

Mitchell, Catherine (2008) The political economy of sustainable energy. London: Palgrave Macmillan (Energy, climate and the environment).

Mitchell, Catherine (2012): Who is in charge of Britain's energy policy? In *the guardian*, 2/28/2012. Available online at https://www.theguardian.com/environment/2012/feb/28/britain-energy-politics, checked on 3/14/2016.

Mitchell, Catherine; Bauknecht, Dierk; Connor, Peter (2006): Effectiveness through risk reduction: a comparison of the renewable obligation in England and Wales and the feed-in system in Germany. In *Energy Policy* 34 (3), pp. 297–305. https://doi.org/10.1016/j.enpol.2004.08.004.

Mitchell, Catherine; Connor, Peter (2004): Renewable energy policy in the UK 1990–2003. In *Energy Policy* 32 (17), pp. 1935–1947. https://doi.org/10.1016/j.enpol.2004.03.016.

Mitchell, Catherine; Woodman, Bridget (2010): Towards trust in regulation—moving to a public value regulation. In *Energy Policy* 38 (6), pp. 2644–2651. https://doi.org/10.1016/j.enpol.2009.05.040.

Mitchell, J. Clyde (1983): Case and Situation Analysis. In *The Sociological Review* 31 (2), pp. 187–211. https://doi.org/10.1111/j.1467-954x.1983.tb00387.x.

Möller, Bernd; Sperling, Karl; Nielsen, Steffen; Smink, Carla; Kerndrup, Søren (2012): Creating consciousness about the opportunities to integrate sustainable energy on islands. In *Energy* 48 (1), pp. 339–345. https://doi.org/10.1016/j.energy.2012.04.008.

Morris, Aldon D.; Mueller, Carol McClurg (Eds.) (1992): Frontiers in Social Movement Theory (1992). New Haven: Yale University Press.

Morris, Aldon D.; Staggenborg, Suzanne (2004): Leadership in Social Movements. In: Snow, David A.; Soule, Sarah A.; Kriesi, Hanspeter (Eds.): The Blackwell Companion to Social Movements. Hoboken, New Jersey: Blackwell Publishing Ltd.

Moss, Timothy; Becker, Sören; Naumann, Matthias (2015): Whose energy transition is it, anyway? Organisation and ownership of the Energiewende in villages, cities and regions. In *Local Environment* 20 (12), pp. 1547–1563. https://doi.org/10.1080/13549839.2014.915799.

Müller, Jakob R. (2018): Narrative im Innovationsgeschehen der Energiewende. In Jannis Hergesell, Arne Maibaum, Clelia Minnetian, Ariane Sept: Innovationsphänomene. Modi und Effekte der Innovationsgesellschaft. Wiesbaden Springer VS research, pp. 225–246.

Muench, Stefan; Thuss, Sebastian; Guenther, Edeltraud (2014): What hampers energy system transformations? The case of smart grids. In *Energy Policy* 73, pp. 80–92. https://doi.org/10.1016/j.enpol.2014.05.051.

MICT Mull and Iona Community Trust (2019a): About. Membership. Craignure, Isle of Mull. Available online at http://www.mict.co.uk/about-us/membership/, checked on 3/29/2019.

MICT Mull and Iona Community Trust (2019b): About. History of Mull and Iona Community Trust. Craignure, Isle of Mull. Available online at http://www.mict.co.uk/about-us/, checked on 3/29/2019.

MICT Mull and Iona Community Trust (2019c): Island Castaways. About us. Craignure, Isle of Mull. Available online at https://www.islandcastaways.org/, checked on 3/29/2019.

MICT Mull and Iona Community Trust (2019d): Mull and Iona Ranger Service. Craignure, Isle of Mull. Available online at http://www.mict.co.uk/ranger-service/, checked on 3/29/2019.

MICT Mull and Iona Community Trust (2019e): Projects and Services. Pontoon Development. Craignure, Isle of Mull. Available online at http://www.mict.co.uk/projects-services/pontoon-development/, checked on 3/29/2019.

MICT Mull and Iona Community Trust (2019f): Projects and Services. Tobermory Lighthouse Path. Craignure, Isle of Mull. Available online at http://www.mict.co.uk/projects-services/tobermory-lighthouse-path/, checked on 3/29/2019.

MICT Mull and Iona Community Trust (2019g): Projects and Services. Community Childcare Facility. Craignure, Isle of Mull. Available online at http://www.mict.co.uk/projects-services/childcare/, checked on 3/29/2019.

MICT Mull and Iona Community Trust (2019h): Projects and Services. ReTHink MESS—Community Fridge. Available online at http://www.mict.co.uk/projects-services/rethink-mess/community-fridge/, checked on 3/29/2019.

MICT Mull and Iona Community Trust (2019i): Ulva Ferry Community Transport. About. Craignure, Isle of Mull. Available online at http://ufct.co.uk/about/, checked on 3/29/2019.

MICT Mull and Iona Community Trust (December 2015): The Community Trust Newsletter. 30.04.2019. Available online at http://www.mict.co.uk/wp-content/uploads/2015/12/December-15-newsletter.pdf, checked on 3/14/2016.

MICT Mull and Iona Community Trust (n.d.): Garmony Hydro under construction (image). Available online at https://www.mict.co.uk/projects-services/garmonyhydro/, checked on 3/14/2016.

Mulugetta, Yacob; Jackson, Tim; van der Horst, Dan (2010): Carbon reduction at community scale. In Energy Policy 38 (12), pp. 7541–7545. https://doi.org/10.1016/j.enpol.2010.05.050.

Munday, Max; Bristow, Gill; Cowell, Richard (2011): Wind farms in rural areas: How far do community benefits from wind farms represent a local economic development opportunity? In Journal of Rural Studies 27 (1), pp. 1–12. https://doi.org/10.1016/j.jrurstud.2010.08.003.

Murphy, Joseph (Ed.) (2012) Governing Technology for Sustainability (2nd edition). Hoboken: Earthscan.

Murphy, Joseph (Ed.) (2009): Framing the Present, Shaping the Future: Contemporary Governance of Sustainable Technologies. London: Earthscan.

Murphy, Joseph (2007): Governing technology for sustainability. London: Earthscan.

Musall, Fabian David; Kuik, Onno (2011): Local acceptance of renewable energy—A case study from southeast Germany. In Energy Policy 39 (6), pp. 3252–3260. https://doi.org/10.1016/j.enpol.2011.03.017.

Nadelmann, Ethan A. (1990): Global prohibition regimes: the evolution of norms in international society. In Int. Org. 44 (04), pp. 479–526. https://doi.org/10.1017/s0020818300035384.

nationalgridESO (2014): Grounds for constraint. Available online at https://www.nationalgrideso.com/news/grounds-constraint, updated on 2019, checked on 3/29/2019.

Smelser, Neil; Baltes, Paul B. (2001) International Encyclopedia of the Social and Behavioral Sciences. Oxford: Pergamon.

Neuman, Michael (2006): Infiltrating infrastructures: On the nature of networked infrastructure. In Journal of Urban Technology 13 (1), pp. 3–31. https://doi.org/10.1080/10630730600752728.

NEWTON, KENNETH (1997): Social Capital and Democracy. In American Behavioral Scientist 40 (5), pp. 575–586. https://doi.org/10.1177/0002764297040005004.

Nicholls, Alex; Ziegler, Rafael (2017(2015)): An extended Social Grid Model for the Study of Marginalization Processes and Social Innovation. Oxford. Available online at http://eur eka.sbs.ox.ac.uk/5947/, checked on 3/21/2019.

Nicholls, Walter J. (2007): The Geographies of Social Movements. In *Geography Compass* 1 (3), pp. 607–622. https://doi.org/10.1111/j.1749-8198.2007.00014.x.

Niederberger, Marlen (Eds.) (2015) Methoden der Experten- und Stakeholdereinbindung in der sozialwissenschaftlichen Forschung. Wiesbaden: Springer.

Nielsen, Søren Nors; Jørgensen, Sven Erik (2015): Sustainability analysis of a society based on exergy studies—a case study of the island of Samsø (Denmark). In *Journal of Cleaner Production* 96, pp. 12–29. https://doi.org/10.1016/j.jclepro.2014.08.035.

Nolden, Colin (2013): Governing community energy—Feed-in tariffs and the development of community wind energy schemes in the United Kingdom and Germany. In *Energy Policy* 63, pp. 543–552. https://doi.org/10.1016/j.enpol.2013.08.050.

North Data GmbH (2017): OLEGENO Oldenburger Energiegenossenschaft eG, Oldenburg. Available online at https://www.northdata.de/Olegeno+Oldenburger+Energie-Genossens chaft+eG.,+Oldenburg/GnR+200036, updated on 12/31/2017, checked on 4/1/2019.

NRGi (2018): Om os. Hvem er vi. (About us. Who are we?). Available online at https://nrgi. dk/privat/om-os/hvem-er-vi/, checked on 11/21/2018.

Nussbaum, Martha Craven (2006): Frontiers of Justice. Cambridge, MA: Harvard University Press.

Ockwell, David; Whitmarsh, Lorraine; O'Neill, Saffron (2009): Reorienting Climate Change Communityation for Effective Mitigation: Forcing People to be Green or Fostering Grass-Roots Engagement? In *Science Communication* 30 (3), pp. 305–327.

O'Keeffe, Aoife; Haggett, Claire (2012): An investigation into the potential barriers facing the development of offshore wind energy in Scotland: Case study—Firth of Forth offshore wind farm. In *Renewable and Sustainable Energy Reviews* 16 (6), pp. 3711–3721. https:// doi.org/10.1016/j.rser.2012.03.018.

Ofgem (2017): Final Decision: Introduction of SLC20A, Transmission Constraint Licence Condition. Effective 16.07.2017. Available online at https://www.ofgem.gov.uk/publicati ons-and-updates/final-decision-introduction-slc20a-transmission-constraint-licence-con dition, updated on 2019, checked on 3/29/2019.

Ofgem (2019): Electricity supply market shares by company: Domestic (GB). Available online at https://www.ofgem.gov.uk/data-portal/electricity-supply-market-shares-company-dom estic-gb, checked on 3/27/2019.

Ohlhorst, Dörte (2009): Windenergie in Deutschland. Wiesbaden: Springer.

Ohlhorst, Dörte; Kröger, Melanie (2015): Konstellationsanalyse: Einbindung von Experten und Stakeholdern in interdisziplinäre Forschungsprojekte. In : Methoden der Experten- und Stakeholdereinbindung in der sozialwissenschaftlichen Forschung. Wiesbaden, pp. 95–116.

Ohlhorst, Dörte; Schön, Susanne (2015): Constellation Analysis as a Means of Interdiscipli-nary Innovation Research—Theory Formation from the Bottom Up. In *Historical Social Research* 40(3), pp 258–278.

Okereke, Chukwumerije (2008): An exploration of motivations, drivers and barriers to carbon management: the UK FTSE 100. In *Strategic Direction* 24 (6). https://doi.org/10.1108/sd. 2008.05624fad.002.

olegeno (2019a): Das Konzessionsverfahren um das Strom- und Gasnetz in Oldenburg. Chronologie. Available online at https://www.olegeno.de/ueber-uns/konzessionsverfahren/chronologie.html, checked on 5/10/2019.

olegeno (2019b): Energie sparen. Available online at https://www.olegeno.de/solar-anlagen/energie-sparen.html, updated on 3/14/2019.

olegeno (2019c): Mini-PV: Solarenergie direkt vom Balkon. Available online at https://www.olegeno.de/solar-anlagen/mini-pv-anlagen.html, checked on 4/23/2019.

olegeno (2019d): Olegeno Bürger-Ökogas. Available online at https://www.olegeno.de/olegeno-buerger-oekogas.html, checked on 4/23/2019.

olegeno (2019e): Olegeno Bürger-Ökostrom. Available online at https://www.olegeno.de/olegeno-buerger-oekostrom.html, checked on 2/12/2019.

olegeno (2019f): Solardächer gesucht. Available online at https://www.olegeno.de/solar-anlagen/solard%C3%A4cher-gesucht.html, checked on 4/23/2019.

olegeno (2019g): Warum Genossenschaft? Available online at https://www.olegeno.de/ueber-uns/genossenschaft.html, checked on 4/12/2016.

olegeno (2014): Satzung in der Fassung von Dezember 2014. Available online at https://www.olegeno.de/assets/uploads/download/olegeno_satzung.pdf, checked on 4/23/2019

olegeno (2013): Angebot für die Stromkonzession der Stadt Oldenburg. Available online at https://www.olegeno.de/assets/uploads/konzession/angebot-konzession-strom_anonymisiert.pdf, checked on 5/10/2019

(8/22/2013): Konzession für das Strom- und Gasnetz: Olegeno bekommt Rückenwind aus Süddeutschland. Oldenburg. Available online at https://www.olegeno.de/presse/pm/unterstuetzung-ews-vom-22082013.html, checked on 3/21/2019.

Olesen, Gunnar Boye; Maegaard, Preben; Kruse, Jane (2002): Danish Experience in Wind Energy - Local Financing. Working Report for the WELFI project (Wind Energy Local Financing). Available online at https://www.umass.edu/windenergy/sites/default/files/downloads/pdfs/Danish_experience_Local_Financing_Wind.pdf, checked on 4/26/2019.

Olsthoorn, Xander; Wieczorek, Anna J. (Eds.) (2006): Understanding industrial transformation. Views from different disciplines. Dordrecht, Great Britain: Springer (Environment & Policy, v. 44).

Onyx, J.; Leonard, R. J. (2011): Complex systems leadership in emergent community projects. In Community Development Journal 46 (4), pp. 493–510. https://doi.org/10.1093/cdj/bsq041.

Ornetzeder, Michael; Rohracher, Harald (2013): Of solar collectors, wind power, and car sharing: Comparing and understanding successful cases of grassroots innovations. In Global Environmental Change 23 (5), pp. 856–867. https://doi.org/10.1016/j.gloenvcha.2012.12.007.

Oteman, Marieke; Wiering, Mark; Helderman, Jan-Kees (2014): The institutional space of community initiatives for renewable energy: a comparative case study of the Netherlands, Germany and Denmark. In Energ Sustain Soc 4 (1), p. 11. https://doi.org/10.1186/2192-0567-4-11.

Ott, Eckhard (2014): Please, in My Backyard - die Bedeutung von Energiegenossenschaften in der Energiewende. In Christian Aichele, Oliver Doleski (Eds.): Smart Market. Vom Smart Grid zum intelligenten Energiemarkt. Wiesbaden: Springer, pp. 829–841.

Owens, Susan; Driffill, Louise (2008): How to change attitudes and behaviours in the context of energy. In *Energy Policy* 36 (12), pp. 4412–4418. https://doi.org/10.1016/j.enpol.2008. 09.031.

Palmgreen, Philip; Clarke, Peter (1977): Agenda-Setting With Local and National Issues. In *Communication Research* 4 (4), pp. 435–452. https://doi.org/10.1177/009365027700 400404.

Papazu, Irina (2018): Storifying Samsø's Renewable Energy Transition. In *Science as Culture* 27 (2), pp. 198–220. https://doi.org/10.1080/09505431.2017.1398224.

Payne, Rodger (2001): Persuasion, Frames and Norm Construction. In *European Journal of International Relations* 7 (1), pp. 37–61. https://doi.org/10.1177/1354066101007001002.

Peters, Michael; Fudge, Shane; Sinclair, Philip (2010): Mobilizing community action towards a low-carbon future: Opportunities and challenges for local government in the UK. In *Energy Policy* 38 (12), pp. 7596–7603.

Phadke, Roopali (2010): Steel forests or smoke stacks: the politics of visualisation in the Cape Wind controversy. In *Environmental Politics* 19 (1), pp. 1–20. https://doi.org/10.1080/096 44010903396051.

Pickel, Susanne; Pickel, Gert (2009): Qualitative Interviews als Verfahren des Länder-vergleichs. In Susanne Pickel, Gert Pickel, Hans-Joachim Lauth, Detlef Jahn (Eds.): Methoden der vergleichenden Politik- und Sozialwissenschaft. Neue Entwicklungen und Anwendungen, vol. 30. 1. Aufl. Wiesbaden: Springer (Lehrbuch), pp. 441–464.

Pickel, Susanne; Pickel, Gert; Lauth, Hans-Joachim; Jahn, Detlev (2009) Methoden der ver-gleichenden Politik- und Sozialwissenschaft. Neue Entwicklungen und Anwendungen. 1. Aufl. Wiesbaden: Springer (Lehrbuch).

Pierson, Chris; Tormey, Simon (Eds.) (2000) Politics at the Edge. London: Palgrave Macmillan (Political Studies Association Yearbook Series).

Pinch, Trevor J.; Bijker, Wiebe E. (1984): The Social Construction of Facts and Artefacts: or How the Sociology of Science and the Sociology of Technology might Benefit Each Other. In *Social studies of science* 14 (3), pp. 399–441. https://doi.org/10.1177/030631 284014003004.

PlanLoCal (2016) Planning for low carbon living. Centre for Sustainable Living. Bristol, UK.

Pohlmann, Angela (forthcoming): Dismantling the relationship between energy innovations and power. In *Energy Policy*.

Pohlmann, Angela (2018): Situating social practices in community energy projects. Three case studies about the contextuality of renewable energy production. Wiesbaden: Springer (Energiepolitik und Klimaschutz).

Pohlmann, Angela (2011): Local climate change governance. Working Paper. Universität Hamburg (Global Transformations towards a Low Carbon Society, 8).

Polletta, Francesca; Jasper, James M. (2001): Collective Identity and Social Movements. In *Annual Review of Sociology* 27 (1), pp. 283–305. https://doi.org/10.1146/annurev.soc.27. 1.283.

Polsby, Nelson W. (1963): Community Power and Political Theory. New Haven: Yale University Press.

Posner, Barry Z. (2015): An investigation into the leadership practices of volunteer leaders. In *Leadership & Org Development J* 36 (7), pp. 885–898. https://doi.org/10.1108/lodj-03-2014-0061.

Potthoff, Antje (n.d.): Die Schwarzwald-Rebellin. BRIGITTE. Available online at https://www.ews-schoenau.de/export/sites/ews/ews/presse/.files/die-schwarzwald-rebellin-bri gitte.pdf, checked on 6/22/2019.

Przeworski, Adam (1987): Methods of Cross-National Research 1970-1983: An Overview. In Meinolf Dierkes, Hans N. Weiler, Ariane Berthoin Antal (Eds.): Comparative Policy Research: Learning from Experience. Aldershot.

Przeworski, Adam; Teune, Henry (1970): The logic of comparative social inquiry. New York: Wiley.

Przyborski, Aglaja; Wohlrab-Sahr, Monika (2014): Qualitative Sozialforschung. Ein Arbeitsbuch. 4., erw. Aufl. München: Oldenbourg Wissenschaftsverlag.

Pulla, Venkat (2014): Grounded Theory Approach in Social Research. In *Space and Culture* 2 (3), pp. 14–23. https://doi.org/10.20896/saci.v2i3.93.

Putnam, Robert (2000): Bowling alone. The collapse and revival of American community. New York, NY: Simon & Schuster.

Putnam, Robert (1993): Making democracy work: Civic traditions in modern Italy. Princeton, NJ: Princeton University Press.

Radtke, Jörg (2016): Bürgerenergie in Deutschland: Partizipation zwischen Gemeinwohl und Rendite. Wiesbaden: Springer.

Rammert, Werner; Windeler, Arnold; Knoblauch, Hubert; Hutter, Michael (2016) Innovationsgesellschaft heute. Wiesbaden: Springer.

Rammert, Werner (2000): National Systems of Innovation, Idea Innovation Networks and Comparative Innovation Biographies (Technology Studies Working Paper, TUTS-WP-5-2000).

Rand, Joseph; Hoen, Ben (2017): Thirty years of North American wind energy acceptance research: What have we learned? In *Energy Research & Social Science* 29, pp. 135–148. https://doi.org/10.1016/j.erss.2017.05.019.

Rathi, Sambhu Singh; Chunekar, Aditya (2015): Not to buy or can be 'nudged' to buy? Exploring behavioral interventions for energy policy in India. In *Energy Research & Social Science* 7, pp. 78–83. https://doi.org/10.1016/j.erss.2015.03.006.

Rau, Irina; Schweizer-Ries, Petra; Hildebrand, Jan (2012): Participation. The Silver Bullet for the Acceptance of Renewable Energies? In : Vulnerability, Risks, and Complexity. Impacts of Global Change on Human Habitats. 1st ed. Ashland (Advances in People-Environment Studies, v.3), pp. 177–192.

Rayner, Steve; Malone, Elizabeth (Eds.) (1998): Human choice and climate change. Columbus, Ohio: Battelle Press.

Reason, Peter (2011(1994)): Three approaches to participative inquiry. In Norman K. Denzin, Yvonna S. Lincoln (Eds.): The Sage handbook of qualitative research. 4th ed. Thousand Oaks, pp. 324–339.

Reiche, Danyel (Ed.) (2012(2005)): Handbook of Renewable Energies in the European Union. Case studies of the EU-15 States. 2nd, revised ed. Frankfurt a.M., Berlin, Bern, Bruxelles, New York, Oxford, Wien: Peter Lang.

Rein, Martin; Schön, Donald (1996): Frame-critical policy analysis and frame-reflective policy practice. In *Knowledge and Policy* 9 (1), pp. 85–104. https://doi.org/10.1007/bf0 2832235.

Ricci, Miriam; Bellaby, Paul; Flynn, Rob (2010): Engaging the public on paths to sustainable energy: Who has to trust whom? In *Energy Policy* 38 (6), pp. 2633–2640. https://doi.org/10.1016/j.enpol.2009.05.038.

Richardson, Sam (2013): The Island of Samsø: From Green to Grey (Jutland Station). Available online at http://www.jutlandstation.dk/the-island-of-samso-from-green-to-grey/, checked on 4/30/2019.

Rip, Arie; Misa, Thomas J.; Schot, Johan (Eds.) (1995): Managing technology in society. The approach of constructive technology assessment. London: Pinter.

Rip, Arie (1998): Technological change. In Steve Rayner, Elizabeth Malone (Eds.): Human choice and climate change. Columbus, Ohio: Batelle Press, pp. 327–399.

Ritchie, Jane; Lewis, Jane; McNaughton Nicholls, Carol; Ormston, Rachel (Eds.) (2003): Qualitative research practice. A guide for social science students and researchers. London, Thousand Oaks: Sage.

Robson, Colin (2002): Real World Research. Hoboken, New Jersey: Blackwell Publishing Ltd.

Röcklinsberg, Helena; Sandin, Per (Ed.) (2013): The ethics of consumption. The citizen, the market and the law : EurSafe 2013, Uppsala, Sweden, 11–14 September 2013. Wageningen, The Netherlands: Wageningen Academic Publishers.

Rödl & Partner (2013): Kriterienkatalog Stadt Oldenburg, Konzessionsvergabe Strom. Available online at https://www.olegeno.de/assets/uploads/konzession/kriterienkatalog-konzessionsvergabe-oldenburg-strom.pdf, checked 5/10/2019.

Rogers, Everett; Dearing, James (1988): Agenda-setting research: Where has it been, where is it going? In *Annals of the International Communication Association* 11 (1), pp. 555–594.

Rogers, Jennifer C.; Simmons, Eunice A.; Convery, Ian; Weatherall, Andrew (2008): Public perceptions of opportunities for community-based renewable energy projects. In *Energy Policy* 36 (11), pp. 4217–4226. https://doi.org/10.1016/j.enpol.2008.07.028.

Rogers, Jennifer C.; Simmons, Eunice A.; Convery, Ian; Weatherall, Andrew (2012): Social impacts of community renewable energy projects: findings from a woodfuel case study. In *Energy Policy* 42, pp. 239–247. https://doi.org/10.1016/j.enpol.2011.11.081.

Romero-Rubio, C.; Andrés Díaz, J. R. de (2015): Sustainable energy communities: A study contrasting Spain and Germany. In *Energy Policy* 85, pp. 397–409.

Rose, Gillian (1997): Situating knowledges: positionality, reflexivities and other tactics. In *Progress in Human Geography* 21 (3), pp. 305–320. https://doi.org/10.1191/030913297673302122.

Rucht, Dieter (2011): Zum Stand der Forschung zu sozialen Bewegungen. In *Forschungsjournal Soziale Bewegungen* 24 (3), p. 37. https://doi.org/10.1515/fjsb-2011-0303.

Rudolph, David; Haggett, Claire; Aitken, Mhairi (2018): Community benefits from offshore renewables: The relationship between different understandings of impact, community, and benefit. In *Environment and Planning C: Politics and Space* 36 (1), pp. 92–117. https://doi.org/10.1177/2399654417699206.

Ruggiero, Salvatore; Martiskainen, Mari; Onkila, Tiina (2018): Understanding the scaling-up of community energy niches through strategic niche management theory: Insights from Finland. In *Journal of Cleaner Production* 170, pp. 581–590. https://doi.org/10.1016/j.jclepro.2017.09.144.

Rut, Monika; Davies, Anna R. (2018): Transitioning without confrontation? Shared food growing niches and sustainable food transitions in Singapore. In *Geoforum* 96, pp. 278–288. https://doi.org/10.1016/j.geoforum.2018.07.016.

Ryghaug, Marianne; Skjølsvold, Tomas Moe; Heidenreich, Sara (2018): Creating energy citizenship through material participation. In *Social studies of science* 48 (2), pp. 283–303. https://doi.org/10.1177/0306312718770286.

Sabel, C.F. (1993): Studied trust: Building new forms of cooperation in a volatile economy. In *Human Relations* 46, pp. 1133–1170.

Sadownik, Bryn; Jaccard, Mark (2002): Shaping Sustainable Energy Use in Chinese Cities. In *disP - The Planning Review* 38 (151), pp. 15–22. https://doi.org/10.1080/02513625.2002.10556819.

Salzborn, Samuel (Ed.) (2016): Klassiker der Sozialwissenschaften. 100 Schlüsselwerke im Portrait. 2. Aufl. Wiesbaden: Springer.

Samsø Energiselskab (1997): Tiarsplan: Første energiplan for Samsø. Project report. Samsø Energiselskab (Ten-year plan: First energy plan for Samsø). Available online at www.ene rgiinstituttet.dk/177, checked on 3/12/2016.

Samsø Posten (1998): Min opgave er at åbne dørene. (My task is to open doors). In *Samsø Posten* 237, 3/10/1998. Available online at http://arkiv.energiinstituttet.dk/187/, checked on 3/21/2016.

Scharpf, Fritz W. (2000): Interaktionsformen. Akteurzentrierter Institutionalismus in der Politikforschung. Opladen (UTB für Wissenschaft Uni-Taschenbücher Politikwissenschaft, 2136).

Schattschneider, E. E. (1975 (1960)): The semi sovereign people. A realists view of democracy in America. Paperback edition. Boston, Mass.: Cengage Learning.

Schön, Donald A.; Rein, Martin (1995 (1994)): Frame reflection. Toward the resolution of intractable policy controversies. 2nd edition. New York, N.Y: Basic Books.

Schön, Susanne; Kruse, Sylvia; Meister, Martin; Nölting, Benjamin; Ohlhorst, Dörte (2007): Handbuch Konstellationsanalyse. Ein interdisziplinäres Brückenkonzept für die Nachhaltigkeits-, Technik- und Innovationsforschung. München: Oekom. Available online at http://deposit.d-nb.de/cgi-bin/dokserv?id=2870496&prov=M&dok_var=1& dok_ext=htm.

Schot, Johan; Geels, Frank W. (2008): Strategic niche management and sustainable innovation journeys. Theory, findings, research agenda, and policy. In *Technology Analysis & Strategic Management* 20 (5), pp. 537–554. https://doi.org/10.1080/09537320802292651.

Schot, Johan; Hoogma, Remco; Elzen, Boelie (1994): Strategies for shifting technological systems. The exemplar of the Automobile System. In *Futures* 26 (10), pp. 1060–1076. https://doi.org/10.1016/0016-3287(94)90073-6.

Schreuer, Anna (2018): Bürgerkraftwerke in Österreich: Ein Phänomen mit vielen Gesichtern. In Lars Holstenkamp, Jörg Radtke: Handbuch Energiewende und Partizipation. Wiesbaden: Springer VS Handbuch, pp. 1081–1092.

Schreuer, Anna (2015): Dealing with the diffusion challenges of grassroots innovations. The case of citizen power plants in Austria and Germany. PhD Dissertation, Interdisziplinäres Forschungszentrum für Technik, Arbeit und Kultur IFZ, Graz.

Schreuer, Anna; Weismeier-Sammer, Daniela (2010): Energy cooperatives and local ownership in the field of renewable energy technologies: a literature review. Vienna (RICC Research Reports).

Schreurs, M. A. (2008): From the Bottom Up. Local and Subnational Climate Change Politics. In *The Journal of Environment & Development* 17 (4), pp. 343–355. https://doi.org/10. 1177/1070496508326432.

Schreurs, Miranda A. (2014): The Ethics of Nuclear Energy: Germany's Energy Politics after Fukushima. In *The Journal of Social Science* 77, pp. 9–29, checked on 5/14/2016.

Schröder, Carolin; Walk, Heike (Eds.) (2014): Genossenschaften und Klimaschutz. Akteure für zukunftsfähige, solidarische Städte. Wiesbaden: Springer (Bürgergesellschaft und Demokratie, Band 11).

Schroeder, Ralph (Ed.) (2016): Global powers. Michael Mann's anatomy of the twentieth century and beyond. Cambridge, New York: Cambridge University Press.

Schumacher, Ernest F. (1993 (1974)): Small is beautiful. A study of economics as if people mattered. London: Vintage.

Scott, Kirsty (2009): Scottish villagers stun developer by demanding extra turbine. In *the guardian*, 5/10/2009. Available online at https://www.theguardian.com/environment/2009/may/10/windpower-energy, checked on 3/21/2019.

Scottish Government (2017a): Scottish energy strategy. The future of energy in Scotland. Edinburgh. Available online at https://www.gov.scot/publications/scottish-energy-strategy-fut ure-energy-scotland-9781788515276/pages/5/, checked on 4/21/2019.

Scottish Government (2017b): A new definition of fuel poverty in Scotland. A review of recent evidence. The 2017 Scottish Fuel Poverty Definition Review Panel. November 2017. Available online at https://www.gov.scot/publications/new-definition-fuel-poverty-scotland-review-recent-evidence/, checked on 4/20/2019.

Scottish Government (2016a): People and Communities Fund: Projects 2012–2015. Available online at https://www.gov.scot/publications/people-and-communities-fund-projects-funded-since-2012/, checked on 3/21/2019.

Scottish Government (2016b): SP Bill 76 Land Reform (Scotland) Bill. Available online at https://www.parliament.scot/S4_Bills/Land%20Reform%20(Scotland)%20Bill/b76s4-introd.pdf, checked on 3/21/2019.

Scottish Government (2015a): Community Energy Policy Statement. September 2015. Available online at https://www.webarchive.org.uk/wayback/archive/20170401055145/http://www.gov.scot/Topics/Business-Industry/Energy/CEPS2015, checked on 3/21/2019.

Scottish Government (2015b): Community Empowerment (Scotland) Act 2015. July 24 2015. Available online at http://www.legislation.gov.uk/asp/2015/6/pdfs/asp_20150006_en.pdf, checked on 4/20/2019.

Scottish Government (2014): Scottish Planning Policy, 23 June 2014. Available online at https://www.gov.scot/publications/scottish-planning-policy/pages/12/, checked on 6/22/2019. ISBN: 9781784125677.

Scottish Government (6/7/2013): A million acres in community ownership by 2020. Available online at https://news.gov.scot/news/a-million-acres-in-community-ownership-by-2020, checked on 3/21/2019.

Scottish Government (2011): 2020 Routemap for Renewable Energy in Scotland. Available online at http://www.gov.scot/Publications/2011/08/04110353/3#renewableenergytar gets, checked on 2/12/2016.

Scottish Government (2010): Scottish Planning Policy, February 2010. Available online at https://www2.gov.scot/resource/doc/300760/0093908.pdf, checked on 6/22/2019.

Scottish Government (2009a) Climate Change (Scotland) Act 2009, asp 12. Source: https://www.legislation.gov.uk/asp/2009/12/pdfs/asp_20090012_en.pdf, checked on 3/28/2019.

Scottish Government (2009b) Scotland's Renewables Action Plan. Available online at https://www.webarchive.org.uk/wayback/archive/20170701074158/www.gov.scot/Publications/2009/07/06095830, checked on 3/28/2019.

Scottish Government (2003: Land Reform (Scotland) Act 2003, asp2. Source: https://www.legislation.gov.uk/asp/2003/2/pdfs/asp_20030002_en.pdf, checked on 3/28/2019.

Scottish Parliament (2/15/2011): The Community and Renewable Energy Scheme (CARES) will begin offering loans for the pre-planning costs of reneables projects in Scotland next financial year. Available online at http://www.gov.scot/Topics/Business-Industry/Energy/Energy-sources/19185/Communities/CARESLF, checked on 2/12/2016.

Scottish Parliament (2003): Public Attitudes to Windfarms: A Survey of Local Residents in Scotland. Executive Summary. Available online at http://www.gov.scot/Publications/2003/08/18049/25580, checked on 2/23/2016.

Scottish Parliament (6/30/2004): Enterprise Committee calls for "fully-fledged" Scottish energy policy. Committee News Release. Available online at https://web.archive.org/web/20061129112301/http://www.scottish.parliament.uk/nmCentre/news/news-comm-04/cent04-005.htm,checked on 6/22/2019.

Scottish Renewables (3/31/2015): Renewable future gets Scots' vote. Available online at https://www.scottishrenewables.com/news/renewable-future-gets-scots-vote/, checked on 2/23/2016.

Scottish Renewables (2014): Employment in Renewable Energy in Scotland in 2013. Available online at https://www.scottishrenewables.com/publications/employment-renewable-energy-scotland-2013/, checked on 6/22/2019.

Sen, Amartya (1999): Development as Freedom. Oxford: Oxford University Press.

Sen, Amartya (1992): Inequality reexamined. New York, Oxford: Clarendon Press.

Setton, Daniela (2015): Energiebuerger, werdet Klimaaktivisten! In *klimaretter.info*, 4/13/2015, pp. ONLINE. Available online at http://www.klimaretter.info/standpunkte/18565-qenergiebuerger-werdet-klimaaktivistenq, checked on 5/12/2015.

Seyfang, Gill; Haxeltine, Alex (2012): Growing Grassroots Innovations: Exploring the Role of Community-Based Initiatives in Governing Sustainable Energy Transitions. In *Environ Plann C Gov Policy* 30 (3), pp. 381–400. https://doi.org/10.1068/c10222.

Seyfang, Gill; Hielscher, Sabine; Hargreaves, Tom; Martiskainen, Mari; Smith, Adrian (2014): A grassroots sustainable energy niche? Reflections on community energy in the UK. In *Environmental Innovation and Societal Transitions* 13, pp. 21–44. https://doi.org/10.1016/j.eist.2014.04.004.

Seyfang, Gill; Longhurst, Noel (2013a): Desperately seeking niches: Grassroots innovations and niche development in the community currency field. In *Global Environmental Change* 23 (5), pp. 881–891. https://doi.org/10.1016/j.gloenvcha.2013.02.007.

Seyfang, Gill; Longhurst, Noel (2013b): Growing green money? Mapping community currencies for sustainable development. In *Ecological Economics* 86, pp. 65–77. https://doi.org/10.1016/j.ecolecon.2012.11.003.

Seyfang, Gill; Park, Jung Jin; Smith, Adrian (2013): A thousand flowers blooming? An examination of community energy in the UK. In *Energy Policy* 61, pp. 977–989. https://doi.org/10.1016/j.enpol.2013.06.030.

Seyfang, Gill; Smith, Adrian (2007): Grassroots innovations for sustainable development: Towards a new research and policy agenda. In *Environmental Politics* 16 (4), pp. 584–603. https://doi.org/10.1080/09644010701419121.

Shackley, Simon; Green, Ken (2007): A conceptual framework for exploring transitions to decarbonised energy systems in the United Kingdom. In *Energy* 32 (3), pp. 221–236. https://doi.org/10.1016/j.energy.2006.04.010.

Shove, Elizabeth; Walker, Gordon (2007): Caution! Transitions Ahead: Politics, Practice, and Sustainable Transition Management. In *Environ Plan A* 39 (4), pp. 763–770. DOI: 10.1068/a39310.

Shove, Elizabeth; Walker, Gordon (2008): Transition Management™ and the Politics of Shape Shifting. In *Environ Plan A* 40 (4), pp. 1012–1014. https://doi.org/10.1068/a4004leb.

Sijm, J.P.M. (2002): The Performance of Feed-In Tariffs to Promote Renewable Electricity in European Countries. Report ECN-C02-083. Petten, NL: Energy Center of the Netherlands.

Silverman, David (2013): Doing qualitative research. Fourth edition. London: Sage.

Simcock, Neil; Walker, Gordon; Day, Rosie (2016): Fuel poverty in the UK: beyond heating? In *PPP* 10 (1), pp. 25–41. https://doi.org/10.3351/ppp.0010.0001.0003.

SMART Fintry (2018): Resources. Available online at http://smartfintry.org.uk/about-smart-fintry/resources/, checked on 3/21/2019.

SMILE, Smart Islands Energy System (2017): Samsoe, Denmark. Funded by the European Union Horizon 2020 Research and Innovation programme. Available online at https://www.h2020smile.eu/the-islands/samso-denmark/, checked on 12/18/2017.

Smith, Adrian (2007a): Emerging in between: The multi-level governance of renewable energy in the English regions. In *Energy Policy* 35 (12), pp. 6266–6280. https://doi.org/10.1016/j.enpol.2007.07.023.

Smith, Adrian (2007b): Translating Sustainabilities between Green Niches and Socio-Technical Regimes. In *Technology Analysis & Strategic Management* 19 (4), pp. 427–450. https://doi.org/10.1080/09537320701403334.

Smith, Adrian; Hargreaves, Tom; Hielscher, Sabine; Martiskainen, Mari; Seyfang, Gill (2016): Making the most of community energies: Three perspectives on grassroots innovation. In *Environ Plan A* 48 (2), pp. 407–432. https://doi.org/10.1177/0308518x15597908.

Smith, Adrian; Raven, Rob (2012): What is protective space? Reconsidering niches in transitions to sustainability. In *Research Policy* 41 (6), pp. 1025–1036. https://doi.org/10.1016/j.respol.2011.12.012.

Smith, Adrian; Stirling, Andrew (2017): Innovation, sustainability and democracy: an analysis of grassroots contributions. In *Journal of Self-Governance and Management Economics* 6 (1), pp. 64–97. Available online at https://steps-centre.org/wp-content/uploads/2017/07/SmithStirling-2017-GI-ID-journal-article.pdf, checked on 4/12/2019.

Smith, Jackie (2018): Smart Fintry—Year 2. Innovation Workstream Findings. Available online at http://smartfintry.org.uk/wp-content/uploads/2018/04/Smart-Fintry-Innovation-Report-final.pdf, updated on 4/12/2018, checked on 3/21/2019.

Snow, David A.; Benford, Robert D. (1992): Master Frames and Cycles of Protest. In Aldon D. Morris, Carol Mueller, Carol McClurg (Eds.): Frontiers in Social Movement Theory, New Haven: Yale University Press, pp. 133–155.

Snow, David A.; Benford, Robert D. (1988): Ideology, Frame Resonance and Participant Mobilization. In *International Social Movement Research* 1, pp. 197–217.

Snow, David A.; Soule, Sarah A.; Kriesi, Hanspeter (Eds.) (2004): The Blackwell Companion to Social Movements. Hoboken, New Jersey: Wiley-Blackwell Publishing Ltd.

Solar Green Point (2019): Achtergrond (Background). Available online at https://www.solarg reenpoint.nl/achtergrond, checked on 4/11/2019.

Solarinitative (n.d.): Die Solarinitiative. 50 DM Solaraktie. Available online at http://www. solarini.de/Solaraktie/solaraktie.html, checked on 3/19/2019.

Soule, Sarah A.; King, Brayden G. (2008): Competition and Resource Partitioning in Three Social Movement Industries. In *American Journal of Sociology* 113 (6), pp. 1568–1610. https://doi.org/10.1086/587152.

Soutar, Iain (2015): From local to global value: The transformational nature of community energy. PhD Dissertation. University of Exeter.

Sovacool, Benjamin K. (2016): How long will it take? Conceptualizing the temporal dynamics of energy transitions. In *Energy Research & Social Science* 13, pp. 202–215. https://doi. org/10.1016/j.erss.2015.12.020.

Sovacool, Benjamin K. (2014): What are we doing here? Analyzing fifteen years of energy scholarship and proposing a social science research agenda. In *Energy Research & Social Science* 1, pp. 1–29. https://doi.org/10.1016/j.erss.2014.02.003.

Sovacool, Benjamin K. (2009): The importance of comprehensiveness in renewable electricity and energy-efficiency policy. In *Energy Policy* 37 (4), pp. 1529–1541. https://doi.org/10. 1016/j.enpol.2008.12.016.

Sovacool, Benjamin K; Noel, Lance; Axsen, Jonn; Kempton, Willett (2018): The neglected social dimensions to a vehicle-to-grid (V2G) transition: A critical and systematic review. In *Environmental Research Letters* 13 (1).

Sovacool, Benjamin K.; Brossmann, Brent (2014): The rhetorical fantasy of energy transitions: implications for energy policy and analysis. In *Technology Analysis & Strategic Management* 26 (7), pp. 837–854. https://doi.org/10.1080/09537325.2014.905674.

Sovacool, Benjamin K.; Brown, Marilyn A. (2015): Deconstructing facts and frames in energy research: Maxims for evaluating contentious problems. In *Energy Policy* 86, pp. 36–42. https://doi.org/10.1016/j.enpol.2015.06.020.

Sovacool, Benjamin K.; Lakshmi Ratan, Pushkala (2012): Conceptualizing the acceptance of wind and solar electricity. In *Renewable and Sustainable Energy Reviews* 16 (7), pp. 5268– 5279. https://doi.org/10.1016/j.rser.2012.04.048.

Sperling, Karl (2017): How does a pioneer community energy project succeed in practice? The case of the Samsø Renewable Energy Island. In *Renewable and Sustainable Energy Reviews* 71, pp. 884–897. https://doi.org/10.1016/j.rser.2016.12.116.

SPIEGEL (1996): Rebellen von Schönau. In *Der SPIEGEL* 21, 5/20/1996, pp. 135– 136. Available online at http://www.spiegel.de/spiegel/print/d-8926364.html, checked on 3/27/2019.

Spreng, Daniel (2014): Transdisciplinary energy research—Reflecting the context. In *Energy Research & Social Science* 1, pp. 65–73. https://doi.org/10.1016/j.erss.2014.02.005.

SSE Ltd (3/30/2015): Isle of Mull to trial innovative grid connection model. Available online at https://sse.com/newsandviews/allarticles/2015/03/isle-of-mull-to-trial-inn ovative-grid-connection-model/, checked on 4/30/2019.

Stadt Oldenburg (12/2/2013): Vergabe der Strom- und Gaskonzessionen. Available online at https://www.olegeno.de/assets/uploads/konzession/vorlage-stadtrat-oldenburg-strom-gaskonzessionen.pdf, checked on 5/10/2019.

Stadt Oldenburg, Der Oberbürgermeister (1/30/2014): Mit Solaraktien regenerative Energien fördern. Available online at https://www.oldenburg.de/startseite/leben-wohnen/umwelt/ lokale-agenda-21/agenda-gruppen/archiv/solarinitiative.html, checked on 3/19/2019.

Stadtwerke Wolfhagen (2019): Unternehmen - Über Uns. Philosophie. Available online at https://www.stadtwerke-wolfhagen.de/index.php/unternehmen/ueber-uns/philosophie, checked on 5/10/2019.

Stephan, Hannes R. (2012): Revisiting the Transatlantic Divergence over GMOs. Toward a Cultural-Political Analysis. In *Global Environmental Politics* 12 (4), pp. 104–124. https:// doi.org/10.1162/glep_a_00142.

Stirling, Andy (2014): Transforming power: Social science and the politics of energy choices. In *Energy Research & Social Science* 1, pp. 83–95. https://doi.org/10.1016/j.erss.2014. 02.001.

Stoecker, Randy (2008): Challenging institutional barriers to community-based research. In *Action Research* 6 (1), pp. 49–67. https://doi.org/10.1177/1476750307083721.

Stoecker, Randy (1999): Making Connections: Community Organizing, Empowerment Planning, and Participatory Research in Participatory Evaluation. In *Sociological Practice* 1 (3), pp. 209–231. https://doi.org/10.1023/a:1022874507194.

Stoecker, Randy (1999): Are Academics Irrelevant? Roles for Scholars in Participatory Research. In *American Behavioral Scientist* 42 (5), pp. 840–854. https://doi.org/10.1177/000 27649921954561.

Stonington, Joel (2013): An Unlikely Effort to Buy the Berlin Power Grid. In *SPIEGEL Online*, 4/3/2013. Available online at http://www.spiegel.de/international/business/an-unl ikely-effort-to-buy-the-berlin-power-grid-a-886426.html, checked on 4/1/2019.

Strachan, Peter A.; Cowell, Richard; Ellis, Geraint; Sherry-Brennan, Fionnguala; Toke, David (2015): Promoting Community Renewable Energy in a Corporate Energy World. In *Sust. Dev.* 1 (1), n/a-n/a. https://doi.org/10.1002/sd.1576.

Strauss, Anselm; Corbin, Juliet (1990): Basics of qualitative research: Grounded theory procedures and techniques. Thousand Oaks, CA: Sage.

Stromauskunft (2019): Echter Ökostrom. Available online at https://www.stromauskunft.de/ oekostrom/echter-oekostrom/, updated on 4/30/2019.

Stromreport (2017): Zusammensetzung Strompreis 2017. Available online at https://1-str omvergleich.com/strom-report/strompreis/#zusammensetzung-strompreis-2017, checked on 4/30/2019.

Strömberg, David (2004): Radio's Impact on Public Spending. In *The Quarterly Journal of Economics* 119 (1), pp. 189–221. https://doi.org/10.1162/003355304772839560.

Strübing, Jörg (2018(2004)): Qualitative Sozialforschung. Eine Komprimierte Einführung Für Studierende. 2nd ed. Berlin/Boston (Soziologie Kompakt Ser). Available online at https://ebookcentral.proquest.com/lib/gbv/detail.action?docID=5156911.

Süddeutsche Zeitung (2011): Anti-Atom Bewegung mobilisiert 250.000 Menschen. In *sueddeutsche.de*, 4/27/2011. Available online at https://www.sueddeutsche.de/politik/bun desweite-proteste-anti-atom-bewegung-mobilisiert-zehntausende-1.1077642, checked on 3/27/2019.

Sullivan, Helen (2007): Interpreting 'community leadership' in English local government. In *policy polit* 35 (1), pp. 141–161. https://doi.org/10.1332/030557307779657775.

Süsser, Diana (2016): People Powered Local Energy Transition. Mitigating Climate Change with Community-Based Renewable Energy in North Frisia. PhD, Hamburg. Department of Mathematics, Information Technologies and Natural Sciences, Institute for Geosciences.

SWR Suedwestrundfunk (2016): Wyhl? "Nai hämmer gsait!". Der Widerstand gegen das Atomkraftwerk am Kaiserstuhl. Available online at https://www.swr.de/geschi chte/wyhl-atomkraft-widerstand/-/id=100754/did=12047138/nid=100754/6854hq/index. html, checked on 3/27/2019.

Sydow, Jörg; Schreyögg, Georg; Koch, Jochen (2009): Organizational Path Dependence: Opening the Black Box. In *Academy of Management Review* 34 (4), pp. 689–709.

Szarka, Joseph; Cowell, Richard; Ellis, Geraint; Strachan, Peter A.; Warren, Charles R. (2016 (2012)): Learning from wind power. Governance, societal and policy perspectives on sustainable energy. Basingstoke: Palgrave Macmillan (Energy, climate and the environment).

Tarrow, Sidney (1994): Power in movements, social movements, collective action and politics. Cambridge, MA: Cambridge University Press.

Tews, Kerstin (2014): Energiearmut – vom politischen Schlagwort zur handlungsleitenden Definition. In Achim Brunnengräber, Maria Rosaria Di Nucci (Eds.): Im Hürdenlauf zur Energiewende. Von Transformationen, Reformen und Innovationen; zum 70. Geburtstag von Achim Mez. Wiesbaden: Springer, pp. 441–449.

The Waterfall Fund (2018): The Waterfall Fund Story. Available online at https://www.the waterfallfund.co.uk/background, checked on 7/24/2018.

The Waterfall Fund (2019a): Eligibility Check List. Craignure, Isle of Mull. Available online at https://www.thewaterfallfund.co.uk/eligibility-checker, updated on 2019, checked on 3/28/2019.

The Waterfall Fund (2019b): Recipients. Available online at https://www.thewaterfallfund. co.uk/recipients, checked on 4/12/2019.

Thøgersen, John; Grønhøj, Alice (2010) Electricity saving in households—A social cognitive approach. In *Energy Policy* 38 (12), pp. 7732–7743.

Tilly, Charles (1978): From Mobilization to Revolution. Reading, MA: Addison-Wesley.

Tobiasson, Wenche; Beestermöller, Christina; Jamasb, Tooraj (2016) Public engagement in electricity network development: the case of the Beauly–Denny project in Scotland. In *Economia e Politica Industriale* 43 (2), pp. 105–126.

Toke, David (2012): Wind Power, Governance and Networks. In Joseph Murphy (Ed.): Governing Technology for Sustainability. Hoboken: Earthscan, pp. 168–181.

Toke, David (2011) The UK offshore wind power programme: A sea-change in UK energy policy? In *Energy Policy* 39 (2), pp. 526–534.

Toke, David (2008) Wind power deployment outcomes: How can we account for the differences?. In *Renewable and Sustainable Energy Reviews* 12 (4), pp. 1129–1147.

Toke, David (2005) Community Wind Power in Europe and in the UK. In *Wind Engineering* 29 (3), pp. 301–308.

Tolkien, J.R.R. (1966 (1954)): The fellowship of the ring. Being the first part of the Lord of the Rings. Second edition, fifth impression. London: George Allen & Unwin Ltd.

Torgler, Benno; Garcia-Valinas, Maria A.; Macintyre, Alison (2011): Participation in environmental organizations: an empirical analysis . In *Environment and Development Econonmics* 16 (05), pp. 591–620.

Tranaes, Flemming (n.d.): Danish Wind Energy. Available online at http://www.spok.dk/con sult/reports/danish_wind_energy.pdf.

Trapence, Gift; Collins, Chris; Avrett, Sam; Carr, Robert; Sanchez, Hugo; Ayala, George; Diouf, Daouda; Beyrer, Chris; Baral, Stefan D. (2012): From personal survival to public health: community leadership by men who have sex with men in the response to HIV. In *The Lancet* 380 (9839), pp. 400–410.

trend:research, Leuphana Universität Lüneburg (2013): Definition und Marktanalyse von Bür gerenergie in Deutschland. Bremen/Lüneburg. Available online at https://www.buendnis-buergerenergie.de/fileadmin/user_upload/downloads/Studien/Studie_Definition_und_ Marktanalyse_von_Buergerenergie_in_Deutschland_BBEn.pdf, checked on 4/10/2019.

Turner, Jonathan H. (Ed.) (2001): Handbook of sociological theory. New York, London: Kluwer Academic/Plenum Pub (Handbooks of Sociology and Social Research).

Turnhout, E.; van Bommel, S.; Aarts, M.N.C. (2010): How Participation Creates Citizens: Participatory Governance as Performative Practice. In *Ecology and Society* 15 (4), pp. 26–41.

Umweltbundesamt (2016): Energie-stationär. Dessau. Available online at https://www.umw eltbundesamt.de/themen/klima-energie/treibhausgas-emissionen/emissionsquellen#tex tpart-7, checked on 12/18/2018.

Upham, Paul (2012): Environmental citizens: climate pledger attitudes and micro-generation installation. In *Local Environment: The International Journal of Justice and Sustainability* 17 (1), pp. 75–91.

Usmani, Leyla (2017): Community and locally owned renewable energy in Scotland at June 2017. A report by the Energy Saving Trust for the Scottish Government. Edinburgh. Available online at https://www.energysavingtrust.org.uk/sites/default/files/Community% 20and%20locally%20owned%20renewable%20energy%20report_2017.pdf, checked on 4/23/2018.

Valdivielso, Joaquín (2005): Social Citizenship and the Environment. In *Environmental Politics* 14 (2), pp. 239–254.

van der Horst, Dan (2008): Social enterprise and renewable energy: emerging initiatives and communities of practice. In *Social Enterprise Journal* 4 (3), pp. 171–185.

van der Laak, W.W.M.; Raven, R.P.J.M.; Verbong, G.P.J. (2007): Strategic niche management for biofuels: Analysing past experiments for developing new biofuel policies. In *Energy Policy* 35 (6), pp. 3213–3225.

van der Schoor, Tineke; Scholtens, Bert (2015): Power to the people. Local community initiati-ves and the transition to sustainable energy. In *Renewable and Sustainable Energy Reviews* 43, pp. 666–675.

van der Stoep, Hetty (2014): Stories becoming sticky. How civic initiatives strive for connec-tion to governmental spatial planning agendas. Wageningen, The Netherlands: PhD thesis Wageningen University. ISBN 9461738293.

van Dijk, Terry (2011): Imagining future places (2011). In *Planning Theory* 10 (2), pp. 124–143.

van Hulst, Merlijn; Yanow, Dvora (2016): From Policy "Frames" to "Framing". In *The American Review of Public Administration* 46 (1), pp. 92–112.

van Oers, Laura M.; Boon, W.P.C.; Moors, Ellen H.M. (2018): The creation of legiti-macy in grassroots organisations: A study of Dutch community-supported agriculture. In *Environmental Innovation and Societal Transitions* 29, pp. 55–67.

van Veelen, Bregje (2018): Negotiating energy democracy in practice: governance processes in community energy projects. In *Environmental Politics* 27 (4), pp. 644–665.

van Veelen, Bregje; Haggett, Claire (2017): Uncommon Ground: The Role of Different Place Attachments in Explaining Community Renewable Energy Projects. In *Sociologia Ruralis* 57 (1), pp. 533–554.

van Vugt, Mark (2002): Central, Individual or Collective Control? Social Dilemma Strategies for Natural Resource Management. In *American Behavioral Scientist* 45, pp. 783–800.

van Vugt, Mark (2001): Community Identification Moderating the Impact of Financial Incentives in a Natural Social Dilemma: Water Conservation. In *Pers Soc Psychol Bull* 27 (11), pp. 1440–1449.

van Vugt, Mark; Cremer, David de (1999) Leadership in social dilemmas: The effects of group identification on collective actions to provide public goods. In *Journal of Personality and Social Psychology* 76 (4), pp. 587–599.

Vandenbroucke, Jan P. (2001): In Defense of Case Reports and Case Series. In *Annals of Internal Medicine* 134 (4), pp. 330–334.

Vaughan, Adam; Macalister, Terry (2015): The nine green policies killed off by the Tory government. In *theguardian.com*, 7/24/2015. Available online at http://www.thegua rdian.com/environment/2015/jul/24/the-9-green-policies-killed-off-by-tory-government, checked on 2/18/2016.

Vaughan, Adam (2015): UK doesn't have right policies to meet renewable energy target, admits Amber Rudd. In *theguardian.com*, 11/10/2015. Available online at http://www.the guardian.com/environment/2015/nov/10/rudd-issues-transport-challenge-to-meet-uk-ren ewables-target, checked on 2/17/2016.

Verba, Sydney (1995): Voice and equality: Civic voluntarism in American politics. With K.L. Schlozman, H. Brady. Cambridge, MA: Harvard University Press.

Verbong, Geert; Geels, Frank (2010): Exploring sustainability transitions in the electricity sector with socio-technical pathways. In *Technological Forecasting and Social Change* 77 (8), pp. 1214–1221.

Verbong, Geert; Geels, Frank (2007): The ongoing energy transition: Lessons from a socio-technical, multi-level analysis of the Dutch electricity system (1960–2004). In *Energy Policy* 35 (2), pp. 1025–1037.

Viardot, Eric (2013): The role of cooperatives in overcoming the barriers to adoption of renewable energy. In *Energy Policy* 63, pp. 756–764.

Visit Samsoe (2019): Kanhave Kanalen. (In Danish.). Available online at https://www.visits amsoe.dk/de/inspiration/kanhave-kanalen/, updated on 1/22/2019, checked on 3/27/2019.

Vorholz, Fritz (1991): Ein Dorf unter Spannung. In *DIE ZEIT* 24, 6/7/1991. Available online at https://www.zeit.de/1991/24/ein-dorf-unter-spannung, checked on 3/27/2019.

Vorkinn, Marit; Riese, Hanne (2001): Environmental Concern in a Local Context. In *Environment and Behavior* 33 (2), pp. 249–263. https://doi.org/10.1177/001391601219 72972.

Wakiyama, Takako; Zusman, Eric; Monogan, James E. (2014): Can a low-carbon-energy transition be sustained in post-Fukushima Japan? Assessing the varying impacts of exogenous shocks. In *Energy Policy* 73, pp. 654–666. https://doi.org/10.1016/j.enpol.2014.06.017.

Walker, Benjamin J.A.; Wiersma, Bouke; Bailey, Etienne (2014): Community benefits, framing and the social acceptance of offshore wind farms: An experimental study in England.

In *Energy Research & Social Science* 3, pp. 46–54. https://doi.org/10.1016/j.erss.2014. 07.003.

Walker, Edward T.; McCarthy, John d. (2010): Legitimacy, Strategy, and Resources in the Survival of Community-Based Organizations. In *Social Problems* 57 (3), pp. 315–340. https://doi.org/10.1525/sp.2010.57.3.315.

Walker, Gordon (2011): The role for 'community' in carbon governance. In *WIREs Clim Change* 2 (5), pp. 777–782. https://doi.org/10.1002/wcc.137.

Walker, Gordon (2008): What are the barriers and incentives for community-owned means of energy production and use? In *Energy Policy* 36 (12), pp. 4401–4405. https://doi.org/10. 1016/j.enpol.2008.09.032.

Walker, Gordon (2007): Ambivalence, Sustainability and the Governance of Socio-Technical Transitions. In *Journal of Environmental Policy & Planning* 9 (3-4), pp. 213–225.

Walker, Gordon; Cass, Noel (2007): Carbon reduction, 'the public' and renewable energy: engaging with socio-technical configurations. In *Area* 39 (4), pp. 458–469. https://doi.org/ 10.1111/j.1475-4762.2007.00772.x.

Walker, Gordon; Devine-Wright, Patrick (2008): Community renewable energy. What should it mean? In *Energy Policy* 36 (2), pp. 497–500. https://doi.org/10.1016/j.enpol.2007. 10.019.

Walker, Gordon; Devine-Wright, Patrick; Hunter, Sue; High, Helen; Evans, Bob (2010a): Trust and community: Exploring the meanings, contexts and dynamics of community renewable energy. In *Energy Policy* 38, pp. 2655–2663.

Walker, Gordon; Devine-Wright, Patrick (2010): Trust and community: Exploring the meanings, contexts and dynamics of community renewable energy. In *Energy Policy* 38 (6), pp. 2655–2663. https://doi.org/10.1016/j.enpol.2009.05.055.

Walker, Gordon; Hunter, Sue; Devine-Wright, Patrick; Evans, Bob (2007): Harnessing Community Energies: Explaining and Evaluating Community-Based Localism in Renewable Energy Policy in the UK. In *Global Environmental Politics* 7 (2), pp. 64–82. https://doi. org/10.1162/glep.2007.7.2.64.

Walsh, Bryan (2008): Heroes of the Environment 2008: Soren Hermansen. In *TIME Magazine*, 9/24/2008. Available online at http://content.time.com/time/specials/packages/printout/ 0,29239,1841778_1841782_1841789,00.html?iid=sr-link1, checked on 4/1/2019.

Warner, Joshua (2018): Are the Big Six suppliers losing their grip on the UK energy market? In *IG*, 3/12/2018. Available online at https://www.ig.com/uk/news-and-trade-ideas/shares-news/are-the-big-six-suppliers-losing-their-grip-on-the-uk-energy-mar-180312, checked on 3/27/2019.

Warren, Charles R.; Lumsden, Carolyn; O'Dowd, Simone; Birnie, Richard v. (2005): 'Green On Green': Public perceptions of wind power in Scotland and Ireland. In *Journal of Environmental Planning and Management* 48 (6), pp. 853–875. https://doi.org/10.1080/ 09640560500294376.

Warren, Charles R.; McFadyen, Malcolm (2010): Does community ownership affect public attitudes to wind energy? A case study from south-west Scotland. In *Land Use Policy* 27 (2), pp. 204–213. https://doi.org/10.1016/j.landusepol.2008.12.010.

Wissenschaftliche Beirat der Bundesregierung Globale Umweltveränderungen (WBGU) (2014): Klimaschutz als Weltbürgerbewegung: Sondergutachten. Berlin.

Weston, David (2017): Danish wind share falls in 2016. DENMARK: Lower wind speeds have caused wind's share in total electricity consumption to fall for the first time in eight years. In *Windpower monthly*, 1/13/2017. Available online at https://www.windpowermon thly.com/article/1420900/danish-wind-share-falls-2016, checked on 4/15/2017.

Wetzel, Daniel (2017): Die schmutzige Trickserei mit der Bürgerenergie. In *WELT online*, 6/22/2017. Available online at https://www.welt.de/wirtschaft/article165807760/Die-sch mutzige-Trickserei-mit-der-Buergerenergie.html, checked on 12/14/2018.

White, Rebecca; Stirling, Andrew (2013): Sustaining trajectories towards Sustainability: Dynamics and diversity in UK communal growing activities. In *Global Environmental Change* 23 (5), pp. 838–846. https://doi.org/10.1016/j.gloenvcha.2013.06.004.

Wiersma, Bouke; Devine-Wright, Patrick (2014): Decentralising energy: comparing the drivers and influencers of projects led by public, private, community and third sector actors. In *Contemporary Social Science* 9 (4), pp. 456–470. https://doi.org/10.1080/21582041. 2014.981757.

Windfang eG (2011): Satzung der Windfang e.G. Frauenenergiegemeinschaft. Oldenburg. Available online at http://s477174211.website-start.de/%C3%BCber-uns/satzung/, checked on 4/11/2019.

Winskel, Mark (2007): Multi-level governance and energy policy. Renewable energy in Scotland. In: Joseph Murphy (ed.): Governing technology for sustainability. London, Sterling, VA: Earthscan: 182–219.

Winterer, Andreas (12/21/2018): Ökostromanbieter: Wie man gute erkennt. Available online at https://www.oekotest.de/bauen-wohnen/Oekostromanbieter-Wie-man-gute-erk ennt-_600763_1.html, checked on 3/21/2019.

Wirth, Steffen (2014): Communities matter: Institutional preconditions for community renewable energy. In *Energy Policy* 70, pp. 236–246. https://doi.org/10.1016/j.enpol.2014. 03.021.

Wolfe, Philip (2016): Embarassingly successful: an obituary for the UK's Feed-in Tariffs. Business Green, 25 April 2016. Available online at https://www.businessgreen.com/ bg/opinion/2455741/embarrassingly-successful-an-obituary-for-the-uk-s-feed-in-tariffs, checked on 6/22/2019.

Wolk, Constanze (2016): Aus Schönau ins Weisse Haus. Available online at https://www. ews-schoenau.de/energiewende-magazin/zur-ews/goldman-environmental-prize-fuer-urs ula-sladek/, checked on 4/1/2019.

Woodman, Bridget; Mitchell, Catherine (2011): Learning from experience? The development of the Renewables Obligation in England and Wales 2002–2010. In *Energy Policy* 39 (7), pp. 3914–3921. https://doi.org/10.1016/j.enpol.2011.03.074.

Wüstenhagen, Rolf; Wolsink, Maarten; Bürer, Mary Jean (2007): Social acceptance of renewable energy innovation: An introduction to the concept. In *Energy Policy* 35 (5), pp. 2683–2691. https://doi.org/10.1016/j.enpol.2006.12.001.

Yildiz, Özgür (2014): Financing renewable energy infrastructures via financial citizen participation—The case of Germany. In *Renewable Energy* 68, pp. 677–685. https://doi.org/ 10.1016/j.renene.2014.02.038.

Yildiz, Özgür; Rommel, Jens; Debor, Sarah; Holstenkamp, Lars; Mey, Franziska; Müller, Jakob R. et al. (2015): Renewable energy cooperatives as gatekeepers or facilitators? Recent developments in Germany and a multidisciplinary research agenda. In *Energy Research & Social Science* 6, pp. 59–73. https://doi.org/10.1016/j.erss.2014.12.001.

Yin, Robert K. (1981): The Case Study as a Serious Research Strategy. In *Knowledge* 3 (1), pp. 97–114. https://doi.org/10.1177/107554708100300106.

Yin, Robert K. (1981): The Case Study Crisis: Some Answers. In *Administrative Science Quarterly* 26 (1), pp. 58–65.

Yin, Robert K. (2014): Case Study Research Design and Methods. 5th edition. Thousand Oaks, CA: Sage.

Young, Ken; Ashby, Deborah; Boaz, Annette; Grayson, Lesley (2002): Social Science and the Evidence-based Policy Movement. In *Soc Pol Soc* 1 (03). https://doi.org/10.1017/s14 74746402003068.

Zeitschrift für kommunale Wirtschaft (2018): Eneco erwirbt Lichtblick komplett. In *Zeitschrift für kommunale Wirtschaft*, 12/19/2018. Available online at https://www.zfk.de/untern ehmen/nachrichten/artikel/eneco-erwirbt-lichtblick-komplett-2018-12-19/, checked on 4/30/2019.

Ziegler, Rafael (2017a): Social innovation as a collaborative concept. In *Innovation: The European Journal of Social Science Research* 30 (4), pp. 388–405. https://doi.org/10. 1080/13511610.2017.1348935.

Ziegler, Rafael (2017b): Wer zur Quelle will, muss gegen den Strom schwimmen. Innovation aus Bürgerhand für eine demokratisch-ökologische Wasserwirtschaft. München: oekom. Available online at https://www.content-select.com/index.php?id=bib%5Fview& ean=9783962384593.

Ziegler, Rafael (2017b): Ziegler, Rafael; Molnár, György; Chiappero-Martinetti, Enrica; Jacobi, Nadia von (2017): Creating (Economic) Space for Social Innovation. In *Journal of Human Development and Capabilities* 18 (2), pp. 293–298. https://doi.org/10.1080/194 52829.2017.1301897.

Ziegler, Rafael; Ott, Konrad (2011): The quality of sustainability science: a philosophical perspective. In *Sustainability: Science, Practice and Policy* 7 (1), pp. 31–44. https://doi. org/10.1080/15487733.2011.11908063.

Zoellner, Jan; Rau, Irina; Schweizer-Ries, Petra (2011): Beteiligungsprozesse und Entwicklungschancen für Kommunen und Regionen. In *ÖW* 26 (3), p. 25. https://doi.org/10.14512/ oew.v26i3.1141.

The manufacturer's authorised representative in the EU is Springer
Nature Customer Service Centre GmbH, Europaplatz 3, 69115 Heidelberg,
Germany. If you have any concerns regarding our products, please
contact ProductSafety@springernature.com

Printed and bound by CPI Group (UK) Ltd, Croydon, CR0 4YY
28/04/2026
02098498-0001